PLANT PHYSIOLOGY

Research Progress in Botany

PLANT PHYSIOLOGY

Edited By
Philip Stewart, PhD
Head, Multinational Plant Breeding Program;
Author; Member, US Rosaceae Genomics,
Genetics and Breeding Executive Committee;
North Central Regional Association of
State Agricultural Experiment Station Directors, U.S.A.

Sabine Globig
Associate Professor of Biology, Hazard Community
and Technical College, Kentucky, U.S.A.

Apple Academic Press

TORONTO   NEW YORK

© 2012 by
Apple Academic Press Inc.
3333 Mistwell Crescent
Oakville, ON L6L 0A2
Canada

Apple Academic Press Inc.
1613 Beaver Dam Road, Suite # 104
Point Pleasant, NJ 08742
USA

First issued in paperback 2021

Exclusive worldwide distribution by CRC Press, a Taylor & Francis Group

ISBN 13: 978-1-77463-189-8 (pbk)
ISBN 13: 978-1-926692-69-2 (hbk)

Library and Archives Canada Cataloguing in Publication

Plant physiology/Philip Stewart, Sabine Globig.

Includes index.
ISBN 978-1-926692-69-2
1. Plant physiology–Textbooks. I. Stewart, Philip, 1974-
II. Globig, Sabine, 1949-

QK711.2.P53 2011   571.2   C2011-905431-0

Apple Academic Press also publishes its books in a variety of electronic formats. Some content that appears in print may not be available in electronic format. For information about Apple Academic Press products, visit our website at **www.appleacademicpress.com**

Preface

The chapters within this book focus on the study of all the internal activities of plants —those chemical and physical processes associated with life as they occur in plants. At the smallest scale, this includes research into the molecular interactions of photosynthesis and the internal diffusion of water, minerals, and nutrients. At the largest scale are investigations into the processes of plant development, seasonality, dormancy, and reproductive control.

Furthermore, these chapters focus on various aspects of plant physiology, including phytochemistry (plant chemistry); the interactions within a plant between cells, tissues, and organs; the ways in which plants regulate their internal functions; and how plants respond to conditions and variations within the environment. Given the environmental crises brought about by pollution and climate change, this is a particularly vital area of study, since stress from water loss, changes in air chemistry, or crowding by other plants can lead to changes in the way a plant functions. These changes may in turn be affected by genetic, chemical, and physical factors. Readers will gain the information they need to stay current with the latest research being done in this essential field of study.

—**Philip Stewart, PhD**

List of Contributors

Pierre Abad
National Institute of Agronomic Research, UMR 1301, Sophia-Antipolis, France.
National Centre for Scientific Research, UMR 6243, Sophia-Antipolis, France.
University of Nice Sophia-Antipolis, UMR 1301, Sophia-Antipolis, France.

Miguel A. Aranda
Department of Biology of Stress and Plant Pathology, Center for Soil Science and Applied Biology of the Segura (CEBAS)-CSIC, PO Box 164, 30100 Espinardo (Murcia), Spain.

Philip Benfey
Department of Biology, Duke University, Durham, North Carolina, USA.

José Blanca
Institute of Conservation and Improvement of agrobiodiversity Valenciana (COMAV-UPV), CPI, Ed.8E, Camino de Vera s/n, 46022 Valencia, Spain.

Hernán G. Bondino
Institute of Molecular and Cellular Biology of Rosario (IBR-CONICET), Faculty of Biochemical and Pharmaceutical Sciences, Universidad Nacional de Rosario, Suipacha 531, Rosario S2002LRK, Argentina.

Nils Braun
Institute of Plant Science, CNRS UPR2355, Université Paris Sud Orsay, Gif sur Yvette, France.
Institute of Biotechnology, University of Cambridge, Cambridge, UK.

Jessica L. Butler
Harvard University, Harvard Forest, Petersham, Massachusetts, USA.

Marie-Cécile Caillaud
National Institute of Agronomic Research, UMR 1301, Sophia-Antipolis, France.
National Centre for Scientific Research, UMR 6243, Sophia-Antipolis, France.
University of Nice Sophia-Antipolis, UMR 1301, Sophia-Antipolis, France.

Carlos Calderón-Vázquez
Center for Research and Advanced Studies, National Laboratory of Genomics for Biodiversity (LANGEBIO) North Bypass Road Irapuato-León, Irapuato, Guanajuato, Mexico.

Steven Cannon
USDA-ARS, Corn Insect and Crop Genetics Research Unit, Iowa State University, Ames, Iowa 50011 USA.

Ana I. Caño-Delgado
Molecular Genetics Department, Centre for Research in Agricultural Genomics CRAG (CSIC-IRTA-UAB), Barcelona (08034), Spain.

Roy G. Cantvell
Monsanto, 1 800 N. Lindbergh Blvd, St Louis, MO 63167, USA.

Shou-Yi Chen
Plant Gene Research Center, National Key Laboratory of Plant Genomics, Institute of Genetics and Developmental Biology, Chinese Academy of Sciences, Beijing, China.

Helene San Clemente
Cell surfaces and signaling in plants, UMR 5546 CNRS - UPS-Université de Toulouse, Pôle de Biotechnologie Végétale, 24 Chemin de Borde-Rouge, BP 42617 Auzeville, 31326 Castanet-Tolosan, France.

Sudhansu Dash
Virtual Reality Applications Center, Iowa State University, Ames, IA 50010, USA.

Wim Deleu
IRTA, Plant Genetics Department, Centre for Research in Agricultural Genomics CRAG (CSIC-IRTA-UAB), Cabrils (08348), Spain.

Laurent Deslandes
National Institute of Agronomic Research Centre, National Scientific Research, UMR 2594, Castanet-Tolosan, France.

Allen Van Deynze
Seed Biotechnology Center, University of California, 1 Shields Ave, Davis, CA, USA.

Julie A. Dickerson
Virtual Reality Applications Center, Iowa State University, Ames, IA 50010, USA.
Department of Electrical and Computer Engineering, Iowa State University, Ames, IA 50010, USA.

Aaron M. Ellison
Harvard University, Harvard Forest, Petersham, Massachusetts, USA.

Bruno Favery
National Institute of Agronomic Research, UMR 1301, Sophia-Antipolis, France.
National Centre for Scientific Research, UMR 6243, Sophia-Antipolis, France.
University of Nice Sophia-Antipolis, UMR 1301, Sophia-Antipolis, France.

Jordi Garcia-Mas
IRTA, Plant Genetics Department, Centre for Research in Agricultural Genomics CRAG (CSIC-IRTA-UAB), Cabrils (08348), Spain.

Daniel Gonzalez-Ibeas
Department of Biology of Stress and Plant Pathology, Center for Soil Science and Applied Biology of the Segura (CEBAS)-CSIC, PO Box 164, 30100 Espinardo (Murcia), Spain.

Michelle A. Graham
USDA-ARS, Corn Insect and Crop Genetics Research Unit, Iowa State University, Ames, Iowa 50011 USA.
Department of Agronomy, Iowa State University, Ames, Iowa 50011 USA.

David Grant
USDA-ARS, Corn Insect and Crop Genetics Research Unit, Iowa State University, Ames, Iowa 50011 USA.
Department of Agronomy, Iowa State University, Ames, Iowa 50011 USA.

Jane Grimwood
Joint Genome Institute—Stanford Human Genome Center, Department of Genetics, Stanford University, School of Medicine, Palo Alto, CA 94304 USA.

Dianjing Guo
Department of Biology and the State Key Lab for Agrobiotechnology, The Chinese University of Hong Kong, Shatin, Hong Kong SAR, PR China.

Yu-Jun Hao
Plant Gene Research Center, National Key Laboratory of Plant Genomics, Institute of Genetics and Developmental Biology, Chinese Academy of Sciences, Beijing, China.

Corina Hayano-Kanashiro
Center for Research and Advanced Studies, Department of Plant Genetic Engineering, North Bypass Road Irapuato-León, Irapuato, Guanajuato, Mexico.

Luis Herrera-Estrella
Center for Research and Advanced Studies, National Laboratory of Genomics for Biodiversity (LANGEBIO)
North Bypass Road Irapuato-León, Irapuato, Guanajuato, Mexico.

Jian Huang
Plant Gene Research Center, National Key Laboratory of Plant Genomics, Institute of Genetics and Developmental Biology, Chinese Academy of Sciences, Beijing, China.

Enrique Ibarra-Laclette
Center for Research and Advanced Studies, National Laboratory of Genomics for Biodiversity (LANGEBIO)
North Bypass Road Irapuato-León, Irapuato, Guanajuato, Mexico.

Muhammad Irshad
Cell surfaces and signaling in plants, UMR 5546 CNRS - UPS-Université de Toulouse, Pôle de Biotechnologie Végétale, 24 Chemin de Borde-Rouge, BP 42617 Auzeville, 31326 Castanet-Tolosan, France.

Elisabeth Jamet
Cell surfaces and signaling in plants, UMR 5546 CNRS - UPS-Université de Toulouse, Pôle de Biotechnologie Végétale, 24 Chemin de Borde-Rouge, BP 42617 Auzeville, 31326 Castanet-Tolosan, France.

Jim D. Karagatzides
Harvard University, Harvard Forest, Petersham, Massachusetts, USA.

Tatyana Khodus
Institute of Biology II—Cell Biology University of Freiburg, Freiburg, Germany.

Russel J. Kohel
USDA-ARS, Southern Plains Agricultural Research Center, 2881 F&B Road, College Station, TX 77845, USA.

Alexander Kozik
Genome and Biomedical Sciences Facility, University of California, 1 Shields Ave, Davis, CA, USA.

Manuel Le Bris
Mediterranean Institute of Ecology and Paleoecology IMEP, UMR-Centre National de la Recherche Scientifique-Institute of Development Research 6116, Université Paul Cézanne, Marseille, France.

Philippe Lecomte
National Institute of Agronomic Research, UMR 1301, Sophia-Antipolis, France.
National Centre for Scientific Research, UMR 6243, Sophia-Antipolis, France.
University of Nice Sophia-Antipolis, UMR 1301, Sophia-Antipolis, France.

Mike Lee
Seed Biotechnology Center, University of California, 1 Shields Ave, Davis, CA, USA.

Ji-Young Lee
Department of Biology, Duke University, Durham, North Carolina, USA.

Xue-Yi Liu
Institute of Economic Crops, Shanxi Academy of Agricultural Sciences, Fenyang, Shanxi, China.

Karin Ljung
Department of Forest Genetics and Plant Physiology, Umeå Plant Science Centre, Sveriges Lantbruksuniversitet, Umeå, Sweden.

Nuria López-Bigas
Research Unit on Biomedical Informatics (GRIB), Experimental and Health Science Department (Universitat Pompeu Fabra) Barcelona (08080), Spain.

Biao Ma
Plant Gene Research Center, National Key Laboratory of Plant Genomics, Institute of Genetics and Developmental Biology, Chinese Academy of Sciences, Beijing, China.

Linyong Mao
Virtual Reality Applications Center, Iowa State University, Ames, IA 50010, USA.

Nicolas Marfaing
National Institute of Agronomic Research, UMR 1301, Sophia-Antipolis, France.
National Centre for Scientific Research, UMR 6243, Sophia-Antipolis, France.
University of Nice Sophia-Antipolis, UMR 1301, Sophia-Antipolis, France.

Albert Mascarell-Creus
Molecular Genetics Department, Centre for Research in Agricultural Genomics CRAG (CSIC-IRTA-UAB), Barcelona (08034), Spain.

Abdul R. Mohammed
Texas AgriLife Research and Extension Center, 1509 Aggie Drive, Beaumont, Texas-77713, USA.

Santiago Mora-García
Molecular Genetics Department, Centre for Research in Agricultural Genomics CRAG (CSIC-IRTA-UAB), Barcelona (08034), Spain.

Philippe Muller
Institute of Plant Science, UPR2355 CNRS, Université Paris Sud Orsay, Gif sur Yvette, France.

James A. H. Murray
Institute of Biotechnology, University of Cambridge, Cambridge, UK.

Rex T. Nelson
USDA-ARS, Corn Insect and Crop Genetics Research Unit, Iowa State University, Ames, Iowa 50011 USA.

Fernando Nuez
Institute of Conservation and Improvement of Agrobiodiversity Valenciana (COMAV-UPV), CPI, Ed.8E, Camino de Vera s / n, 46022 Valencia, Spain.

Jamie A. O'Rourke
Department of Genetics, Developmental and Cellular Biology, Iowa State University, Ames, Iowa 50011 USA.

Laetitia Paganelli
National Institute of Agronomic Research, UMR 1301, Sophia-Antipolis, France.
National Centre for Scientific Research, UMR 6243, Sophia-Antipolis, France.
University of Nice Sophia-Antipolis, UMR 1301, Sophia-Antipolis, France.

Klaus Palme
Institute of Biology II - Biology University of Freiburg, Freiburg, Germany.

Ivan A. Paponov
Institute of Biology II - Biology University of Freiburg, Freiburg, Germany.

Yann Pecrix
Mediterranean Institute of Ecology and Paleoecology IMEP, UMR-Centre National de la Recherche Scientifique-Institute of Development Research 6116, Université Paul Cézanne, Marseille, France.

Catherine Perrot-Rechenmann
Institute of Plant Science, UPR2355 CNRS, Université Paris Sud Orsay, Gif sur Yvette, France.

Belén Picó-Silvent
Institute of Conservation and Improvement of Agrobiodiversity Valenciana (COMAV-UPV), CPI, Ed.8E, Camino de Vera s / n, 46022 Valencia, Spain.

Rafael Pont-Lezica
Cell surfaces and signaling in plants, UMR 5546 CNRS - UPS - University of Toulouse, Pole Plant Biotechnology, 24 Chemin de Borde-Rouge, BP 42617 Auzeville, 31326 Castanet-Tolosan, France.

Pere Puigdomènech
Molecular Genetics Department, Centre for Research in Agricultural Genomics CRAG (CSIC-IRTA-UAB), Barcelona (08034), Spain.

Yan Qi
Department of Biology and the State Key Lab for Agrobiotechnology, The Chinese University of Hong Kong, Shatin, Hong Kong SAR, PR China.

Michaël Quentin
National Institute of Agronomic Research, UMR 1301, Sophia-Antipolis, France.
National Centre for Scientific Research, UMR 6243, Sophia-Antipolis, France.
University of Nice Sophia-Antipolis, UMR 1301, Sophia-Antipolis, France.

Jean-Pierre Renou
Cell surfaces and signaling in plants, UMR 5546 CNRS - UPS - University of Toulouse, Pole Plant Biotechnology, 24 Chemin de Borde-Rouge, BP 42617 Auzeville, 31326 Castanet-Tolosan, France.

Cristina Roig
Institute of Conservation and Improvement of Agrobiodiversity Valenciana (COMAV-UPV), CPI, Ed.8E, Camino de Vera s / n, 46022 Valencia, Spain.

David Roujol
Cell surfaces and signaling in plants, UMR 5546 CNRS - UPS - University of Toulouse, Pole Plant Biotechnology, 24 Chemin de Borde-Rouge, BP 42617 Auzeville, 31326 Castanet-Tolosan, France.

Montserrat Saladié
IRTA, Plant Genetics Department, Centre for Research in Agricultural Genomics CRAG (CSIC-IRTA-UAB), Cabrils (08348), Spain.

Ben Scheres
Department of Molecular Cell Biology, Utrecht University, Utrecht, The Netherlands.

Jeremy Schmutz
Joint Genome Institute—Stanford Human Genome Center, Department of Genetics, Stanford University School of Medicine, Palo Alto, CA 94304 USA.

Randy C. Shoemaker
USDA-ARS, Corn Insect and Crop Genetics Research Unit, Iowa State University, Ames, Iowa 50011 USA.
Department of Agronomy, Iowa State University, Ames, Iowa 50011 USA.

June Simpson
Center for Research and Advanced Studies, Department of Plant Genetic Engineering, North Bypass Road Irapuato-León, Irapuato, Guanajuato, Mexico.

Ludivine Soubigou-Taconnat
Cell surfaces and signaling in plants, UMR 5546 CNRS - UPS - University of Toulouse, Pole Plant Biotechnology, 24 Chemin de Borde-Rouge, BP 42617 Auzeville, 31326 Castanet-Tolosan, France.

David M. Stelly
Department of Soil and Crop Sciences, Texas A and M University, College Station, TX 77843, USA.

Kevin Stoffel
Seed Biotechnology Center, University of California, 1 Shields Ave, Davis, CA, USA.

Lee Tarpley
Texas AgriLife Research and Extension Center, 1509 Aggie Drive, Beaumont, Texas-77713, USA.

Alexandre Tromas
Institute of Plant Science, CNRS UPR2355, Université Paris Sud Orsay, Gif sur Yvette, France.

Estela M. Valle
Institute of Molecular and Cellular Biology of Rosario (IBR-CONICET), Faculty of Biochemical and Pharmaceutical Sciences, Universidad Nacional de Rosario, Suipacha 531, Rosario S2002LRK, Argentina.

John L. Van Hemert
Program of Bioinformatics and Computational Biology, Iowa State University, Ames, IA 50010, USA.

Carroll P. Vance
USDA-ARS, Plant Science Research Unit, University of Minnesota, St. Paul, MN 55108 USA.

Josep Vilarrasa-Blasi
Molecular Genetics Department, Centre for Research in Agricultural Genomics CRAG (CSIC-IRTA-UAB), Barcelona (08034), Spain.

Hui-Wen Wang
Plant Gene Research Center, National Key Laboratory of Plant Genomics, Institute of Genetics and Developmental Biology, Chinese Academy of Sciences, Beijing, China.

Wei Wang
Department of Biology and the State Key Lab for Agrobiotechnology, The Chinese University of Hong Kong, Shatin, Hong Kong SAR, PR China.

Yejun Wang
Department of Biology and the State Key Lab for Agrobiotechnology, The Chinese University of Hong Kong, Shatin, Hong Kong SAR, PR China.

Wei Wei
Plant Gene Research Center, National Key Laboratory of Plant Genomics, Institute of Genetics and Developmental Biology, Chinese Academy of Sciences, Beijing, China.

Thea A. Wilkins
Department of Plant and Soil Science, Texas Tech University, Experimental Sciences Building, Room 215, Mail Stop 3132, Lubbock, TX 79409-3132, USA.

John Z. Yu
USDA-ARS, Southern Plains Agricultural Research Center, 2881 F&B Road, College Station, TX 77845, USA.

Jing-Yun Zhao
Institute of Economic Crops, Shanxi Academy of Agricultural Sciences, Fenyang, Shanxi, China.

Jin-Song Zhang
Plant Gene Research Center, National Key Laboratory of Plant Genomics, Institute of Genetics and Developmental Biology, Chinese Academy of Sciences, Beijing, China.

Qing Zhang
Department of Biology and the State Key Lab for Agrobiotechnology, The Chinese University of Hong Kong, Shatin, Hong Kong SAR, PR China.

Wan-Ke Zhang
Plant Gene Research Center, National Key Laboratory of Plant Genomics, Institute of Genetics and Developmental Biology, Chinese Academy of Sciences, Beijing, China.

Hong-Feng Zou
Plant Gene Research Center, National Key Laboratory of Plant Genomics, Institute of Genetics and Developmental Biology, Chinese Academy of Sciences, Beijing, China.

List of Abbreviations

2CPB	2-Cys peroxiredoxin B
ABA	Abscisic acid
ABP1	Auxin binding protein 1
AFLPs	Amplified fragment length polymorphisms
AGCN	*Arabidopsis* gene co-expression network
AGPs	Arabinogalactan proteins
APC/C	Anaphase-promoting complex/cyclosome
ARF	Auxin response factor
BiFC	Bimolecular fluorescence complementation
BUB	Budding uninhibited by benzymidazol
CAF-1	Chromatin assembly factor-1
CaMV	Cauliflower mosaic virus
CC	Cajete Criollo
CDK	Cyclin-dependent protein kinases
CDPKs	Calcium dependent protein kinases
CEs	Carbohydrate esterases
CESAs	Cellulose synthases
CFUs	Colony-forming units
Ck	Clustering coefficient
CMD	Cotton Microsatellite Database
CMV	Cucumber mosaic virus
COB	Chip on board
COS	Conserved orthologous set
CSLs	Cellulose synthases-like
Ct	Cycle thresholds
CWGs	Cell wall genes
DAB	3'3'-diaminobenzidine
DAC	Data acquisition and control
DAE	Days after emergence
DART	Data analysis for real-time
DCS	Distributed control system
DHCP	Dynamic Host Configuration Protocol
DIN	Deutsche industrial norm
dpg	Days post-germination
DRE	Dehydration response elements

DUF	Domains of unknown function
ERF	Ethylene response factor
ESTs	Expressed sequence tags
F6PK	Fructose 6 phosphate kinase
FATI	Free air temperature increase
FBPase	Fructose-1,6-bisphosphatase
FC	Fold-change
FDR	False Discovery Rate
FLAs	Fasciclin-like arabinogalactan proteins
FWER	Family-wise error rate
G3PD	Gyceraldehydes 3 phosphate dehydrogenase
GASAs	Gibberellic acid-stimulated *Arabidopsis* proteins
GGM	Graphical gaussian model
GHs	Glycoside hydrolases
GO	Gene ontology
GRP	Glycine-rich protein
GSP	Genome-specific primers
GSTs	Gene-specific tag
GTs	Glandular trichomes
GUS	β-glucuronidase
HATs	Histone acetyltransferases
HCL	Hierarchical clustering method
HMI	Human machine interface
HSP	Heat shock protein
HSV	Herpes simplex virus
HXK	Hexokinases
IAA	Indole 3 acetic acid
IDC	Iron deficiency chlorosis
IN	Inorganic nitrogen
IP	Internet Protocol
IPP	Isopentenyl diphosphate
IR	Infrared
ISS	Inferred from sequence similarity
kMT	Kinetochore-microtubule
KRPs	Kinase inhibitors
LAN	Local area network
LEA	Late embryogenesis abundant protein
LEAs	Late abundant embryogenesis
M21	Michoacán 21

MAD	Mitotic arrest-deficient
MAS	Marker-assisted selection
MBEI	Model based expression index
MCC	Mitotic checkpoint complex
MCL	Markov clustering
MEP	2C-methyl-D-erythritol 4-phosphate
MIPS	Munich Information Center for Protein Sequences
MOA	Maize oligonucleotide array
MTs	Microtubules
MU	4-methylumbelliferone
MV	Methyl viologen
MVA	Mevalonate
N	Nitrogen
NGTs	Non-glandular trichomes
NIPs	Nodulin-like intrinsic proteins
NPA	1-N-naphthylphthalamic acid
NR	Non-redundant
NTCs	No template controls
ON	Organic nitrogen
ORC	Origin recognition complex
OTC	Open-top chambers
P5C5	Pyrroline-5-carboxylate synthetase
P5CDH	Pyrroline-5-carboxylate dehydrogenase
PCA	Principal component analysis
Pcc	Pearson correlation coefficient
PEG	Polyethylene glycol
PID	Proportional integral derivative
PK	Pyruvate kinase
PLs	Polysaccharide lyases
PMEIs	Pectin methylesterase inhibitors
PMEs	Pectin methylesterases
POD	Peroxidase
Ppdb	Plant Promoter Database
PSI	Photosystem I
PSII	Photosystem II
PSKs	Phytosulfokines
PUF	Protein of unknown function
QLT	Quantitative trait locus
qRT-PCR	Quantitative real-time polymerase chain reaction

QTL	Quantitative trait loci
RBR	Retinoblastoma-related protein
REGs	Regulatory element groups
RFLPs	Restriction fragment length polymorphisms
RIL	Recombinant inbred line
RMA	Robust multichip analysis
RNR	Ribonucleotide reductase
ROS	Reactive oxygen species
RTD	Resistance temperature detector
RT-PCR	Reverse transcription-polymerase chain reaction
SAC	Spindle assembly checkpoint
SAMs	Significance analysis of microarrays
SCADA	Supervisory control and data acquisition
SCR	Silicon controlled rectifier
SDs	Standard deviations
SFPs	Single feature polymorphisms
SHR	Short-root
SMT	Surface mount technology
SNPs	Single nucleotide polymorphisms
SPAR	Soil-plant-atmosphere-research
SPGs	Secretory pathway genes
sqRT-PCR	Semi quantitative real time reverse transcriptase polymerase chain reaction
SSRs	Simple sequence repeats
ST	Support trees
STS	Sequence-tagged site
TF	Transcription factor
TFBS	Transcription factor binding sites
TIPs	Tonoplast intrinsic proteins
TIR1	Transport inhibitor response 1
TPI	Triosephosphate isomerase
TSS	Transcription start site
UPF	Uncharacterized protein families
VPD	Vapor pressure deficit
XTHs	Xyloglucan endotransglucosylase/hydrolases

Contents

Chapter 1

Sampling Nucleotide Diversity in Cotton

Allen Van Deynze, Kevin Stoffel, Mike Lee, Thea A. Wilkins,
Alexander Kozik, Roy G. Cantrell, John Z. Yu, Russel J. Kohel,
and David M. Stelly

INTRODUCTION

Cultivated cotton is an annual fiber crop derived mainly from two perennial species, *Gossypium hirsutum* L. or upland cotton, and *G. barbadense* L., extra long-staple fiber Pima or Egyptian cotton. These two cultivated species are among five allotetraploid species presumably derived monophyletically between *G. arboreum* and *G. raimondii*. Genomic-based approaches have been hindered by the limited variation within species. Yet, population-based methods are being used for genome-wide introgression of novel alleles from *G. mustelinum* and *G. tomentosum* into *G. hirsutum* using combinations of backcrossing, selfing, and inter-mating. Recombinant inbred line (RIL) populations between genetics standards TM-1, (*G. hirsutum*) × 3–79 (*G. barbadense*) have been developed to allow high-density genetic mapping of traits.

This chapter describes a strategy to efficiently characterize genomic variation (SNPs and indels) within and among cotton species. Over 1,000 SNPs from 270 loci and 279 indels from 92 loci segregating in *G. hirsutum* and *G. barbadense* were genotyped across a standard panel of 24 lines, 16 of which are elite cotton breeding lines and eight mapping parents of populations from six cotton species. Over 200 loci were genetically mapped in a core mapping population derived from TM-1 and 3-79 and in *G. hirsutum* breeding germplasm.

In this research, SNP and indel diversity is characterized for 270 single-copy polymorphic loci in cotton. A strategy for SNP discovery is defined to pre-screen loci for copy number and polymorphism. Our data indicate that the A and D genomes in both diploid and tetraploid cotton remain distinct from each such that paralogs can be distinguished. This research provides mapped DNA markers for intra-specific crosses and introgression of exotic germplasm in cotton.

The cotton family consists of 45 diploid species (2n = 2x = 26) representing eight genome groups (A, B, C, D, E, F, G, K) and five AD allotetraploid species (2n = 4x = 52) that are inter-crossable to various degrees. Cultivated cotton is an annual fiber crop derived mainly from two perennial species, *Gossypium hirsutum* L. or upland cotton, and *G. barbadense* L., extra long-staple fiber Pima or Egyptian cotton. These two cultivated species are among five allotetraploid species presumably derived monophyletically from a single polyploidization event that occurred 1-2 MYA between ancestors most closely represented today by *G. arboreum* (A2 genome) and *G. raimondii* (D5 genome). Breeding of cotton is primarily focused on intra-specific crosses to introduce transgenic traits and to improve baseline lint yield and quality (Bowman, 2000), although significant advances can be made through inter-specific introgression (He et al., 2007; Robinson et al., 2007; Zhang et al., 2003). The relatively recent speciation of tetraploid cotton affords opportunities to transfer novel traits between species, but also amplifies the challenge of maintaining the high yields and quality requisite of commercial products. *Gossypium tomentosum* (*AD3*) and *G. mustelinum* (*AD4*) are rich sources of novel traits that are currently being mined to improve cotton agronomy and fiber. The DNA markers can provide means of detecting, manipulating, and identifying genes associated with desirable agronomic and quality traits within breeding programs, as well as novel alleles from wide crosses. They have shown to be useful in accelerating the transfer of novel traits into elite backgrounds, particularly when these markers have been placed on genetic maps (Frary et al., 2004).

The most extensive genetic maps in cotton have been derived mainly from wide crosses between the two AD-genome species *G. hirsutum* and *G. barbadense* (Lacape et al., 2003; Reinisch et al., 1994; Rong et al., 2004). Inter-specific maps also exist between *G. hirsutum* and *G. tomentosum* (Waghmare et al., 2005). The DNA markers have been oriented to chromosomes and used to establish co-linearity among genomes and species using radiation hybrids and hypo-aneuploid F1 hybrids, available for most chromosomes. In total, approximately 5,000 DNA markers have been mapped. These were derived from approximately 3,300 restriction fragment length polymorphisms (RFLPs), 700 amplified fragment length polymorphisms (AFLPs), 1,000 simple sequence repeats (SSRs), and 100 single nucleotide polymorphisms (SNPs) (Chen et al., 2007). Furthermore, 2,584 sequence-tagged site (STS) loci are mapped in an AD genome and 1,014 in and the D genome (Rong et al., 2004, 2005). An EST-SSR map with 1,017 loci is also available (Guo et al., 2007). Only a few low resolution intra-specific maps that focus on specific traits exist due to the low level of polymorphism within a species and the paucity of cost-efficient markers available to be used in breeding programs.

The vast majority of markers in cotton that are useful in breeding are as SSRs. Over 8,000 pairs of SSR primers are identified in cotton from *G. arboreum*, *G. raimondii*, and *G. hirsutum* (Blenda et al., 2006). The frequency of polymorphism within species has been reported to be limited to 11% (Frelichowski et al., 2006). Informative, abundant, high-throughput markers associated with genes such as SNPs or insertion/deletions (indels) are desirable for both breeding and genetic analyses. Expressed genes are available as templates to study variation. In, *G. arboreum*, 24,597 non-redundant transcripts are available; with 27,355 in *G. raimondii*; and 63,138 in *G. hirsutum* (Volker et al.). The goal of the current project was to design a strategy to efficiently identify and characterize SNP markers that are useful to manipulate and transfer novel alleles to breeding germplasm in cotton. Different DNA templates were evaluated for their specificity to amplify single-copy loci, and polymorphism within and among species, with emphasis on cultivated cotton. We show that single-copy loci can be efficiently amplified in cotton despite redundancy conferred by its allopolyploid origin, and that they can be mapped to specific genomes. The information can be queried in the Cotton Marker Database which has been modified for presentation of SNP data (Blenda et al., 2005).

MATERIALS AND METHODS

Plant Materials

The 24 cotton lines and species screened for this study were chosen based on an expansion of the Cotton Marker Database (Table 2, (Blenda et al., 2006)) standard germplasm panel assembled to represent the breadth of US cotton breeding germplasm and genetic standards. Consequently, the specific polymorphisms are expected to be relevant to these applications. The Cotton Microsatellite Database (CMD) panel has been genotyped with thousands of SSR markers and serves as a resource to assess the utility of genetic markers in cotton. To address the low polymorphism in cotton breeding germplasm (11%) (Frelichowski et al., 2006) the CMD panel was expanded to include 12 additional elite breeding lines (Table 1). The panel contains crossing parents from *G. tomentosum* and *G. mustelinum*, and representation of the diploid species, *G. arboreum* and *G. raimondii*. As our main goal is to develop markers relevant to breeding germplasm 16 *G. hirsutum* and three *G. barbadense* (Pima) lines were selected. An additional landrace, TX2094, was added to represent *G. hirsutum* race yucatanense (Liu et al., (2000b)). All accessions represent self-pollinating or inbred lines, thus should be homozygous for the majority of the loci.

Primer Design and Screening

To determine the optimum target template to identify SNPs specific to single-copy sequences in tetraploid cotton, we tested primer pairs from available BAC-end sequences; sequences flanking SSRs or predicted introns in ESTs; and sequences in the 3' UTR or 5' terminus of ESTs (Table 1). All sequences were downloaded from GenBank (Sept, 2004) except for SSR sequences (kindly provided by Dr. Ben Burr, Brookhaven National Laboratories, NY). Based on the results, a database of primer sets with predicted product sizes of 600–800 bp was generated using Primer 3 (Rozen and Skaletsky, 2000) to design primer pairs in the 3' or 5' termini of unigenes from *G. arboreum* ESTs (Arpat et al., 2004). As more ESTs became available, *G. arboreum* and *G. raimondii* ESTs were trimmed and assembled into 7,666 contigs with SNPs and deletions between these genomes. To evaluate the usefulness of cotton genome-specific primers (GSP), 48 primer pairs were tested with a deletion in the forward primer and 48 with a SNP at the 3' end of the forward primer. A Conserved Orthologous Set (COS) for cotton of 2,390 contigs based on 27,878 *G. arboreum*, 35,509 *G. raimondii*, and 14,354 *G. hirsutum* ESTs using the procedures described in Van Deynze et al. (Van Becelaere et al., 2007; Van Deynze et al., 2007) was also generated as template for SNP discovery. A set of 576 primers (labeled with prefix of COT) was designed to amplify across predicted intron sites (based on *Arabidopsis*) with primers positioned 50–100 bp from the predicted introns.

Two near-homozygous lines (*G. arboreum* and *G. hirsutum* breeding lines) were amplified and tested on agarose gels for amplification (Table 5). The amplified products were subsequently run on SSCP gels to screen for the presence of duplicated loci in amplification products. The SSCP is a highly sensitive technique that detects variation in the nucleotide sequences of single-stranded molecules. Single-copy loci display two bands (the sense and antisense DNA strands). Loci displaying greater than two bands were not sequenced and were assumed to be not from a single locus. Primer pairs that successfully amplified a product and showed SSCP patterns of single-copy loci were tested for polymorphism using sequencing in a series of three pools representing different degrees of diversity in breeding germplasm: within elite *G. hirsutum*, a genetically diverse *G. hirsutum*, and between *G. hirsutum* and *G. barbadense* (Van Becelaere et al., 2007). Each pool had three similar lines and one complementary genetically-distant line to maximize the chance of detecting a polymorphism within or among pools. Using a series of empirical tests with lines with known SNPs in ratios of 1:7, 1:5, 1:3, and 1:1, we determined that an unknown polymorphism can be reliably detected with sequencing with a 1:3 dilution. Pool 1 consisted of DPL458BR, Fibermax 832, Stoneville 4892BR, Sealand 5 42; Pool 2 consisted of PD1, Maxxa, Tamcot Sphinx, TX2094; and Pool 3 consisted of TM-1, 3-79, Pima S-7, Maxxa (Table 2). These pools represent increasing diversity in breeding germplasm. DNA was extracted from each line using Qiagen DNEASY (Qiagen, Valencia, USA) and was combined in equi-molar concentrations.

For all sequencing reactions, forward and reverse primers were tailed with M13 sequences and sequenced using standard protocols for Sanger sequencing (Applied Biosystems, Foster City, CA) in forward and reverse directions using a ABI 3730 (Applied Biosystems, Foster City, CA). Trace files were trimmed with Phred options–"trim_cutoff 0.02" which translates to Phred 17 score (Ewing and Green, 1998). Assembly was achieved with Phrap/Consed and options were set at "-retainduplicates and -forcelevel 5." These options were optimized to give the best trim and assembly parameters for calling SNPs. Stringent trim parameters are favored in this case to minimize the high number of false SNPs associated with poor sequence on the ends.

The SNPs were first identified semi-manually using Polyphred as heterozygotes within pools or homozygous differences among pools. The line, HS200 (*G. hirsutum*), was used as reference to confirm that amplicons of single-copy loci were represented. Amplicons with putative SNPs were then amplified in the individual 24 lines (Table 2) and sequenced as described above. Data was extracted from Polyphred using custom scripts (Tearse). Similarly, data for indels were extracted from Polyphred. Total polymorphism was calculated among genotypes and species as total number of SNPs, bases per SNP (SNP frequency), and percent polymorphic loci. Sequences are available through the Cotton Marker database (Blenda et al., 2005).

A set of 384 SNPs was selected to develop an Illumina Golden Gate® oligonucleotide pooled assay. In order of priority, SNPs were selected to maximize the number of loci represented that were polymorphic between TM-1 and 3-79, or were polymorphic within *G. hirsutum* with moderate minor allele frequencies (>15%) in *G. hirsutum* germplasm sequenced (Table 2). The SNPs were genotyped in 186 RILs, the parents, and the F_1 as per manufacturer protocols at the University of California, Genome and Biosciences Facility, Davis, CA. Data were extracted and exported and mapped using JoinMap 4.0 (Van Ooijen, 2006) with the Kosambi mapping function (Kosambi, 1944). A LOD score of at least 6.0 was used to determine the linkage groups, of which the marker orders were verified at LOD score 3.0. Individual linkage groups were assigned to respective chromosomes by use of the TM-1 × 3–79 base map (Yu et al., in preparation).

DISCUSSION AND RESULTS

Defining Optimum Regions to Sample for SNPs in Cotton

Our goal was to define and optimize a SNP discovery strategy in cotton that exploits current genomic resources. The disomic polyploid nature of cotton poses a particular challenge in that most loci are duplicated, and breeding germplasm is derived from a relatively narrow genetic base (Rungis et al., 2005; Van Becelaere et al., 2005). A re-sequencing strategy developed by our lab was modified to anchor and screen DNA primer pairs that amplify single-copy, polymorphic regions of the genome relevant to current breeding germplasm that can also be used for introgression of novel alleles from exotic germplasm (See Materials and Methods and (Van Becelaere et al., 2007)). To address this, we empirically evaluated the proportion of single-copy sequences obtained from amplicons originating from different genomic regions, namely: BAC-end sequences, sequences flanking SSRs or predicted introns in ESTs, and sequences in the 3' or 5' untranslated region of ESTs (Table 1). Although the majority (93%, data not shown) of amplifiable primer sets showed single bands on agarose gels, SSCP analysis revealed that only 51% and 40% of the primer pairs amplified single-copy sequences in *G. arboreum* and *G. hirsutum*, respectively. With SSCP analysis of single varieties, an amplicon from a single-copy locus in homozygous state, whether from a diploid or tetraploid is expected to display only two bands, one from sense and one anti-sense strands. As the lines being assayed were near-homozygous, only loci with two SSCP bands were carried forward.

The amplicon templates exhibited a range of amplification, copy number, and polymorphism (Table 1). The low rate of amplification (29%) achieved when primers were designed from BAC-end genomic sequence deterred us from using this as template for primer design. The relatively poor rate of success might have resulted from inadequate quality of BAC-end genomic sequence available at the time (Sept, 2004). Cotton GSP amplicons, those derived from primers spanning a SNP between *G. arboreum* and *G. raimondii*, resulted in the highest percentage (48%) of single-copy sequences (Table 1). Primer pairs neighboring genomic SSRs also resulted in a high percentage (46%) of single-copy sequences, but the relatively small amount of sequence flanking each SSR, 150–200 bp, limited the number of nucleotides available within an amplicon for SNP discovery. Amplicons that encompass introns or are derived from a COS of sequences that are single copy in *Arabidopsis* and cotton ESTs showed excellent amplification. However, as their primers are anchored in coding regions, they are likely to be the most conserved across paralogs, as indicated by the relatively low proportion (38% for introns and 45% for COS primers) of single-copy sequences. The 5' UTR of ESTs are the most abundant in the EST dataset and resulted in the highest proportion of amplicons with single-copy loci overall (42%), next to the GSPs that became available only late in the project. It is important to note that the ESTs used were created by capturing sequences from the 3' UTR and sequencing from either the 3' or 5' end. Consequently, the 5' sequences may not necessarily represent the 5' terminus of genes if clones were not full length. As a comparison, Chee et al. (2004) used sequences from *G. arboreum* to amplify *G. hirsutum* (Chee et al., 2004). The authors reported that 33% (16/89) of primer pairs yielded amplicons from single-copy loci, which is similar to the current

Table 1. Summary statistics for SNPs and indels among 24 lines of cotton derived from different DNA templates.

DNA Template	Primer pairs designed	Amplified Primer pairs (%)	Single-copy loci (%)[1]	Polymorphic SNP loci (%)[2]	SNPs	SNPs/locus[1]	bp/SNP[2,3]	Polymorphic indel loci (%)[2]	Indels	Indels/Locus	bp/indel[2,3]
EST 3' end	160	102 (64)	45 (28,44)	30 (67)	144	3.2	252	14 (31)	49	1.1	741
EST 5' end	988	802 (73)	417 (42,52)	142 (34)	417	1.0	807	37 (9)	109	0.3	3,087
COS	576	523 (91)	201 (35,38)	56 (28)	296	1.5	548	20 (10)	70	0.3	2,317
Intron	126	98 (78)	44 (35,45)	17 (39)	46	1.0	772	11 (25)	32	0.7	1,110
GSP	48	29 (60)	23 (48,79)	11 (48)	40	1.7	464	5 (22)	12	0.5	1,547
SSR	52	46 (88)	21 (40,46)	14 (67)	62	3.0	273	5 (24)	7	0.3	2,421
BAC	24	7 (29)	0 (0,0)	n/a	n/a	n/a	n/a	n/a	n/a	n/a	n/a
Total	1,974	1,607 (81)	751 (47)	270 (36)	1,005	1.3	603	92 (12)	279	0.4	2,172

[1]First nuber is percent of designed; second is percent of amplified.
[2]Relative to single-copy loci.
[3]Adjusted to average contig length of 807 bp.

results (average 38% across EST-derived amplicons, Table 1). This indicates the high-level of homology of exons between homoeologous genomes in tetraploid cotton. The results of this study indicate that although genomic regions harbor genetic diversity, strategies must be developed to ensure that allelic diversity and not diversity between homoeologs and paralogs are being assayed.

Diversity of Genomic Regions

To study the diversity of cotton, we tested different DNA templates to select the optimum regions that will yield single-copy, yet polymorphic amplicons from PCR. Of the 1,974 primer pairs designed, 8% were from the 3' end; 50% 5' end; 29% COS; 6% intron; 3% SSR; 2% GSPs; and 1% genomic (BAC-end) primers. Eighty-one percent successfully amplified DNA in a single *G. arboreum* and *G. hirsutum* line and 47% produced single-copy amplicons based on SSCP gels (Table 2). The relatively high amplification rate across species from primers designed mainly from diploid species confirms the transferability of markers across species as indicated in previous studies (An et al., 2007,2008; Park et al., 2005; Wang et al., 2006; Zhang et al., 2007). To pre-screen primer pairs for polymorphism, DNA pools representing increasing diversity within *G. hirsutum* and *G. barbadense* germplasm and among these two species were amplified and sequenced. Pool 1 represents *G. hirsutum*; Pool 2, *G. hirsutum*/*G. hirsutum* race yucatenense; and Pool 3, *G. hirsutum* and *G. barbadense* (Table 2). Polymorphism within pools was identified as heterozygotes, whereas polymorphism among pools was identified as differences in homozygous alleles. Polymorphic loci were then sequenced individually in the forward and reverse directions in 24 lines (Table 2). Forty-nine percent of primers with single-copy loci were polymorphic within or among pools. Overall, the pools represent the two cultivated species in the United States, *G. hirsutum* and *G. barbadense*, which explains the high polymorphism (percent polymorphic loci) and SNP frequency (bp/SNP) compared to within species polymorphism (Tables 2, 3, and 4). The SSCP pre-screening and pooling strategy saved 77% of the resources compared to direct sequencing of individuals with 1,607 amplifiable primers (Table 5).

Only amplicons showing polymorphism in pools were sequenced in individual lines. The germplasm panel included 16 *G. hirsutum* and three *G. barbadense* genotypes, plus tetraploids *G. mustelinum* and *G. tomentosum* and diploids *G. arboreum* and *G. raimondii* (Table 2). Sequencing of DNA templates resulted in different frequencies of SNPs and indels (Table 1). Sequences from 3' UTRs (252 bp/SNP), and SSR-associated (273 bp/SNP) sequences yielded 2–3 fold the frequency of SNPs than 5' ends, COS, GSP and introns (464–807 bp/SNP). It is important to note that COS and GSP sequences contained introns. Similarly the frequency of indels was least in 5' end sequences and greatest in 3' UTRs (Table 1). These results are consistent with those in maize showing that 3' UTRs are a rich source of nucleotide diversity (Ching et al., 2002; Eveland et al., 2008). The lowest diversity in the 5' ends is consistent with representation of more highly conserved coding sequences (Van Becelaere et al., 2007). In comparison, the vacuolar H+-ATPase subunit (Wilkins et al., 1994) and Myb transcription factor families (Hsu et al., 2008) were examined for diversity in diploid and tetraploid cotton species. The, 3' UTRs were 10-fold more polymorphic than coding sequences for Myb transcription factors (Hsu et al., 2008).

Table 2. Germplasm panel sequenced for SNP or indel discovery.

Line[1]	Genome	Description	CMD Panel	Pool[2]
Acala Maxxa	[AD]₁	California Upland cotton and BAC donor	Yes	Pool 2, 3
AHA 6-1-4	[AD]₁	Upland cotton		
Auburn 623RNR	[AD]₁	Upland cotton		
Coker 312	[AD]₁	Upland cotton		
Deltatype Webber	[AD]₁	Upland cotton		
DPL 458BR	[AD]₁	Upland cotton	Yes	Pool 1
Fibermax 832	[AD]₁	Upland cotton	Yes	Pool 1
Paymaster 1218BR	[AD]₁	Upland cotton	Yes	
PD-1	[AD]₁	Upland cotton		Pool 2
Sealand 542	[AD]₁	Upland cotton		Pool 1
Stoneville 20	[AD]₁	Upland cotton		
Stoneville 4892BR	[AD]₁	Upland cotton	Yes	Pool 1
Tamcot Sphinx	[AD]₁	Upland cotton		Pool 2
Tidewater Seabrooks	[AD]₁	Upland cotton		
TM-1	[AD]₁	Genetic standard (BAC donor/RI parent)	Yes	Pool 3
Wilt Acala 1517	[AD]₁	California Upland cotton		
TX 2094	[AD]₁	*G. hirsutum* race *yucatanense*		Pool 2
3-79	[AD]₂	Genetic standard (fiber QTLs/RI parent)	Yes	Pool 3
Pima S-6	[AD]₂	Pima germplasm breeding source	Yes	
Pima S-7	[AD]₂	Pima germplasm breeding source		Pool 3
G. tomentosum	[AD]₃	Introgression breeding source	Yes	
G. mustelinum	[AD]₄	Introgression breeding source	Yes	
G. arboreum	A₂₋₈	A-genome species representative	Yes	
G. raimondii	D₅₋₃	D-genome species representative	Yes	

[1]All 12 members of the Cotton Marker Database Panel were included [13].
[2]Pools of 4 lines representing cotton germplasm were used to screen primer sets (see Materials and Methods).

Table 3. Species-specific statistics for Gossypium SNPs and indels.

Species	N	Loci with SNPs	Number of SNPs	Loci with SNPs (%)[1]	SNPs. Locus[1]	bp/ SNP[1,2]	Loci with Indels	Number of Indels	Loci with Indels (%)[1]	Indels/ Locus[1]	Bases/ Indel[1,2]
G. arboreum	1	149	238	19.8	0.3	2,546	34	67	2.1	0.1	9,046
G. raimondii	1	142	379	18.9	0.5	1,599	29	66	1.8	0.1	9,183
G. hirsutum[3]	16	124	245	16.5	0.3	2,474	70	161	4.4	0.2	3,764
G. barbadense	3	208	439	27.7	0.6	1,381	48	117	3.0	0.2	5,180
G. mustelinum	1	182	432	24.2	0.6	1,403	42	94	2.6	0.1	6,447
G. tomentosum	1	156	382	20.8	0.5	1,587	34	84	2.1	0.1	7,215
Total	16	270	1,005	36.0	1.3	603	92	279	5.8	0.4	2,172

[1]Relative to 751 single-copy loci.
[2]Adjusted to average contig length of 807 bp.
[3]Excluding *G. hirsutum* race *yucatenense*.

Table 4. Summary statistics for SNPs and indels between pairs of *Gossypium* species. The percentage is calculated out of total of 1005 SNPs and 279 indels.

Number of SNPs	Gr[1]	Gh	Gb	Gm	Gt	Number of Indels	Gr	Gh	Gb	Gm	Gt
G. arboreum	317	303	201	334	261	G. arboreum	111	140	122	93	94
G. raimondii		452	430	266	317	G. raimondii		169	153	84	115
G. hirsutum			396	427	451	G. irsutum			140	92	133
G. barbadense				396	448	G. barbadense				91	113
G. mustelinum					410	G. mustelinum					96
SNP Polymorphism (%)	Gr	Gh	Gb	Gm	Gt	Indel Polymorphism (%)	Gr	Gh	Gb	Gm	Gt
G. arboreum	32	30	20	33	26	G. arboreum	40	50	44	33	34
G. raimondii		45	43	26	32	G. raimondii		61	55	3	41
G. hirsutum			39	42	45	G. hirsutum			50	33	48
G. barbadense				39	45	G. barbadense				33	41
G. mustelinum					41	G. mustelinum					34

[1] Gr = *G. raimondii*, Gh = *G. hirsutum*, Gb = *G. barbadense*, Gm = *G. mustelinum*, Gt = *G. tomentosum*.

Table 5. Summary of results for primer pairs tested in pools.

Primer Results[1]	Number of Primers	Percentage of Preceding Primer Pool	Percentage of all Primers Tested
Tested	1974	-	-
Amplified	1607	81	81
Single-copy	751	47	38
Polymorphic in Pools	365	49	18
Successfully sequenced	351	96	18

[1]Thirty-four single-copy loci were sequenced without pre-screeing in pools.

Alternatively, designing primers from conserved single-copy sequences (COS), or GSPs is an efficient method to access variation in introns for single loci in cotton. A closer examination at the different strategies indicates that GSP-based primers are less likely to amplify informative products than COS (60 vs. 91%), but more likely to target single-locus sequences when they do (79 vs. 45%). The majority of the COS sequences in our dataset were derived from diploid progenitors (see Materials and Methods). A COS-derived from tetraploid cotton only is likely to yield a larger proportion of single-copy loci in tetraploids., cotton intron regions were found to be less conserved than exons in the H+-ATPase subunit family (Wilkins et al., 1994). Intron sequences have shown to be 3.7-fold more polymorphic than exons in cotton (Chee et al., 2004). The COS-based amplicons have also yielded a very high proportion of single-copy sequences (>95%) in Solanaceae (Van Becelaere et al., 2007; Wu et al., 2006).

Another template target is to use predicted SNPs between diploid progenitor species to anchor primers to specific genomes (GSPs). Yang et al. (2006) predicted 32,229 genome-specific SNPs from EST databases with 31% showing perfect concordance to the A or D genomes in the genotypes examined. In the current study, 3,000 SNPs were identified between only the diploid species and not within the tetraploid species (data not shown). As we sampled introns, many of these would be novel to EST-mined SNPs. Only 317 SNPs between *G. arboreum* and *G. raimondii* (Table 4) are polymorphic in *G. hirsutum* and *G. barbadense*. In our study, the estimates of diversity within and among species may not be representative of other species than *G. hirsutum* and *G. barbadense*, as only sequences that were polymorphic within and among these species were characterized. To ensure that single-copy loci were being assayed, only SNPs that had clear homozygotes in at least one tetraploid were called.

Diversity of Cotton Germplasm

Within Species Diversity

The SNPs were recorded as a base substitution relative to the consensus sequence derived from all genotypes. Although the number of individuals sampled varied, the species-specific diversity (as measured by the number of bases per SNP) was similar with two-fold lower frequency of SNPs for *G. arboreum* and *G. hirsutum* than the other species tested (Table 3). *Gossypium barbadense* showed the highest level of diversity of the species tested with 1 SNP per 1,381 bases and 28% of its loci being polymorphic. Conversely, *G. hirsutum* has the highest frequency of indels and the diploid progenitor species the least (Table 3). The average frequency of SNPs among 16 *G. hirsutum* lines was 1/2474 bp (0.04%). In *Adh* genes, within *G. hirsutum* and *G. barbadense* diversity ranged from 1 to 3 SNPs in 983 bases (0.1–0.3%) (Small et al., 1999). With the few reports of SNP frequencies in cotton, within *G. hirsutum* (3–4 lines), SNP frequency ranged from 0/1,000 bp (Hsu et al., 2008; Small et al., 1999), 1/5,000 bp (0.02%) (An et al., 2007). to 1/947 bp (1%) (An et al., 2008) in *Adh*, *Myb*, and *expansin* genes, respectively. Estimates of SNP divergence within and among species in the referenced studies were greater than those reported in our study for over 1,900 loci.

Among Species Diversity

The SNP diversity in cotton has a large range that is both genome and gene specific. Compared to *G. hirsutum*, SNP frequency in *G. arboreum*, *G. barbadense*, *G. mustelinum*, *G. tomentosum*, and *G. raimondii* ranged from 1/2,000 bp (0.05%) to 1/1,341 bp/SNP (0.075%) in order of increasing divergence (Table 4). Estimates among these species ranged from 1/51 bp per SNP (1.96%) in *Myb* and *expansin* genes between *G. raimondii* and *G. hirsutum* (An et al., 2007; Hsu et al., 2008) to 1/714 bp per SNP (0.14%) between *G. hirsutum* and *G. barbadense* in R2R3 Myb transcription factors (An et al., 2008; Senchina et al., 2003). The discrepancy in estimates is likely due to gene-specific estimates, number of genes, germplasm used and the regions of the genes being sampled (coding vs non-coding). Non-coding sequence was reported to be 2–3 fold more polymorphic than coding regions (An et al., 2007, 2008).

Comparisons of tetraploid species with *G. raimondii* (D-genome) were consistently more divergent than those with *G. arboreum* (A-genome) except for *G. mustelinum* (Table 4). This agrees with observations using SSRs. The EST-derived SSRs from diploid species tended to map more often to their orthologous genomes in corresponding tetraploid species (Park et al., 2005), with *G. raimondii*-derived SSRs (43% polymorphism) being more polymorphic than *G. arboreum* derived SSRs (18%) in the same cross (Han et al., 2004; Wang et al., 2006). Furthermore, in comparisons between A-genome diploids, a D-genome diploid, and tetraploids (*G. hirsutum* and *G. barbadense*), the rate of divergence in 48 genes was significantly higher in D-genomes than A-genomes, although the rate of divergence was gene-specific (Senchina et al., 2003). This was verified for specific gene families of transcription factors, Adh and expansins (An et al., 2007,2008; Small et al., 1999) in cotton.

Several methods have been proposed for SNP discovery in allopolyploids and highly duplicated genomes including *in silico* analysis (Udall et al., 2006; Yang et al., 2006), genome specific-PCR (Caldwell et al., 2004), cloning and sequencing (Hsu et al., 2008). We have evaluated genome-specific amplification and high-throughput direct sequencing using M13-tailed amplicons combined with targeted primer design and pre-screening of primer pairs. We show that several options are feasible for high-throughput SNP discovery in cotton, each with their own advantages and disadvantages. The current research agrees with current literature that although tetraploid cottons were derived from a single polyploidization event from their diploid progenitors only 1–2 MYA (Wendel and Cronn, 2003), the genomes remain distinct and have sufficient diversity for breeding.

Linkage Mapping

The SNPs were evaluated as markers by designing a 384-SNP array for the Illumina Golden Gate assay®. Of the 384 SNPs on the array, 268 putative SNPs representing 240 contigs were expected to be polymorphic between the parents of our mapping population, TM-1 (*G. hirsutum*) and 3–79 (*G. barbadense*) and validated by assessing their segregation and amenability to linkage mapping. Of the 268 expected parental SNPs, 247 polymorphisms were detected using the Illumina assays on a population of 186 RILs (Park et al., 2005). Segregation was as expected (1:1) for 188 markers, whereas 59 had skewed segregation resulting in 223 SNPs being placed on the linkage map (Figure 1). Markers that were not mapped had missing data or skewed segregation. The SNP markers are added to the TM-1/3–79 base map (Figure 1, and Yu et al., in preparation) providing new tools for high-throughput genotyping in cotton. The SNP data are thus cross-referenced to several genetic (Frelichowski et al., 2006; Lacape et al., 2003; Rong et al., 2004) and physical maps (Gao et al., 2004; Liu et al., (2000a)) via common SSRs (Yu et al., in preparation). All SNPs, indels and flanking sequences can be accessed through the CMD database(Blenda et al., 2005).

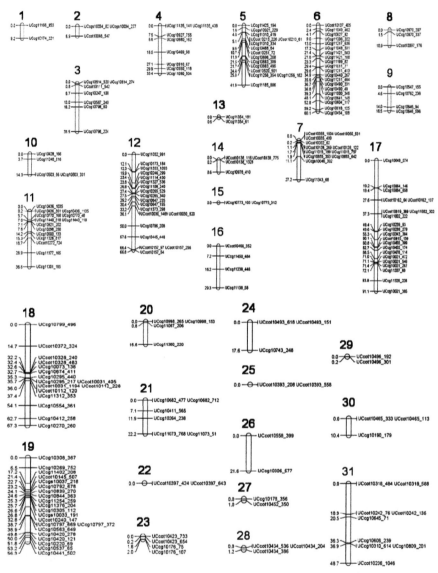

Figure 1. Genetic map of 223 SNP markers in 186 recombinant inbred lines from a cross between TM-1 and 3–79. Genetic distances in cM (Kosambi, 1944).

Although markers were evenly distributed within linkage groups, there was a disproportionate number of markers on linkage groups in the A subgenome (191 markers, 86%) versus the D subgenome (32 markers, 14%; Figure 1, data not shown). At least 70% of the primers including those designed from 3'-end, 5'-end, intron and GSP, were derived exclusively from *G. arboreum* sequences, whereas the remainder were from *G. arboreum*, *G. hirsutum*, and *G. raimondii* assemblies (see Materials and

Methods). The fact that the majority of loci were derived from *G. arboreum* sequences and that these loci were in turn mapped to A-subgenome chromosomes in the TM-1 × 3–79 cross suggests that (a) our approach for identifying single-locus markers was very effective; (b) there is strong sequence conservation between the A2 genome of *G. arboreum*, and the A subgenomes of *G. hirsutum* (AD_1) and *G. barbadense* (AD_2), c) the A and D subgenomes of contemporary AD genomes are distinct from each other; and d) the A subgenomes of *G. hirsutum* and *G. barbadense* are moderately divergent. The above conclusions are emphasized in the present study first by amplifying and sequencing single-copy loci and being able to assay the same sequences using an independent assay, Illumina Golden Gate. Because our template sequences were primarily drawn from *G. arboreum* and selected for both PCR function and single-locus attributes using SSCP, it is quite possible that the populations of selected loci in the two subgenomes would have been differentially affected. The potential for bias across the two subgenomes precludes using these amplicon sequence or genotyping data to infer relative diversity of the A versus D subgenomes of *G. hirsutum* and *G. barbadense*. Our findings are not discordant with the theory that these two extant tetraploids originated from a common allopolyploid ancestor and have evolved independently long enough to create the detected variation and a significant degree of diploidization.

The Use of SNPs in Cotton

The SNP diversity depends on population size sampled and the natural evolutionary and directed selection within those populations. The present study indicates that a moderate amount of variation associated with genes exists in breeding germplasm (1 SNP in 2,474 bp in *G. hirsutum*) at the nucleotide level. *Gossypium barbadense* germplasm sampled was 1.8 times as diverse (1 SNP in 1,381 bp) as *G. hirsutum* even though less than one-fifth as many *G. barbadense* lines were sampled. The A and D genomes of tetraploid cotton are significantly divergent such that individual SNP loci can be assayed with high specificity. Although the above strategy for SNP identification and validation was fruitful, new high-throughput sequencing technologies such as 454 (Roche Biosciences, Branford, USA), Illumina Genome Analyzer (Hayward, USA), and SOLiD (Applied Biosystems, Foster City, USA) offer an opportunity to complement the current strategy to rapidly uncover nucleotide diversity at the whole genome level in multiple breeding lines. The present study demonstrates that sufficient SNP diversity exists in tetraploid cotton populations for genetic and breeding studies and it can be efficiently assayed.

CONCLUSION

In this research, SNP and indel diversity is characterized for 270 single-copy polymorphic loci in cotton. A strategy for SNP discovery is described to pre-screen loci for copy number and polymorphism. Diversity was characterized in a broad set of breeding lines and exotic lines representing a standard germplasm panel indicating that *G. barbadense* is much more diverse than *G. hirsutum*. Our data also indicate that the A and D genomes in both diploid and tetraploid cotton remain distinct from each other such that homoeologs can be distinguished. All marker data and flanking sequences have been submitted to Cotton Marker database http://www.cottonmarker. org/Downloads.shtml.

KEYWORDS

- **Conserved orthologous set**
- **Restriction fragment length polymorphisms**
- **Simple sequence repeats**
- **Single nucleotide polymorphisms**

AUTHORS' CONTRIBUTIONS

Allen Van Deynze conceived, supervised, and wrote the manuscript; Kevin Stoffel carried out the research, contributed to methods and research and edited the chapter; Mike Lee carried out the research, contributed to methods and research; Alexander Kozik helped conceive, analyze and edit the chapter; Thea A Wilkins aided in initiating the research, the EST database and edited the chapter; Roy G. Cantrell initiated the research and developed the germplasm panel; John Z Yu contributed the mapping population, analyzed the mapping data, and edited the chapter; Russel J. Kohel contributed to the mapping population and analysis; David M Stelly conceived the research, developed the germplasm panel, analyzed the data and edited the chapter.

All authors have read and contributed to the writing of the chapter.

ACKNOWLEDGMENTS

The authors would like to thank Mr. Brandon Tearse for converting raw polymorphism data to spreadsheet (Tearse) format and Ms. Jessica Lund for technical assistance for generating the genotyping data. We would also like to thank Dr. Ben Burr (Brookhaven National Laboratory, NY, USA) for providing sequence data for SSRs. The authors would like to thank Cotton Incorporated and the University of California Discovery Grant for funding this research.

Chapter 2

Pitcher Plant *Sarracenia purpurea* and the Inorganic Nitrogen Cycle

Jim D. Karagatzides, Jessica L. Butler, and Aaron M. Ellison

INTRODUCTION

Despite the large stocks of organic nitrogen (ON) in soil, nitrogen availability limits plant growth in many terrestrial ecosystems because most plants take up only inorganic nitrogen (IN), not ON. Although some vascular plants can assimilate ON directly, only recently has ON been found to contribute significantly to the nutrient budget of any plant. Carnivorous plants grow in extremely nutrient-poor environments and carnivory has evolved in these plants as an alternative pathway for obtaining nutrients. We tested if the carnivorous pitcher plant *Sarracenia purpurea* could directly take up intact amino acids in the field and compared uptake of organic and inorganic forms of nitrogen across a gradient of nitrogen deposition. We hypothesized that the contribution of ON to the nitrogen budget of the pitcher plant would decline with increasing nitrogen deposition.

At sites in Canada (low nitrogen deposition) and the United States (high nitrogen deposition), individual pitchers were fed two amino acids, glycine and phenylalanine, and IN (as ammonium nitrate), individually and in mixture. Plants took up intact amino acids. Acquisition of each form of nitrogen provided in isolation exceeded uptake of the same form in mixture. At the high deposition site, uptake of ON was higher than uptake of IN. At the low deposition site, uptake of all three forms of nitrogen was similar. Completeness of the associated detritus-based food web that inhabits pitcher-plant leaves and breaks down captured prey had no effect on nitrogen uptake.

By taking up intact amino acids, *Sarracenia purpurea* can short-circuit the IN cycle, thus minimizing potential bottlenecks in nitrogen availability that result from the plant's reliance for nitrogen mineralization on a seasonally reconstructed food web operating on infrequent and irregular prey capture.

Nitrogen (N) limits plant growth in most terrestrial ecosystems (Vitousek et al., 1997) yet many ecosystems, including arctic tundra (Kielland et al., 1994), coastal salt marshes (Henry and Jefferies, 2002), alpine meadows (Miller and Bowman, 2003), boreal forests (Persson and Nasholm, 2001), and bogs (Bridgham et al., 1996) have large stocks of ON. More than 90% of soil N is bound in an organic form (humus), 20–40% of this as amino acids (Senwo and Tabatabai, 1998). The availability of amino acids may drive ecosystem function in N-limited environments such as arctic tundra because of the very high turnover rates (2–24 hr) of amino acids that result from microbial uptake and release (Jones et al., 2004). Plants that can use amino acids as an N source may be able to co-exist with, or even outgrow, plants that only use IN, especially in environments where N mineralization rates are low and N limits plant growth (McKane et al., 2002; Miller et al., 2007; Wiegelt et al., 2005). In Arctic tundra, for example, the most productive species used the most abundant N forms and less productive species used less abundant forms (McKane et al., 2002).

It has been known for decades that vascular plants can assimilate ON directly when grown in culture (Virtanen and Linkola, 1946), with mycorrhizae (Melin and Nilsson, 1953), or in the absence of microbial competition (Miller and Schmidt, 1965), but only in the last decade has ON been shown to be a significant N source for a wide range of plant species in different N-limited systems (Lipson and Nasholm, 2001). Standard theory of N cycling with respect to plant uptake (Stark, 2000) assumes that ON has to be mineralized to IN before it can be assimilated, but direct ON uptake by plants has been proposed to "short-circuit" the N cycle as plants bypass microbial mineralization of ON (Neff et al., 2003). This short-circuit is thought to be energetically favorable to plants because ON immediately provides amino acids whereas NH_4^+ and NO_3^- (after reduction to NH_4^+) must be synthesized into amino acids (Clarkson, 1985). Because most ecosystems are N limited and plants can potentially access multiple forms of N in the environment, more information is needed on the generality of direct acquisition of amino acids by plants to fully assess current models of N cycling for a wider range of environments.

Carnivorous plants are generally restricted to extremely N-limited habitats, such as bogs, outwash sand plains, and inselbergs (Ellison and Gotelli, 2001; Givnish, 1989; Porembski and Barthlott, 2000). In North America, carnivorous pitcher plants (*Sarracenia* spp. and *Darlingtonia californica* Torrey (both in the Sarraceniaceae)) acquire little N from root uptake; up to 80% of their N is obtained from prey captured in their pitcher-shaped leaves (Butler and Ellison, 2007; Chapin and Pastor, 1995; Ellison and Gotelli, 2001; Schulze et al., 1997). Most North American pitcher plants secrete chitinases and proteases that directly break down the prey (Hepburn et al., 1927), but *S. purpurea* L. secretes digestive enzymes only at very low levels (Gallie and Chang, 1997) and enzyme secretions have not been observed in *D. californica*. Instead, these two species rely on a food web of aquatic insect larvae, protozoa, and bacteria that inhabits the pitchers (Addicott, 1974; Naeem, 1988) to break down the captured prey, mineralize the available ON to IN, and release it for absorption by the pitchers (Bradshaw and Creelman, 1984; Butler et al., 2008). In northeastern North America where N deposition rates are relatively high, *Sarracenia purpurea* also acquires IN directly from rainfall that collects in its pitchers (Ellison and Gotelli, 2002; Gotelli and Ellison, 2002).

This "*Sarracenia* microecosystem" (*S. purpurea* plus its resident food web) has been developed as a model system in which we have examined N cycling of an entire detritus-based food web (Butler and Ellison, 2007; Butler et al., 2008). In northeastern North America, *S. purpurea* grows in peat bogs and poor fens where plant growth is predominantly N-limited (Bedford et al., 1999). These bogs have massive stores of ON in peat that is generally assumed to be unavailable for uptake and use by vascular plants, but high nutrient flux and organic production occurs in bogs (Bridgham et al., 1996), and many non-carnivorous plants in these habitats have been shown to be able to take up ON (as amino acids) directly through their roots (Clemmensen et al., 2008; Kielland et al., 1994; Raab et al., 1999). Carnivorous plants such as *S. purpurea* have weakly developed root systems (root:shoot ratio ≈0.2) (Butler and Ellison, 2007), and although carnivorous plants take up some nutrients from their roots, they obtain most of their nutrients from prey captured by modified leaves (Darwin, 1875; Ellison and Gotelli, 2009; Lloyd, 1942).

Recent research on the N budget of *S. purpurea* has focused on the relative importance of bacteria and the macroinvertebrates in its food web (larvae of the midge *Metriocnemus knabi* Coq., the mosquito *Wyeomyia smithii* (Coq.), and the sarcophagid fly *Fletcherimyia fletcheri* (Aldrich)) in the nutrient mineralization and excretion process (Butler et al., 2008). This work has demonstrated that bacteria are the primary agents of N mineralization, although the mosquito and fly larvae regulate both the abundance and the diversity of the bacteria (Gotelli and Ellison, 2006; Peterson et al., 2008). Inorganic N derived from atmospheric deposition is directly assimilated by plants (Butler and Ellison, 2007; Ellison and Gotelli, 2002). However, neither the ability of pitcher plants to assimilate ON directly, nor the role of the food web in modulating such ON uptake has been investigated experimentally.

Here, we report the results of a 72-hr pulse-chase experiment conducted in the field in which we fed two isotopically enriched amino acids (glycine and phenylalanine) and ammonium nitrate, singly and in combination, to pitcher plants in the field. Our factorial experimental design also assessed whether the macroinvertebrate component of the pitcher-plant's associated food web altered the observed patterns of nitrogen uptake. Finally, we determined if ON uptake by *S. purpurea* and its associated food web differed between sites with different background levels of atmospheric nitrogen deposition.

MATERIALS AND METHODS

Study Species
Sarracenia purpurea grows in ombrotrophic (rain-fed) bogs (Damman, 1990), poor fens, and seepage swamps throughout Canada east of the Rocky Mountains and in the eastern United States from Maine to Georgia (Schnell, 2002). This long-lived (>50 yr), perennial carnivorous plant grows as a rosette of leaves from a small rhizome crown (Figure 1). In the northeastern United States and Canada where we studied *S. purpurea*, it produces 6–10 new leaves each year; the leaves last 1–2 yr and then senesce. These leaves are modified into pitfall traps ("pitchers") that fill with rainwater in which captured arthropod prey drowns. The pitchers also are inhabited by an aquatic, detritus-based food web consisting of bacteria, protozoa, and invertebrates (Addicott, 1974; Butler et al., 2008; Gotelli and Ellison, 2006). Prey captured by *S. purpurea* is shredded by aquatic larvae and mineralized by bacteria that inhabit the pitchers; the mineralized nutrients are released for uptake by the plant (Bradshaw and Creelman, 1984; Butler et al., 2008; Cochran-Stafira and von Ende, 1998; Heard, 1994). *Sarracenia purpurea* is a somewhat inefficient predator—<3% of insect visitors are actually captured (Newell and Nastase, 1998) and insects and other prey account for 10–80% of their nutrient budget (Chapin and Pastor, 1995); it obtains the remainder of its

nutrients from stored reserves (Butler and Ellison, 2007), remobilization and excretion of N and P by rotifers (Błędzki and Ellison, 1998), and increasingly, atmospheric deposition (Ellison and Gotelli, 2002). As pitchers generally account for ≈80% of the total plant mass with roots and rhizome crowns accounting for the remaining ≈20% (Butler and Ellison, 2007), *S. purpurea* derives <5% of its nutrients from the pore water in the peat where it grows (Błędzki and Ellison, 1998, 2002).

Figure 1. The pitcher plant *Sarracenia purpurea*. This carnivorous plant grows as a rosette of leaves modified into pitchers that act as pitfall traps in which rainfall is collected and prey are captured.

Field Sites

We measured nitrogen acquisition by *S. purpurea* under field conditions at Fort Albany, northern Ontario, Canada (52°15′ N, 81°35′ W) and at Tom Swamp, adjacent to Harvard Pond in Petersham, Massachusetts, U.S.A. (42°30′ N, 72°11′ W). Fort Albany is in the James Bay Lowlands of the Hudson Plains Ecoregion (Environment Canada, 2004), and the dominant vegetation at the study site consists of sedges, mosses, and lichens with or without stunted black spruce (*Picea mariana* (Mill.) Britton, Sterns, and Poggenb.) and tamarack (*Larix laricina* (Du Roi) K. Koch). Tom Swamp is a ≈50 ha bog at the north end of Harvard Pond, an artificial pond created in the 1800s by the construction of two dams on Riceville Creek (Swan and Gill, 1970). The bog vegetation is dominated by leatherleaf (*Chamaedaphne calyculata* (L.) Moench.). Fort Albany is near the northern limit of *S. purpurea* in Ontario, but pitcher plants are abundant there as well as at Tom Swamp (densities >5/m²). Annual wet IN deposition for Chapais, Quebec (the nearest data available for Fort Albany and 600 km to the southeast) was ≈2.5 kg/ha in 2002 (Environment Canada, 2004) compared to ≈4.5 kg/

ha for central Massachusetts (National Atmospheric Deposition Program (NRSP-3), 2009).

Experimental Design

We used a 72-hr pulse-chase experiment with isotopically enriched amino acids as our ON source and ammonium nitrate ($^{15}NH_4^{15}NO_3$) as our IN source to determine if pitcher plants can acquire ON directly and to compare ON and IN uptake under different conditions. We focused on uptake of N by pitchers because our previous research showed that pitchers acquired ≈70% of added IN while roots acquired less than 2.5% of added IN (Butler and Ellison, 2007).

At each site, 125 mature individuals were selected with at least three live (no sign of senescence) mature pitchers (firm and open). Five of these plants served as untreated controls and were harvested at the end of the experiment to determine baseline ^{15}N and ^{13}C natural abundances. The remaining 120 plants were randomly assigned to one of six treatment groups: uniformly-labeled (U-) glycine (U-Gly: 98 atom% U-^{13}C-^{15}N-glycine), uniformly-labeled phenylalanine (U-Phe: 98 atom% U-^{13}C-^{15}N-phenylalanine), I^{15}N (98 atom% $^{15}NH_4^{15}NO_3$), U-Gly plus unlabeled phenylalanine and unlabeled NH_4NO_3 (hereafter U-Gly+), U-Phe plus unlabeled glycine and unlabeled NH_4NO_3 (hereafter U-Phe+), or I^{15}N plus unlabeled glycine and phenylalanine (hereafter I^{15}N+). Plants within treatment groups were assigned randomly to one of two harvests (3- or 72-hr) and one of two food web treatments (complete food webs or partial food webs, which lacked the macroinvertebrate larvae of the detritivorous midge *Metriocnemus knabi* and the keystone predator, the mosquito *Wyeomyia smithii*). Larvae of the sarcophagid fly *Fletcherimyia fletcheri*, which are found more commonly in pitchers in Massachusetts than in Canada, were excluded from all experimental pitchers. There were N = 5 pitchers in each treatment at each site.

Pitchers were significantly larger at Tom Swamp than at Fort Albany (Table 1). Control (unfed) plants at both sites had similar concentrations of N and natural abundance levels of $\delta^{15}N$ in leaf tissues (Table 1). These control plants also had similar C concentrations and similar C:N ratios, but background natural abundance of δ13C was slightly lower at Tom Swamp than at Fort Albany (Table 1).

Any liquid in the pitchers, along with the food web, was removed from all experimental pitchers the day before the pulse-chase experiment began; the liquid removed (pitcher "liquor") was kept for the food web manipulations. Following food web removal in the field, pitchers were rinsed with distilled water to remove as much detritus and as many microbes as possible and the pitcher opening was blocked with a fine nylon mesh to limit subsequent entry of animals and prey. In the laboratory, all living midge and mosquito larvae were removed from liquid collected from each pitcher and kept alive overnight in a solution of pitcher liquor.

The next day, the largest pitcher on each plant was fed with one of the ^{15}N treatments. We fed each manipulated pitcher with a 0.8 mM ^{15}N solution (2 ml for Fort Albany and 9 ml for the larger pitchers at Tom Swamp) and an equal amount of pitcher liquor, resulting in pitchers filled to approximately three-quarters of their volume. Thus, all experimental pitchers contained an enriched (^{15}N) nutrient solution along with the microbial component of the food web (supplied in the pitcher liquor). Pitchers at Fort Albany were fed 0.022 mg N, whereas the larger pitchers at Tom Swamp were fed 0.101 mg N. The amount of N added to pitchers represented <1.0% of N content of pitcher tissue at Fort Albany and <1.9% of N content of pitcher tissues at Tom Swamp.

Table 1. Comparison of traits for pitchers of untreated plants (N = 5) harvested at Fort Albany, James Bay, Ontario, Canada (FA) and Tom Swamp, central Massachusetts, USA (TS) with significance level of unpaired t-test comparing sites.

	Site	Mean	SD	P
Dry Mass (mg)	FA	232	57	0.00003
	TS	754	142	
Length (cm)	FA	8.4	1.9	0.00008
	TS	18.7	1.9	
$\delta^{15}N$ (%)	FA	1.8	0.8	0.373
	TS	1.1	1.3	
$\delta^{15}C$ (%)	FA	−26.8	0.7	0.0013
	TS	−29.4	1.0	
Nitrogen (%)	FA	1.08	0.13	0.609
	TS	1.02	0.28	
Carbon (%)	FA	46.4	3.3	0.408
	TS	47.7	0.3	
C:N	FA	43	4	0.282
	TS	49	10	

When we added only single forms of N (i.e., the U-Gly, U-Phe, and I^{15}N treatments), all N added to the pitchers was enriched in 15N. When we added three forms of N (the U-Gly+, U-Phe+, and I^{15}N+ treatments), only one-third of the N added to each pitcher was enriched in ^{15}N; the remaining two-thirds was comprised of equal amounts of the other two forms as unlabeled N. We are confident that we minimized potential effects of excess N availability on pitcher N uptake, which could have been particularly important when only one form of N was added in a single feeding event. The total amount of N supplied and the actual concentration of N were both substantially lower than that used in other studies of *Sarracenia*: 1.2–3.6 mg N/plant as a mixture of amino acids to pitchers of *S. flava* (Plummer and Kethley, 1964); 1–10 mM alanine fed to Nepenthes pitchers (Lüttge, 1965); 20 ml/pitcher of 6.8–8.6 mM NH_4-N (1.9–2.4 mg N/pitcher) (Butler and Ellison, 2007; Bradshaw and Creelman, 1984) or 6–8.7 mg N/pitcher as NH_4Cl to *S. purpurea* (Chapin and Pastor, 1995; Ellison and Gotelli, 2002), which is comparable to the mass of prey-N captured by *S. purpurea* in a growing season (Heard, 1998).

Finally, for the complete food web treatments, we put invertebrate larvae into the pitchers immediately after we added the ^{15}N solution. We added two midge and two mosquito larvae in each pitcher in the complete food web treatment at Fort Albany and nine midge and nine mosquito larvae in each pitcher in the complete food web treatment at Tom Swamp (i.e., 1 midge+1 mosquito larva per ml of pitcher liquor). Unfed (control) pitchers for which we measured natural abundance of ^{13}C and ^{15}N also had complete food webs (pitcher liquor+midge+mosquito larvae).

Because of the dramatic size differences between plants at the two sites (Table 1), total N fed to each plant and food web manipulations (numbers of midge and mosquito larvae added to pitchers) differed at the two sites. Therefore, statistical analyses were conducted separately for each site.

Harvest

Target pitchers were removed from the rest of the plant 3 or 72 hr after feeding with a stainless steel razor blade that was rinsed in 50% ethanol between cuttings. Pitcher liquor was transferred to a sealed sterile plastic tube and the pitcher was placed in a zip-lock plastic bag. Both were stored in a cooler with cold packs and taken immediately to the laboratory for processing. Pitchers were cut open longitudinally, washed thoroughly with tap water, rinsed with 0.5 mM $CaCl_2$ to remove any amino acids from the surface (Persson et al., 2003), and finally rinsed three times with distilled-deionized water before being transferred to paper bags. Midge and mosquito larvae were removed from the pitchers with an eye dropper, transferred through three sequential baths of distilled-deionized water and stored in new sterile vials. Because of the small mass of larvae in each pitcher, larvae from the five replicates of each harvest × treatment combination were pooled into one composite larval sample. Plant and invertebrate samples were then oven-dried at 65°C for 48 hr and then weighed.

Isotopic Analyses

Each pitcher and composite larval sample was ground to a fine powder in a stainless steel capsule with a stainless steel ball using a Wig-L-Bug mixer (Bratt Technologies, LLC., East Orange, New Jersey, USA). A 4-mg subsample of plant tissue or a 1-mg subsample of larvae was then placed into an 8 × 5 mm tin capsule (Elemental Microanalysis Mason, Ohio, USA) and combusted in a Costech ECS4010 Elemental Analyzer and DeltaPlus XP mass spectrometer at the University of New Hampshire to measure $^{13}C/^{12}C$, %C, $^{15}N/^{14}N$, and %N concurrently. A reference standard (NIST 1515, NIST 1575a, or an internal tuna standard) was included after every five samples.

Recovery of added tracer in pitchers was calculated as:

$$^{15}N_{rec} = m_{pool} \frac{\left(atom\%^{15}N_{pool} - atom\%^{15}N_{ref}\right)}{\left(atom\%^{15}N_{tracer} - atom\%^{15}N_{ref}\right)}$$

where $^{15}N_{rec}$ = *mass of* ^{15}N *tracer recovered in the labeled* N_{pool}, m_{pool} = N *mass of the total* N_{pool}, $atom\%^{15}N_{pool}$ = *atom%* ^{15}N *in the labeled* N_{pool}, $atom\%^{15}N_{ref}$ = *atom%* ^{15}N *in the reference* N_{pool} (non-labeled plants harvested at the end of the experiment), and $atom\%^{15}N_{tracer}$ = *atom%* ^{15}N *of the applied tracer* (Nadelhoffer and Fry, 1994).

Statistical Analysis

First, we measured ^{15}N and ^{13}C in pitchers to determine if *S. purpurea* could acquire intact amino acids. We emphasize that observing enrichment in ^{13}C and ^{15}N by itself does not provide evidence for acquisition of intact glycine or phenylalanine because both ^{13}C and ^{15}N may be acquired in products of microbial mineralization of amino acids. Rather, a comparison of the slope of excess ^{13}C versus ^{15}N (per gram dry mass of plant tissue) to the slope of the ^{15}N source provides a conservative estimate of N acquired as amino acid (Nasholm et al., 1998). Ratios below the expected slope (= $^{13}C:^{15}N$ ratio of the amino acids) indicate loss of ^{13}C (e.g., respiration) and/or ^{15}N acquisition after mineralization of labeled amino acid. Such an analysis should be undertaken within a few hours of the application of dual-labeled amino acids (hence our 3-hr harvest). We used ordinary least-squares linear regression analysis to compare the slope of $^{13}C:^{15}N$ in pitchers fed U-Gly or U-Phe as their only N source with the expected slopes of $^{13}C:^{15}N$ if glycine (expected slope = 2) or phenylalanine (expected slope = 9) were taken up intact within the first 3 hr after feeding.

For all other analyses, we analyzed nitrogen uptake by pitchers as µg ^{15}N per gram dry mass. Data were arcsin-square-root transformed to reduce heteroscedasticity. Although ^{15}N enrichment was measured in midge and mosquito larvae (Figure 2), analysis of variance with a main effect for the invertebrate food web manipulation

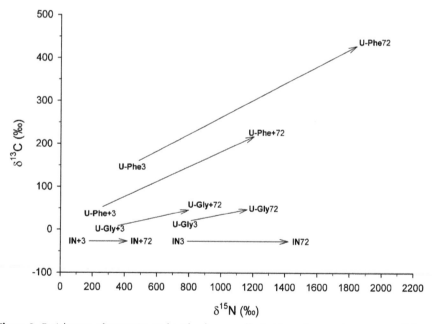

Figure 2. Enrichment of mosquito and midge larvae collected from *Sarracenia purpurea* pitchers. The arrows connect points showing enrichment after 3 and 72 hr of plants at Tom Swamp in Massachusetts that were fed ^{15}N as ammonium nitrate ($^{15}NH_4{}^{15}NO_3$), uniformly-labeled (U-) ^{13}C-^{15}N-glycine or U-^{13}C-^{15}N-phenylalanine alone (IN, U-Gly, and U-Phe, respectively) or in combination (IN+, Gly+, and Phe+, respectively). Larvae in pitchers fed Phe or Phe+ had the highest ^{13}C and ^{15}N enrichment after 72 hr.

revealed no significant differences in ^{15}N uptake at each site for pitchers with and without the invertebrate food web (Fort Albany $F_{1,96} = 0.137$, $P = 0.71$; Tom Swamp $F_{1,94} = 1.68$, $P = 0.20$). Therefore, ^{15}N uptake by pitchers was analyzed for each site separately using a fixed-effect two-way analysis of variance to test for differences in ^{15}N uptake only as a function of the form of N fed to pitchers and harvest time. In these analyses, pitchers in the two food-web treatments were pooled within each N addition treatment. *A priori* contrasts of the N-form treatment at the 72-hr harvest were used to compare plant N uptake across treatments and between sites at the end of the experiment.

DISCUSSION

Uptake of intact amino acids has been demonstrated for species in ecosystems ranging from alpine and arctic tundra to a subtropical rainforest and in ephemeral pools in the Namibian desert (Lipson and Nasholm, 2001). The use of a variety of forms of different nutrients could provide selective advantages to plants inhabiting nutrient-limited environments. Carnivorous plants, whose growth and reproduction are strongly limited by nutrient availability and that grow in extremely nutrient-poor habitats (Ellison and Gotelli, 2001; Porembski and Barthlott, 2000) acquire organic nutrients as prey, but whether or not they can directly take up intact amino acids had not been studied previously.

Our data are consistent with the hypothesis that *S. purpurea* takes up amino acids directly. However, while this was clearly demonstrated for phenylalanine, *S. purpurea* either did not assimilate glycine directly or else it assimilated glycine but metabolized it more rapidly than our 3-hr sample could detect (Figure 3). All pitchers fed U-Gly and U-Phe were highly enriched in ^{13}C and ^{15}N. Rapid enrichment of pitcher tissue with both ^{13}C and ^{15}N at a $^{13}C:^{15}N$ ratio similar to that of the amino acid fed to the plant would suggest uptake of intact amino acids. Whereas pitchers fed U-Phe had $^{13}C:^{15}N$ ratios similar to the expected value of nine, pitchers fed U-Gly were highly enriched in both ^{13}C and ^{15}N but their $^{13}C:^{15}N$ ratio was significantly less than the expected value of two. The results for acquisition of intact glycine, and comparison to phenylalanine, however, must be interpreted with caution (Harrison et al., 2007; von Felton et al., 2008). First, carbon respiration following acquisition of phenylalanine will have less effect on the $^{13}C:^{15}N$ relationship than it would for glycine because of the greater amount of ^{13}C acquired per unit phenylalanine (C:N = 9) compared to glycine (C:N = 2). Carbon acquired from glycine is rapidly catabolized (Hodge et al., 2000), which can lead to slopes of $^{13}C:^{15}N$ substantially below the expected value of 2 (Figures 3A, B). This may have been exacerbated by the relatively small amounts of ^{13}C- and ^{15}N-enriched amino acids provided to pitchers; in order to avoid potential effects of excess N availability on the plant, added N represented <2% of the total N in the pitchers.

Furthermore, the substantial enrichment of pitchers (Figure 4) and the differences in ^{13}C and ^{15}N enrichment of invertebrate larvae (Figure 2) between pitchers fed U-Gly and U-Phe provides some additional support for intact glycine acquisition by *S. purpurea* pitchers under field conditions. As expected if plants preferentially acquire amino acids with low C:N (Lipson et al., 1999), the ^{15}N enrichment of *S. purpurea* pitchers was generally greater for plants fed U-Gly than U-Phe (Figure 4).

More detailed comparisons of enrichment of ^{15}N in plant tissues of the different treatment groups suggests that *Sarracenia* should show preferential uptake of amino acids with lower C:N ratios (such as glycine) than those with higher C:N ratios (such as phenylalanine). After 72-hr, ^{15}N concentration in plants fed glycine alone was significantly greater than ^{15}N concentration in plants fed phenylalanine alone at Fort Albany but not at Tom Swamp (Figure 4). At both sites, uptake of ^{15}N from glycine tended to be higher than uptake of 15N from phenylalanine when all three forms of N were available to pitchers (Figure 4). Because Fort Albany has cooler air temperatures, a shorter growing season, and lower atmospheric N deposition than Tom Swamp, we interpret these results to suggest that pitchers at the Canadian site can maximize N uptake by taking up relative more amino acids with low C:N ratios (e.g., glycine) and avoid the energetically costly synthesis of new amino acids from IN plus a carbon skeleton (Clarkson, 1985). In contrast, at Tom Swamp in Massachusetts, the climate is warmer, the growing season is longer, and IN is more readily available because of higher atmospheric deposition rates. Thus, direct uptake of low C:N amino acids such as glycine may not be as important at Tom Swamp because suitable environmental conditions exist to synthesize amino acids from readily available IN plus available carbon.

These inferences are supported by our results showing greater ^{15}N uptake from ON than from IN (energetic benefit) when each form was provided in isolation at Tom Swamp (Figure 4). At the more N-limited Fort Albany site, however, ^{15}N uptake by *S. purpurea* pitchers was similar for all three forms of ^{15}N when each was provided in isolation. *Sarracenia purpurea* pitchers are open to the atmosphere, collecting rainwater as well as prey, and the difference in ON uptake between sites may represent a response to the higher atmospheric IN deposition at Tom Swamp. Similar preferential acquisition of the predominant forms of N in the local environment has been observed for plants in boreal forests (Nordin et al., 2001), arctic tundra (Chapin et al., 1993; Nordin et al., 2004), alpine meadows (Raab et al., 1999), and cold-temperate forests (Finzi and Berthrong, 2005).

Finally, our results illustrate that the acquisition of any one form of N provided in isolation will exceed uptake of this form when multiple forms of N are made available to the plant simultaneously. At both sites and for each form of ^{15}N supplied, uptake of ^{15}N was significantly greater when only one form was made available compared to that same form of ^{15}N provided in mixture. However, there were no significant differences in ^{15}N uptake among the three forms when all forms of ^{15}N were made available simultaneously (Figure 4). This result highlights the versatility of N acquisition by *S. purpurea* because N uptake of any one particular form decreases when all three forms are available.

Sarracenia purpurea is one of only two species of North American pitcher plants (the other being *Darlingtonia californica*) in which a food web mineralizes captured prey and simultaneously competes with *S. purpurea* for N. Our results revealed no differences in ^{15}N uptake between pitchers with and without higher trophic levels in their associated food webs. This result is consistent with previous results showing that the upper trophic levels in the *S. purpurea* microecosystem actively process detritus but that the activity of the microbial component of the food web ultimately determines N availability for *Sarracenia* (Butler et al., 2008).

Taken as a whole, our field experiments indicate versatility of N acquisition by this carnivorous plant and variability in N acquisition across a gradient of atmospheric deposition. These results are consistent with data reported from other N-limited environments (Atkin, 1996; Persson et al., 2003). Similarly, there is pronounced spatiotemporal variation in the availability, form, quantity, and proportions of each form of N in bogs in general and in the *Sarracenia* microecosystem in particular. The growth of plants in eastern North American bogs is predominantly N-limited (Bedford et al., 1999) but bogs have massive stores of ON in peat with high nutrient flux and organic production (Bridgham et al., 1996) and receive variable inputs of atmospheric IN deposition throughout the growing season. Additionally, pitcher plants collect varying amounts of organic and inorganic N via trapped prey. We suggest that the energetic benefits of direct and rapid (<3 hr) acquisition of ON as intact amino acids allow *Sarracenia* to short-circuit the inorganic N cycle and to minimize potential bottlenecks in N availability because of the plant's reliance for N mineralization on a seasonally reconstructed food web (Ellison et al., 2003) operating on irregular and infrequent seasonal pulses of prey capture (Fish and Hall, 1978; Newell and Nastase, 1998). Experiments employing a greater range of N concentrations for a longer duration would improve our ability to determine the upper limit of N acquisition by *Sarracenia* and characterize the importance of the acquisition of intact amino acids to the N budget of this carnivorous plant.

RESULTS

Uptake of Intact Amino Acids

Pitcher plants rapidly assimilated the two amino acids we fed to them. There was a significant positive relationship between tissue ^{13}C and ^{15}N for pitchers harvested 3-hr after receiving only U-Gly at Fort Albany ($P = 3.83 \times 10^{-4}$; Figure 3A) and Tom Swamp ($P = 6.84 \times 10^{-6}$; Figure 3B). The slope of the [13]C:[15]N line at both sites was significantly less than 2 the value expected if intact glycine was taken up directlyat both Fort Albany (slope = 0.98, 95% CI = 0.6–1.4) and at Tom Swamp (slope = 1.1, 95% CI = 0.8–1.3).

There was a highly significant positive relationship between tissue ^{13}C and ^{15}N for pitchers harvested 3-hr after receiving U-Phe at both Fort Albany ($P = 1.28 \times 10^{-8}$; Figure 3C) and Tom Swamp ($P = 7.17 \times 10^{-8}$; Figure 3D). The slope of the [13]C:[15]N relationship for pitcher tissue at both sites did not differ from 9 the value expected if intact phenylalanine was taken up directlyat both Fort Albany (slope = 8.6 with a 95% CI = 7.7–9.5; Figure 3C) and at Tom Swamp (slope = 8.9 with a 95% CI = 7.8–10.0; Figure 3D).

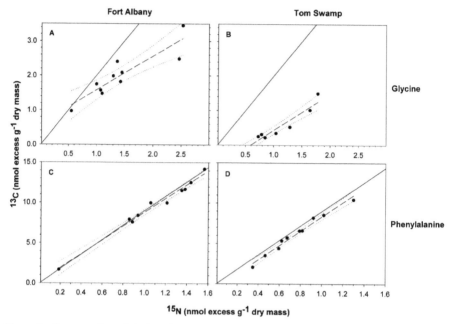

Figure 3. Concentration of 13C and 15N recovered in *Sarracenia purpurea* pitchers 3 hr after feeding. Nitrogen solutions were applied into the pitcher leaves at Fort Albany in Ontario, Canada (left, panels A, C) and at Tom Swamp in Massachusetts, USA (right, panels B, D) as uniformly-labeled (U-) ^{13}C-^{15}N-glycine (top row, panels A, B) or as U-^{13}C-^{15}N-phenylalanine (bottom row, panels C, D). Solid lines represent the ^{13}C:^{15}N ratio of glycine (2:1) or phenylalanine (9:1); dashed-lines represent ordinary least squares regression (with 95% confidence intervals as dotted lines) for plant uptake.

Uptake of Different Forms of Nitrogen

At each site, there were highly significant differences among treatments (Fort Albany: $F_{5,108} = 41.8$, $P < 0.0001$; Tom Swamp: $F_{5,106} = 47.0$, $P<0.0001$) and harvest times (Fort Albany: $F_{1,108} = 62.3$, $P < 0.0001$; Tom Swamp: $F_{1,106} = 158.3$, $P < 0.0001$), and significant treatment × harvest time interactions (Fort Albany: $F_{5,108} = 2.3$, $P = 0.05$; Tom Swamp: $F_{5,106} = 4.5$, $P = 0.0009$; Figure 4). At Fort Albany, pitchers acquired significantly more ($P = 0.004$) 15N from glycine than from phenylalanine when those forms were provided in isolation (Figure 4A), and showed a similar trend when all three forms of N were available ($P = 0.07$; Figure 4C). At Tom Swamp pitchers acquired similar amounts ($P = 0.56$) of N from glycine and phenylalanine when ON was provided in isolation (Figure 4B) but tended to favor glycine when all forms of N were available ($P = 0.08$; Figure 4D).

At Tom Swamp, pitchers took up significantly more N from either glycine or phenylalanine than from ammonium nitrate ($P = 0.034$ and $P = 0.009$, respectively) when only one form of ^{15}N was provided (Figure 4D). In contrast, at Fort Albany, pitchers took up similar amounts of ^{15}N either the amino acids or the ammonium nitrate when only one form of ^{15}N was available ($P = 0.22$ and $P = 0.09$, respective contrasts; Fig. 4C). At both sites, pitchers acquired similar amounts of ^{15}N from the amino acids and ammonium nitrate when all three forms of N were provided simultaneously ($P \geq 0.29$). Finally, highly significant differences of ^{15}N uptake were found for each form of N when it was provided in isolation relative to when it was provided in mixture (Figures 4C, D).

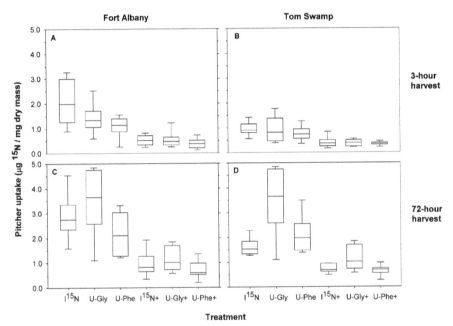

Figure 4. 15N acquisition by *Sarracenia purpurea* pitchers supplied with multiple forms of nitrogen. Nitrogen was provided to pitcher leaves as ammonium nitrate ($^{15}NH_4^{15}NO_3$), uniformly-labeled (U-) ^{13}C-^{15}N-glycine or U-^{13}C-^{15}N-phenylalanine (IN, U-Gly, and U-Phe, respectively) or in combination (IN+, U-Gly+, U-Phe+, respectively) at Fort Albany, Ontario, Canada (left, panels A, C) and at Tom Swamp, Massachusetts, USA (right, panels B, D). Uptake was measured after 3 hr. (top, panels A, B) and 72 hr (bottom, panels C, D) in a pulse-chase experiment. The total amount of N provided to pitchers at Fort Albany was 0.022 mg, and was 0.101 mg N at Tom Swamp. Pitchers with and without the invertebrate food webs were pooled for these analyses because there were no significant differences in ^{15}N uptake in the two food-web treatments (Fort Albany $F_{1,96} = 0.137$, $P = 0.71$; Tom Swamp $F_{1,94} = 1.68$, $P = 0.20$).

KEYWORDS

- **Inorganic nitrogen cycle**
- **Organic nitrogen**
- ***Sarracenia* microecosystem**
- ***Sarracenia purpurea***

AUTHORS' CONTRIBUTIONS

Conceived and designed the experiments: Jim D. Karagatzides, Jessica L. Butler, and Aaron M. Ellison. Performed the experiments: Jim D. Karagatzides. Analyzed the data: Jim D. Karagatzides and Aaron M. Ellison. Wrote the paper: Jim D. Karagatzides, Jessica L. Butler, and Aaron M. Ellison.

ACKNOWLEDGMENTS

We thank Erik Hobbie and Andew Ouimette at the University of New Hampshire Stable Isotopes Laboratory for analyzing our samples. L. Tsuji, E. Liberda, C. Hart, X. Wen, D. Kozlovic, and J. Liedtke assisted in the field. Paul Grogan and Steve Heard provided helpful comments on an earlier draft.

Chapter 3

Analysis of Pathogen Resistance and Fruit Quality Traits in Melon (*Cucumis melo* L.)

Albert Mascarell-Creus, Joaquin Cañizares, Josep Vilarrasa-Blasi,
Santiago Mora-García, José Blanca, Daniel Gonzalez-Ibeas,
Montserrat Saladié, Cristina Roig, Wim Deleu, Belén Picó-Silvent,
Nuria López-Bigas, Miguel A. Aranda, Jordi Garcia-Mas,
Fernando Nuez, Pere Puigdoménech, and Ana I. Caño-Delgado

INTRODUCTION

Melon (*Cucumis melo*) is a horticultural specie of significant nutritional value, which belongs to the Cucurbitaceae family, whose economic importance is second only to the Solanaceae. Its small genome of approximate 450 Mb coupled to the high genetic diversity has prompted the development of genetic tools in the last decade. However, the unprecedented existence of a transcriptomic approaches in melon, highlight the importance of designing new tools for high-throughput analysis of gene expression.

We report the construction of an oligo-based microarray using a total of 17,510 unigenes derived from 33,418 high-quality melon expressed sequence tags (ESTs). This chip is particularly enriched with genes that are expressed in fruit and during interaction with pathogens. Hybridizations for three independent experiments allowed the characterization of global gene expression profiles during fruit ripening, as well as in response to viral and fungal infections in plant cotyledons and roots, respectively. Microarray construction, statistical analyses, and validation together with functional-enrichment analysis are presented in this study.

The platform validation and enrichment analyses shown in our study indicate that this oligo-based microarray is amenable for future genetic and functional genomic studies of a wide range of experimental conditions in melon.

Cucurbits, comprising up to 750 different species distributed in 90 genera, are among the most important horticultural crops worldwide (Gonzalez-Ibeas et al., 2007). Species of this family have been independently and repeatedly domesticated by different cultures in both the Old and New World, either for food or as materials for a range of products. As a result, they display an enormous, and mostly untapped, genetic diversity.

Melon (*Cucumis melo*) belongs to this family and is one of the most important fleshy fruits for fresh consumption in temperate, subtropical, and tropical regions (Gonzalez-Ibeas et al., 2007). It has been classified into two subspecies, *C. melo* sp. *agrestis* and *C. melo* sp., with India and Africa as their centers of origin, respectively (Garcia-Mas et al., 2004; Kirkbride, 1993). Melon is a diploid species, with a basic number of chromosomes x = 12 (2x = 2n = 24) and an estimated genome size of 450–500 M (Arumuganathan,1991), similar in size to the rice genome (419 Mb) (Goff et al., 2002; Yu et al., 2002) and about four times the size of the *Arabidopsis* genome (125 Mb) (Huala et al., 2001).

Based on its relatively small genome, wide morphological, physiological and biochemical diversity and the ability to produce hybrids between unrelated cultivars, *C. melo* has a great potential for becoming a genetic model species (Kirkbride, 1993; Liu et al., 2004). Classically, melon and other cucurbits have been used to analyze the development of plant vasculature and its role in the transport of macromolecules (Gomez et al., 2005; Haritatos et al., 1996). Recently, the development of novel genetic and genomics tools, such as a high quality genetic map (Fernandez-Silva et al., 2008) a draft of a physical map (P. Puigdomènech, unpublished), a TILLING platform (J. Garcia-Mas, unpublished) and an EST collection (Gonzalez-Ibeas et al., 2007) have enabled the identification and study of genes with agronomic interest in this species. Together, these tools fostered the use of melon as an experimental system to dissect biological processes of economic relevance, such as flavor and textural changes that take place during fruit ripening (Moreno et al., 2008), and the interactions between melon and its pathogens (Chen et al., 1997; Nieto et al., 2006). So far, these studies have followed a gene-by-gene approach. As an example, the recent discovery of an ethylene biosynthetic enzyme, CmACS-7, that is required for the development of female flowers in monoecious lines, revealed a novel mechanism involved in organ development (Boualem et al., 2008) and brought up the necessity for developing new tools for high-throughput analysis of gene expression in melon.

Partial sequencing of cDNA inserts of ESTs have been used as an effective method for gene discovery (Gonzalez-Ibeas et al., 2007). In the last decade, the development of several EST collections has opened the way to functional genomic studies in several plant species (Alba et al., 2004; Rudd, 2003). In addition, EST collections are good sources of simple sequence repeats (SSRs) and single-nucleotide polymorphisms (SNPs) that can be used for creating saturated genetic maps (Morgante et al., 2002; Rafalski, 2002). Microarray technology has demonstrated the power of high-throughput approaches to unravel key biological processes (Aharoni and Vorst,2002; Clarke and Zhu, 2006). The cDNA-based microarrays are specially relevant for crop species where little genome information is available both to study a particular biological process or to identify candidate target genes for breeding (Rensink and Buell, 2005). Whereas full-length cDNA microarrays were the first choice for the generation of microarray platforms, oligonucleotide-based chips are gradually gaining importance due to the reduction of manipulation steps and their ability to differentiate similar members of gene families (Larkin et al., 2005). Although cDNA arrays are known to have higher precision across technical replicates, oligonucleotide-based platforms show a greater dynamic range in the evaluation of expression levels (Woo et al., 2004). Microarray platforms can be classified with respect to their accuracy and precision (Cope et al., 2004; He et al., 2003). The precision of the microarray can be assessed by comparison between different replicas. Accuracy, on the other hand, requires knowledge of the expression of particular genes in the biological system under study (Cope et al., 2004; He et al., 2003).

In this study, a dataset of 33,418 high-quality melon ESTs obtained from nine MELOGEN normalized cDNA libraries and another available collections from different melon cultivars, corresponding to various tissues in different physiological conditions have been used to generate the first oligonucleotide-based microarray platform in melon. Similar microarray platforms have been generated in other important crop species such as watermelon using 832 EST-unigenes (Wechter et al., 2008), citrus with 21,081 putative unigenes (Martinez-Godoy et al., 2008), pea from 2,735 ESTs (Wong et al., 2008), and canola using 10,642 unigenes (Xiang et al., 2008). However, this high-density microarray is estimated to represent a significant portion of the whole melon transcriptome (Gonzalez-Ibeas et al., 2007), offering new possibilities for the study of multiple transcripts in a variety of conditions.

In this study, we describe the design and construction of the melon microarray platform, and we validate our tool by means of the analysis of global changes in gene expression profiles in response to pathogen infection and during fruit ripening, taken as model experiments. Our results show that this melon microarray is suitable for the study of global changes in gene expression in different scenarios, until the full genome-sequencing project becomes a reality.

MATERIALS AND METHODS

Plant Material

The material used for the transcriptomic profile analyses came from three different *C. melo* accessions: the line T-111, which corresponds to a Piel de Sapo breeding line, the Tendral variety (Semillas Fitó, Barcelona, Spain), and the pat81 accession of *C. melo* L. sp. *agrestis* (Naud) Pangalo maintained at the germplasm bank of CO-MAV (COMAV-UPV, Valencia, Spain). Seeds of T-111 were germinated at 30°C for 2 days and plants were grown in a greenhouse in peat bags, drip irrigated, with 0.25-m spacing between plants. Fruits were collected 15 and 46 days after pollination and mesocarp tissues were used for RNA extractions. Roots from pat81 plants were mock treated or inoculated with 50 colony-forming units (CFUs) of *Monosporascus cannonballus* per gram of sterile soil and harvested 14 days after inoculation. Cotyledons from var. Tendral, mock treated or mechanically inoculated with Cucumber Mosaic Virus (CMV), were harvested 3 days post inoculation. Plant growth and infections were done as described in González-Ibeas et al. (2007).

RNA Isolation and cDNA Preparation

Total RNA from each of the tissues was extracted using Plant RNeasy Mini Kit (Qiagen, Hilden, Germany) following the manufacturer protocols. The resulting RNA is enriched in molecules longer than 200 nucleotides. The DNA contaminations were removed using the DNA-free™ Kit (Ambion, Austin, TX) according to the manufacturer instructions. The RNA quantity and purity were determined by Nanodrop ND-1000 spectrophotometer (Nano Drop Technologies, Wilmington, Delaware). Integrity and quality of the RNA were checked by agarose electrophoresis. Only those high-quality RNA samples were reverse transcribed into cDNA using the SuperScript™ Double-Stranded cDNA Synthesis Kit (Invitrogen, Carlsbad, CA). The cDNA integrity was assessed with an Agilent Bioanalyzer 2100 and RNA 6000 NanoLabChip Kit (Agilent Technologies, Palo Alto, CA).

Design and Production of the Melon Nimblegen® Custom Array

A dataset of 17,510 quality-filtered unigenes from different normalized cDNA libraries were submitted for probe design and production of the high-density microarray (Table 1). The Nimblegen Maskless Array Synthesis (MAS) technology was used to fabricate the microarray, combining photo-deposition chemistry with digital light projection. Our design is based on a single chip containing two internal replicate probe sets (2×) of 11 probes per unigene and controls such as random GC oligonucleotides, covering the whole 385K 1-plex platform spots. Thus, every unigene is represented by an average of eleven 60-mer probes, synthesized *in situ* by photolithography on glass slides using a random positional pattern. The melon microarrays are now commercially available at Nimblegen®.

Microarray Probe Preparations

Long oligonucleotide probes generated for this novel *C. melo* microarray provide superior signal-to-noise ratio, increased sensitivity, specificity, and discrimination. Probes were designed taking into account characteristics such as non-repetitive elements, frequency, uniqueness, and melting temperature (Tm). Highly repetitive elements of the transcript set were excluded using a method developed by Morgulis (2006) that identifies these regions and exclude them from probe selection (Morgulis et al., 2006). Each oligonucleotide is then compared to the transcriptome and the similarity is reduced to a Boolean value based on the weighted mismatch score. This score is compared to a set threshold (usually 10) and if the score is higher, it is considered as unique. The Tm and the self-complementarities of the 60 mer oligos are also ranked and those with higher scores were selected as probes.

Microarray Hybridization and Quality Control

The synthesized cDNA (1 µg) from each sample was transcribed to cRNA and labelled with cyanine 3 (Cy3)-labeled nucleotides following Nimblegen® specifications. Hybridization signal intensity (HSI) was calculated using a GenePix 4000B Scanner and associated software. Quality controls such as visualization of hybridization images for finding artefacts or high HSI and background zones were visually performed. The RNA digestion plots to know the amount of RNA degradation that occurred during its preparation and the quality of the second strand synthesis were performed using Perl scripts.

Normalization and Statistical Analysis

The intensity values obtained from the array scanning were normalized using the oligo package for the R statistics software. The workflow used to normalize our data was followed as explained by the package provided for Nimblegen® expression microarrays. An automatic pipeline using Perl and R scripts was generated to map every probe to its gene and normalize the data using the Robust Multichip Average (RMA) algorithm (Irizarry et al., 2003). It consists of three steps: a background adjustment, quantile normalization, and finally log transformed PM values summarization (Bolstad et al., 2003). It has been reported that RMA gives the most reproducible results and shows the highest correlation coefficients with qRT-PCR data on genes identified as differentially expressed (Millenaar et al., 2006).

After normalization, a hierarchical clustering support tree method was performed using TMeV 4.0 (Tusher et al., 2006) software from TIGR. The Euclidean distance was used as a measure of similarity or distance between samples. As a rule for the linkage of clusters, the average linkage method was used. In this method, distance calculation is based on the average distance between objects from the first cluster and objects from the second cluster. The averaging is performed over all pairs, determining the actual distance between two clusters.

To statistically infer the deregulated genes in the microarray, the Significance Analysis of Microarrays (SAMs) algorithm was run using the TMeV 4.0 software. Multiple testing adjustments for False Discovery Rate (FDR) method, based in Benjamini and Hochberg's method (Benjamini and Hochberg, 1995) were performed, forcing it to be FDR ≤ 0, therefore allowing for a highly stringent analysis with no false positive identification of differentially regulated genes. This methodological approach resulted in a variable SAM delta value (δ)—which defines the threshold of false positive in the validated dataset—depending on the experiment. The Venn Diagram was generated using the online Venny tool (Oliveros, 2007). The correlation between the microarray and qRT-PCR data was assessed through Pearson's moment correlation analysis. The R software package was used to systematically test all the expression changes of the qRT-PCR validated genes. Correlation analyses were also performed in order to check for similarities between the biological and technical replicates.

Gene Ontology Functional-Enrichment Analysis

Functional annotation of genes based on Gene Ontology (GO) was provided by González-Ibeas et al. (2007). We used an inclusive analysis for differentially expressed genes, taking into account the GO hierarchy, in which genes annotated with terms that are descendent of the correspondent term in a specific level, obtain the annotation from the parent.

To infer the statistical enrichment we used a similar method and visualization approach as in Lopez-Bigas et al., (2008) In particular, enrichment was assessed using binomial distribution and p-value was calculated as

$$P\left(X \geq x\right) = \sum_{i=0}^{|x|} \binom{n}{i} p^i \left(1-p\right)^{n-i}$$

where: x = number of differentially expressed genes in the category, n = total number of genes in the category, and p = frequency of upregulated or downregulated genes. Resulting p-values were adjusted for multiple testing using the Benjamin and Hochberg's method of FDR (Benjamini and Hochberg, 1995). Statistical results were visualized in a scale-based color gradient depending on the p-value. Red indicates enrichment and gray indicates no statistical significance after FDR adjustment.

Quantitative Real-Time PCR

To validate the expression changes found in the microarray experiments, transcript levels of the 12 selected genes transcripts were quantified by the ABI Prism 7700 (Applied Biosystems, Foster City, CA, U.S.A). The oligonucleotides chosen to amplify the selected genes were designed using the Primer Express Software (Applied Biosystems). Detection of the PCR products was performed as explained by the manufacturer using the Power SYBR green dye (Applied Biosystems) and ROX as passive reference. To calculate the relative change in expression levels, we used the Data Analysis for Real-Time PCR (DART) software with three technical replicates for statistical analysis (Peirson et al., 2003). Melting curves analyses at the end of the process and No Template Controls (NTCs) were carried out to ensure product-specific amplification and no primer-dimer quantification. A control reaction without reverse transcriptase was performed to evaluate genomic DNA contamination. Endogenous controls were performed using relative quantitative accumulation of Cyclophilin (cCL1375) levels in every condition.

Availability of the Microarray Data

Microarray data are publically available at Arrayexpress http://www.ebi.ac.uk/microarray-as/ae/. The experiment name for to the reported hybridizations is "melogen_melo_v1," ArrayExpress accession E-MEXP-2334.

DISCUSSION AND RESULTS

Microarray Design and Experimental Datasets

A dataset of 17,510 unigenes from different melon cDNA libraries was used to generate the probes for the microarray (Table 1). The unigenes used to generate each probe are primarily expressed in fruits (24%), in tissues infected with pathogens (40.6%), and derived from a set of different healthy melon tissues (Gonzalez-Ibeas et al., 2007). Probes for the microarray were designed as described in Nimblegen® protocols (NimbleGen Arrays User's Guide v3.0; Roche NimbleGen Probe Design Fundaments) following quality rules such as length (60 mers), non-repetitiveness thus uniqueness, frequency in the transcriptome, and Tm. Nimblegen® technology provides long oligonucletides which has been shown to improve sensitivity and discrimination in microarray experiments (NimbleGen Arrays User's Guide v3.0; Roche NimbleGen Probe Design Fundaments). In turn, fewer probes are required per gene to achieve consistent expression results. The chip design is based in a single chip containing two internal replicate probe-sets (2×) of eleven probes per unigene, covering completely the 350K spotted platform. Thereby, we generated a high-density oligonucleotide-based microarray platform for transcriptome studies in melon.

To test the microarray platform, we designed three different experimental set ups. First, we compared transcriptional changes of (1) fruit development, using immature fruits (15 days after pollination, namely 15d) versus mature fruits (46 days after pollination, 46d) in the non-climacteric Piel de Sapo T111 cultivar; (2) *C. melo* ssp *agrestis* Pat81 (resistant to the fungus responsible for melon root/vine decline, *Monosporascus cannonballus*, henceforth *M.c.*) after inoculation with *Monosporascus cannonballus*

(AGI) versus non-infected roots (S) and, (3) cotyledons infected with CMV inoculated in the Piel de Sapo tendral cultivar (S) versus a mock inoculation (M).

Table 1. Melon expressed sequence TAG (EST) dataset used for microarray construction.

Library	Subspecies/ Cultivar Accession	Tissue/ Physiological condition	High Quality ESTs	Mean EST Length (pb) ESTs ± SD	Unigenes	Redundancy (%)	Library specific EST
15D	Ssp. melo cv. "Piel de Sapo" T111	Fruit 15 days after pollination	3582	608.1 ± 175.2	2939	18	1100
46d	Ssp. melo cv. "Piel de Sapo" T111	Fruit 46 days after pollination	3493	583.0 ± 161.1	2854	18	1063
A	Ssp. Agrestis accession pat81	Healthy roots	3666	700.0 ± 185.4	3189	13	1365
Al	Ssp. Agrestis accession pat81	Roots infected with M. Cannonballus	3255	756.3 ± 137.1	2616	20	1005
Cl	Ssp. melo var. Cantaloupe accession C-35	Cotyledons infected with CMV	5664	651.4 ± 205.7	4679	17	2264
HS	Ssp. melo var. Cantaloupe accession C-35	Healthy leaves	3012	669.3 ± 171.1	2548	15	998
cm	Ssp. melo var. charentais	Leaves	11	597.5 ± 74.1	11	0	8
f	Ssp. melo cv. "Piel de Sapo" T111	Fruit immature	206	412.4 ± 174.4	190	8	70
mc_fi	Ssp. melo cv. "Piel de Sapo" T111	Fruit immature	106	610.8 ± 119.6	99	7	49
mc_p	Ssp. Agrestis accesion P1 161375	Seedlings	748	565.2 ± 135.8	623	17	268
PS	Ssp. Melo cv. "Piel de Sapo" Piñonet torpedo torpedo	Healthy roots	3377	679.9 ± 198.7	2945	13	1258
PSI	Ssp. Melo cv. "Piel de Sapo" Piñonet torpedo	Roots infected with M. Cannonballus	3555	749.3 ± 156.2	3105	13	1363
cornell_fr	Ssp. Melo var. "tam_Dew	Mature fruits	2783	529.7 ± 159.1	1922	31	733

Microarray Quality Testing and Normalization

To evaluate the microarray quality, three biological replicates were hybridized for each experiment. Technical replicates, as well as internal replicates, were also carried out in order to check the reliability of the hybridization and the precision of the microarray. After hybridization and array scanning, images of the physical hybridization were visually inspected for artefacts such as scratches, bubbles, and high regional or overall background. Detected pixels or other non-uniformities in a scanned image were determined as outliers and thus excluded from downstream analyses (data not shown).

Replicates were further taken into a normalization step using the R statistical software (2008) and the oligo package (Carvalho et al., 2007) applied for high-density oligonucleotide Nimblegen® one-color microarrays within the Bioconductor project of open source software (Gentleman et al., 2004). The expression intensities were calculated from scanned images and normalized using the RMA method. Box plots of both pre- (Figure 1A) and post-normalization (Figure 1B) confirmed that our data were successfully normalized. Data quality was assessed comparing the signal intensity data from each array to that obtained from the technical or biological replicates. Pearson correlation between replicates was calculated for every gene in all the arrays, resulting in a very high correlation level, with a coefficient >0.95 for every independent experiment in a pairwise comparison (Figure 1D). This high coefficient is indicative of the precision level in which the microarray is able to process transcriptomics data reliably.

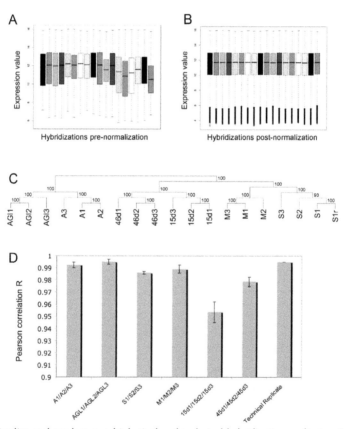

Figure 1. Quality analyses between biological and technical hybridization replicates. Box-plots of the samples before (A) and after (B) the normalization step. The baseline is set to a similar raw expression level, allowing the inter-chips comparison. (C) Hierarchical clustering (HC) of the samples performed using the whole expression data after normalization. Different conditions are separated and replicates cluster together. (D) Pearson correlation at gene-level for all the probes in the replicates of the microarray. All the replicates showed a correlation value greater than $p > 0.95$ thus showing a high level of similarity.

A support tree clustering method with bootstrapping (ST) and a principal compo-nent analysis (PCA) using expression data were performed to statistically validate the tool. The ST is an improvement of the classical hierarchical clustering method (HCL) (Eisen et al., 1998) implementing statistical support for the nodes of the trees based on resampling the data, in our case by bootstrapping from 100 iterations. By performing this statistical analysis, which is used to evaluate the reliability of a tree, we assessed a high level of similarity between the replicates (Figure 1C).

The application of PCA to expression data (where the experimental conditions are the variables, and the gene expression measurements are the observations) let us summarize gene set expression variation in the different conditions (Raychaudhuri et al., 2000). The PCA analysis results together with the ST clustering results showed that the three different experimental populations showed differential expression pat-terns (Figure 2A). Overall, the statistical analyses showed significant separation of the three experiments, with the biological replicates clustering together (Figure 1C; 2A). These results were consistent with the high precision level observed by cor-relation analysis, supporting the high quality of the hybridizations. Together, these studies corroborate the ability of our tool for comparing different experimental conditions.

Taken together the statistical analyses, we conclude that the microarray platform, was efficiently generated, hybridized, and normalized.

Detection of Differentially Regulated Genes

After normalizing the microarray data, we identified the statistically deregulated genes by using the SAM method (Tusher et al., 2001) and its implementation in the Multi Experiment Viewer (MeV v4.2) software suite (Tusher et al., 2006). A total of 937, 198, and 1,182 genes appeared significantly and differentially regulated during fruit ripening, *Monosporascus cannonballus* and CMV infection experiments respectively. Surprisingly, we observed a high variation on differential gene expression inference between the two-pathogen infections (Figure 2C, D). A putative EST density bias due to higher unigene representation for CMV infection is an unlikely explanation, as a similar magnitude of differentially expressed genes was found both in our analysis and previous studies on CMV infection (Whitham et al., 2003). Besides, our microarray has a high percentage of pathogen-interaction EST probes. Rather, the low differen-tially gene expression found in the *Monosporascus cannonballus* experiment could be due to biological issues such as the semi-resistant phenotype of the pat81 line, as suggested in González-Ibeas et al. (2007).

Different fold-change (FC) thresholds were determined according to the SAM analyses for each experiment: FC > 2.1 and FC < 0.53 for 46d versus 15d; FC > 1.3 and FC < 0.75 for AGI versus A; FC > 1.67 and FC < 0.56 for S versus M samples (see Materials and Methods). These results further support the evidence that *Monosporascus cannonballus* infection on the particular cultivar tested induces less and more subtle changes compared to the other conditions.

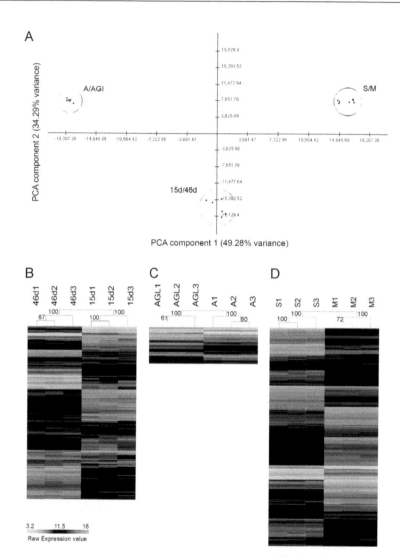

Figure 2. Analyses of differential transcription for the three different data set used in the microarray. (A) Principal Component Analysis (PCA) of the transcript profiles from all the melon samples in the microarray, showing a high separation for each experimental set up. The array data used for this analysis were normalized using the RMA algorithm implemented in the oligo R package for Nimblegen® arrays. Pannels (B), (C), and (D) display support trees (ST) of the deregulated genes for all the conditions. (B) ST of the 937 deregulated genes with a q < 0.01 using the SAM in the 46d vs 15d melon fruit. (C) ST of the 198 deregulated genes with a q < 0.01 using the SAM in the *M. cannonbalus* infection. (D) ST of the 1182 deregulated genes with a q < 0.01 using the SAM in the CMV infection.

Interestingly, a Venn Diagram for all experiments showed that 83.8% of differentially regulated genes during fruit ripening were specific for that condition (809 genes) when comparing it with the other microarray hybridization experiments done in this study (Figure 3). The same was observed in the other two experiments, namely 86.7%

in *Monosporascus cannonballus* infection (166 genes) and 88.5% in CMV infection (1,046 genes) (Figure 3). This high specificity indicates that the microarray platform is able to detect changes in expression profiles both during fruit development or pathogen interaction with high precision with no bias effect due to the probe representation design.

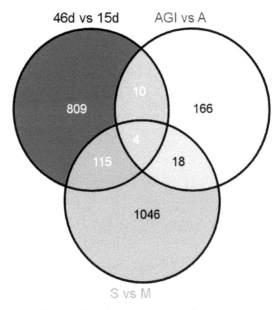

Figure 3. Venn Diagram addresses the high experiment-specificity of the microarray. Venn Diagram using all deregulated genes for all the experiments showed that little overlapping exists. Only four genes were shared between the three experiments, and overlapped genes in pairwise comparison did not exceed 10% of total deregulated genes.

Next, to determine the trend in gene expression for each experiment and to visually display it in two dimensions, we generated a heat map taking into account only genes that showed significant changes in gene expression. The Support Tree (ST) HCL applied to the heat map separated all genes by experiment and condition (Figure 2B–D). This analysis showed a clear trend to down-regulation of gene expression during fruit ripening (Figure 2B) with 74.6% of down-regulated genes and 25.4% of upregulated genes. In the case of *Monosporascus cannonballus* infection, we observed that 31.3% of the genes showed down-regulation and 69.19% were upregulated (Figure 2C). After CMV infection, 58.46% of the genes were upregulated and 41.62% were downregulated (Figure 2C).

Quantitative Real-Time PCR Validation
To validate the microarray data and evaluate the accuracy of the platform, we performed quantitative real-time PCR (qRT-PCR) on 12 genes. Four genes for every microarray experiment were selected based on previous bibliographic reports and deregulation

levels. Then their expression profiles were obtained (Table 2). The oligonucleotide sequences used are summarized in Table 3. The correlation between the microarray results, and those obtained by qRT-PCR was assessed by calculating the Pearson's product moment correlation coefficient (Coppack, 1990; Liu et al., 2004) at replicate level (Table 2). This test statistically assigns a correlation coefficient to the difference in the fold-change from both microarray and qRT-PCR techniques. A global correlation coefficient of 0.868 calculated by the average of every gene was observed. These results indicate that our microarrays are able to detect both low and high fold-changes with high accuracy in different experimental conditions (p-value < 0.01) (Table 2).

Table 2. Microarray and qRT-PCR results of the 12 selected genes with their replicate-level Pearson correlation.

Gene Identifier	Microarray	qRT-PCR	Correlation	p-value	Samples
cCL1715Contig1	10,807	13,279	0.999	1.23E-06	S vs M
cCL3206Contig1	18,273	25,442	0.969	1.35E-03	S vs M
c15d_08-H10-M13R	47,246	138.50	0.988	2.00E-04	S vs M
cPS1_02-F08-M13R	15,915	37,523	0.998	2.84E-06	S vs M
cCL2301Contig1	2,316	2,140	0.889	1.77E-02	AGI vs A
cPSI_33-C12-M13R	2,564	2,370	0.888	1.79E-02	AGI vs A
cCL3647Contig 1	-3,330	-2,941	0.696	1.24E-01	AGI vs A
cCL1700Contig1	-1,581	-1,563	0.781	6.66E-02	AGI vs A
cCL3137Contig1	-57.16	-105.80	0.823	4.43E-02	46d vs 15d
cC15879Contig1	373.92	947.39	0.843	4.58E-04	46d vs 15d
cCL451Contig1	-32.75	-12,367	0.916	1.04E-02	46d vs 15d
c46d_34-C03-M13R	12.02	23,314	0.633	3.07E-03	46d vs 15d

Table 3. Genes and primers used for quantitative RT-PCR.

Gene	Forward Primer (5" -3")	Reverse Primer (5-3')
cCL5879Contig1	AACTTTTTGTGAGTGTGTAATCGTTTTATA	CCGAACATAATGTTACGAATCGATAT
cCL451Contig1	ATAGTAATAAGGAATATTAGAGGGCTT-GTGT	ACCCACTTAAAAAGGGCAAACA
46d_34-C03-M13	CGAAGGGATGAAATTTGTTTGTAA-GAACTAAT	CCATTTTTGGTTCATATATAGAAA
cCL3137Contig1	ATGATATTATTATTCGAAATTGGGAAGTG	AGCAGTCTTGTCTTTTGCTTCTCA
cCL1715Contig1	TAGTTGGTGTGGACCGTGTAGAA	CAGTGTCGGTGTTGAGCACAA
cCL3206Contig1	GCCTTTCGCCCTTCACTTAA	GGAGAAGAAGGCAGCTTATGCTT
c15d_08-HIG-M13R_c	TTATCGTCTTTATGCCCCGAGT	GGTTTCGTTGTCCACTTGATTTT
cPSI_02-F08-M13R_c	TCTTCGAATGTGGTGGGTTCA	CAAAGGCGGTGAATCGAGAA
cCL1700Contig1	TAATCGGTAAGGACGGTTCTG	TAATCGGTAACGACGGTTCTG
cCL2301Contig1	TCGCTCGACTTGATGAAAGAT	AGGTG AAATTCCCTCCTTCAA
cCL3647Contig1	GAGTGGATGGATGAGGAAATG	AAGTTCCAGGCTTAACCCAAA
cPSI 33-C12-M13R	ACTCGATCAACTTCGAGCAAA	TCCCACTGAAGAATACGCATC

Functional Analysis of Microarray Results

In order to shed light into the processes involved in the studied conditions, we analyzed for enrichment of GO terms among genes differentially expressed. The GO annotations were provided by González-Ibeas et al. (2007) (see Materials and Methods). This analysis highlights biological processes that have statistically significant higher number of genes differentially expressed in the studied conditions. Results for GO enrichment analysis and validation by qRT-PCR of selected genes for each independent experiment are explained below.

Fruit Ripening Involves Deregulation in Ethylene Signaling, Sugar Metabolism and Cell Wall-Loosening Enzymes

Several GO terms related to cell wall metabolism were enriched between samples 15d and 46d (Figure 4A). These included "cell wall," "cellulose and pectin-containing metabolic process," and both polygalacturonase and pectinesterase activities (Figure 4A). These are indicative of the important cell-wall modifications associated to the fruit ripening process.

We further corroborated these changes in gene expression during fruit ripening at gene-level and validated the microarray results by qRT-PCR for four selected individual genes that appeared enriched in these categories. Three genes putatively involved in carbohydrate metabolism, cell-wall softening and ethylene production with differential expression between immature (15d) and mature fruit (46d) of the non-climacteric melon Piel de Sapo T111 were selected for qRT-PCR validation. The *cCL-3137Contig1*, is similar to a proline-rich cell wall protein (AT3G22120, 56% similarity, E<1E-15) which was dramatically shut-down in mature fruits (FC = -105.8) (Table 2). In fact, in fleshy-fruit species the flesh softening process implies disassembly and reorganization of the cell walls during the fruit ripening process (Nunez-Palenius et al., 2008).

As determined by the functional enrichment, "carbohydrate metabolism" related genes were differentially expressed during ripening (Figure 4A). Thus, a number of cell wall invertases appeared deregulated (Figure 4A), while others appear upregulated. We chose c46d_34-C03-M13R, which is highly similar to an invertase (UniRef90_Q9ZR55, 64% similarity, E<3E-48) for qRT-PCR validation (FC = 23.314) (Table 2), again confirming our overall observations.

Interestingly, although Piel de Sapo T111 is a non-climacteric variety, we found enrichment for ethylene-related GO terms in the fruit ripening process, showing an over-representation of negative regulation of ethylene-mediated signaling in the up-regulated genes, whereas "ethylene biosynthesis" and "response to ethylene categories" were enriched in down-regulation (Figure 4A). These results provide a molecular insight for the non-climacteric fruit ripening nature of the melon cultivar PS used in our study.

Moreover, the *cCL451Contig1* was selected for validation due to its homology to the CmACO1 (1-aminocyclopropane-1-carboxylate oxidase I) (Balague et al., 1993), involved in ethylene synthesis (UniRef90_Q04644, 88% similarity, E<1E-158). It was shown that CmACO1 antisense expressing plants showed no ethylene production and extended shelf-life in the climacteric cantaloupe cultivar, which is of high economic importance (Ayub et al., 1996). Interestingly, we found that CmACO1 was under-expressed in mature fruits 46d (FC = -12.36) (Table 2), supporting a differential role of this gene controlling ethylene production in climacteric and non-climacteric fruits.

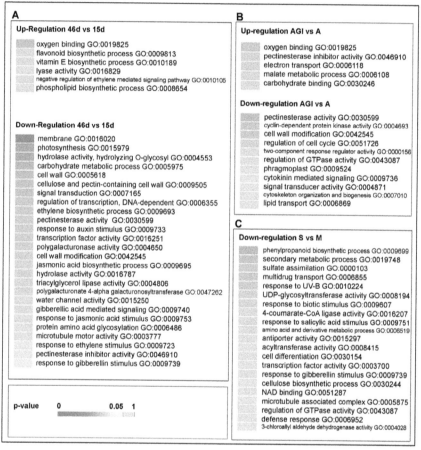

Figure 4. Gene Ontology functional enrichment. (A) Fruit ripening experiment was analyzed for functional enrichment using GO terms and Z-score statistics calculation (see Materials and Methods). Resulting p-values after FDR multiple testing adjustments are visualized in a color code scale; colors toward red signify enrichment for a particular GO term. Gray means no statistically significant enrichment. (B) Enrichment analysis, using Z-score statistics, of the *M. cannonbalus* infection in the *Cucumis melo agrestis* accession roots. (C) Enrichment analysis of the CMV infection in photosynthetic cotyledons in *C. melo* var. tendral.

Finally, to address the detection range of the microarray we chose the most de-regulated gene during fruit ripening in our experiment for qRT-PCR validation. The *cCL5879Contig1*, similar to a gene encoding for a serine-threonine kinase like protein (AT1G01140, 52% similarity, E<1E-12) that showed a FC = 947.39 (Table 2). The homologue of this gene was described as an interactor of calcineurin B-like calcium sensor response proteins (AtCBLs) in *Arabidopsis* (Albrecht et al., 2001). The CBLs are implicated in signaling pathways in response to stress, hormones, and environmental cues (Albrecht et al., 2001), together supporting an important hormonal regulation of fruit maturation in melon.

Overall, our results indicate a high reproducibility level of the dynamic range between the microarray data and the expression profiles by qRT-PCR.

Responses to Fungi Involve Deregulation of Signal Transduction Pathways and Cell Wall and Cytoskeleton Rearrangements

The GO enrichment analyses showed the preferential over-expression of genes related to terms such as "pectinestarase inhibitor activity" and "oxygen binding," related to cytochrome P450 (Figure 4B). It was reported that over-expression of pectin methylesterase inhibitors in *Arabidopsis* restricts fungal infection (Lionetti et al., 2007). It is also well known that pathogen infection usually causes oxidative bursts, triggering the accumulation of reactive oxygen species (ROS) as an early response (Gozzo, 2003; Kotchoni and Gachomo, 2006). Accordingly, GO statistical analyses showed enriched over-expression of genes involved in electron transport (Figure 4B), as expected from known responses (Gozzo, 2003).

Moreover, "cell wall modifications," "cytoskeleton organization and biogenesis," and cell cycle related GO terms such as "cyclin-dependent protein kinase activity," were statistically enriched in downregulated genes, indicating specific changes in these transcriptional responses (Figure 4B).

The expression of four differentially regulated genes involved in "cell wall modifications," "actin cytoskeleton modification," and "oxidative stress" was validated by RT-PCR. The cCL1700Contig1, a putative profilin protein homologue to *Arabidopsis AT2G19760* (PRO1, 74% similarity, E<1E-53) was verified by qRT-PCR (FC = -1.536) (Table 2). Profilin is the main monomer actin binding protein in plant cells (Vidali et al., 2007). Recent studies suggested that profilin is a multifunctional protein with antagonistic effects on actin polymerization. Previous studies related changes in profilin expression in response to Phytophtora infestans (Schutz et al., 2006) in parsley, and to *Tilletia tritici* or *Fusarium gramineum* in wheat (Lu et al., 2005; Zhou et al., 2006). In agreement, the down-regulation of *cCL1700Contig1* in *Monosporascus cannonballus* infected roots support the role for profilins in cytoskeleton reorganization during the defense response in melon. These results were consistent with the determined key role of cell wall and cytoskeleton rearrangement in fungal defense (Schmelzer, 2002).

The unigene *cCL2301Contig1* is a putative melon orthologue of PcCMPG1 from parsley (*Petroselinum crispum*) (76% similarity, E<1E-129), *Arabidopsis AT5G37490* (65% similarity, E<8E-99), (Libault et al., 2007) tobacco NtCMPG1 (71% similarity, E<1E-114), and tomato SlCMPG1 (70% similarity, E<1E-106) (Gonzalez-Lamothe et al., 2006). These have been reported as fast-response genes to elicitor treatment and hence are considered mediators of the immediate early response in *Petroselinum crispum* (Kirsch et al., 2001). Our qRT-PCR results validated the observed up-regulation after inoculation in melon *Cucumis melo agrestis Pat81* accession (FC = 2.140) (Table 2), indicating that this gene might be involved in the response to fungal attack in this resistant melon line (Dias et al., 2004). This result was consistent with a previously reported effect of increased resistance in CMPG1 over-expressing tobacco, while silencing of this gene led to a decreased resistance (Gonzalez-Lamothe et al., 2006).

The *cCL3647Contig1* is a putative orthologue to *Arabidopsis AT2G21610* (81% similarity, E<1E-121), a member of the pectinesterase protein family (Louvet et al., 2006). These enzymes carry demethyl-esterification of pectins, a main component of the primary cell wall. Changes in the degree and pattern of methyl esterification of pectins are associated with cell wall modifications and can influence root development and responses to biotic and abiotic stresses (Camacho-Cristobal et al., 2008). We found a decreased expression of *cCL3647Contig1* (FC = –2.941) (Table 2) after *Monosporascus cannonballus* infection suggesting that pectin biosynthesis might be regulated during defense response in melon.

CMV Infection Induces Structural and Cell-Cycle Deregulation

It has been previously described that plant viruses transiently suppress host gene expression (Aranda and Maule, 1998). In accordance, our analysis showed that 58.46% of deregulated genes were under-expressed in response to CMV infection. Indeed, our GO analysis did not show any category statistically upregulated (Figure 4C), though less stringent analyses pointed to several GO categories potentially upregulated (data not shown). Perhaps strikingly, down-regulated categories include defense-related terms, such as "phenylpropanoid biosynthetic processes" and "defense response" (Figure 4C). Despite this apparent general host gene suppression effect, specific transcripts related to defense responses after pathogen attack increased significantly in their accumulation. These include, for example, WRKY transcription factors and beta-glucanases. To validate these data, we performed qRT-PCR with four CMV-infection related genes. The unigene c15d_08-H10-M13R_c, highly similar to PR-1-like protein in *Arabidopsis* (*AT4G25780*, 83% similarity, E<3E-64), showed a strong induction (FC = 138.50) (Table 2). The unigene cPSI_02-F08-M13R_c, which is highly similar to the elicitor-inducible gene EIG-I24 of tobacco (UniRef90_Q9FXS7, 73% similarity, E<1E-72) and is upregulated after treatment with hyphal wall components (Takemoto et al., 2001), was induced in CMV infected cotyledons (FC = 37.523) (Table 2). Interestingly, *c15d_08-H10-M13R_c* and *cPSI_02-F08-M13R_c* genes were not upregulated at significant level after *Monosporascus cannonballus* infection in melon pat_81 (an accession resistant to this fungus).

Microarray analyses revealed significant changes in gene expression of at least four circadian clock genes. In particular, the *cCL3206Contig1*, highly similar to a pseudo-response regulator (*APRR5*) whose mutation in *Arabidopsis* (*AT5G24470*, 78% similarity, E<5E-21) exhibited circadian-clock associated phenotypes (Yamamoto et al., 2003), was confirmed by qRT-PCR to decrease its accumulation in virus-infected plants. Inhibition of photosynthesis, chloroplast damage, and reduction of chloroplast proteins levels are usual alterations in many virus infections, and oxidative stress has been described to be involved in several plant-virus interactions, including CMV infections (Irizarry et al., 2003). The unigen *cCL1715Contig1* is highly similar to *AT2G15570* gene (67% similarity, E<7E-43), a chloroplast type 3 thioredoxin with thiol-disulfide exchange activity and involved in cell redox homeostasis

(Gelhaye et al.,2005). Downregulation of this gene was also validated by qRT-PCR (FC = 13.279) (Table 2).

CONCLUSION

We report the construction and validation of a high-density oligonucleotide-based microarray tool for functional genomics studies in melon. Our results reveal that this new tool is amenable for high-throughput gene expression analyses in different melon cultivars. Furthermore, the use of the array has proven to be valid for genomic studies on different plant tissues, developmental stages as well as in a range of biotic-stress responses. As a proof of principle, we report changes in gene expression generated by this microarray in three independent experimental set ups. The statistical analyses of our microarray data and the functional analyses using the GO annotations served to characterize the gene expression changes in the validated experiments. These results not only validate our chip, but also provide an important molecular view in yet undocumented ripening and defense responses in melon.

Overall the novel melon microarray platform offers the possibility to carry functional genomics studies of fruit-quality traits in melon, and will be very valuable for those researchers interested in *Cucumis melo* transcriptomics.

KEYWORDS

- **Cucumber mosaic virus**
- **Expressed sequence tags**
- **Hybridization signal intensity**
- **Principal component analysis**
- **Reactive oxygen species**

AUTHORS' CONTRIBUTIONS

Albert Mascarell-Creus, Josep Vilarrasa-Blasi, Santiago Mora-García, Wim Deleu, Daniel Gonzalez-Ibeas, and Ana I Caño-Delgado prepared RNAs, cDNAs for microarray, and Josep Vilarrasa-Blasi. Cristina Roig, Montserrat Saladié, and Daniel Gonzalez-Ibeas carried out the Real-Time-qPCR experiments. Albert Mascarell-Creus, José Blanca, and Joaquin Cañizares carried out the bioinformatics analyses. Pere Puigdomènech is the co-ordinator of The MELOGEN initiative and participated in the conception of the study together with Jordi Garcia-Mas, Miguel A. Aranda, Belén Picó-Silvent, Fernando Nuez, Nuria López-Bigas, and Ana I Caño-Delgado. Ana I Caño-Delgado is the principal investigator of this study and responsible for the delivery of a melon Microarray tool. Ana I Caño-Delgado and Albert Mascarell-Creus processed all the data and wrote the manuscript. All authors read and approved the final chapter.

ACKNOWLEDGMENTS

We acknowledge A. Alibés for critical reading of the manuscript. This work was supported by grants from the Spanish Ministry of Education and Science (BIO2005/007 to Ana I Caño-Delgado; GEN2003-20237-C06 and AGL2006-08069/AGR to Miguel A Aranda). Wim Deleu, Cristina Roig, and Daniel Gonzalez-Ibeas are recipients of CRAG and "Juan de la Cierva" postdoctoral contracts from the Spanish Ministry of Education and Science. Santiago Mora-García was funded by PIV2 program (AGAUR; Generalitat de Catalunya, Spain). Albert Mascarell-Creus and Josep Vilarrasa-Blasi are PhD students funded by I3P program (CSIC; Spain) in Ana I Caño-Delgado laboratory. Ana I Caño-Delgado is funded by a HFSPO/Career Development Award (CDA2004/007). Ana I Caño-Delgado is a "Ramón y Cajal" researcher from the Spanish Ministry of Education and Science.

Chapter 4

Arabidopsis Gene Co-Expression Network

Linyong Mao, John L. Van Hemert, Sudhansu Dash, and Julie A. Dickerson

INTRODUCTION

Biological networks characterize the interactions of biomolecules at a systems-level. One important property of biological networks is the modular structure, in which nodes are densely connected with each other, but between which there are only sparse connections. In this chapter, we attempted to find the relationship between the network topology and formation of modular structure by comparing gene co-expression networks with random networks. The organization of gene functional modules was also investigated.

We constructed a genome-wide *Arabidopsis* gene co-expression network (AGCN) by using 1,094 microarrays. We then analyzed the topological properties of AGCN and partitioned the network into modules by using an efficient graph clustering algorithm. In the AGCN, 382 hub genes formed a clique, and they were densely connected only to a small subset of the network. At the module level, the network clustering results provide a systems-level understanding of the gene modules that coordinate multiple biological processes to carry out specific biological functions. For instance, the photosynthesis module in AGCN involves a very large number (>1,000) of genes which participate in various biological processes including photosynthesis, electron transport, pigment metabolism, chloroplast organization and biogenesis, cofactor metabolism, protein biosynthesis, and vitamin metabolism. The cell cycle module orchestrated the coordinated expression of hundreds of genes involved in cell cycle, DNA metabolism, and cytoskeleton organization and biogenesis. We also compared the AGCN constructed in this study with a graphical Gaussian model (GGM) based *Arabidopsis* gene network. The photosynthesis, protein biosynthesis, and cell cycle modules identified from the GGM network had much smaller module sizes compared with the modules found in the AGCN, respectively.

This study reveals new insight into the topological properties of biological networks. The preferential hub–hub connections might be necessary for the formation of modular structure in gene co-expression networks. The study also reveals new insight into the organization of gene functional modules.

Biological networks characterize the interactions of biomolecules at a systems-level and can help us better understand how biomolecules interact with each other to carry out biological functions in living cells. In the representation of biological networks, it is natural to use graph to describe the interactions between biomolecules. A node in a graph represents a biomolecule such as a gene, a protein or a metabolite, and an edge (or link) indicates the interaction between these two biomolecules. Such interactions could be physical interactions, metabolite flow, regulatory relationships, co-expression relationships, and so forth (Zhang Shihua et al., 2007). One important property of networks is the modular structure, in which nodes are densely connected with each other, but between which there are only sparse connections (Girvan and Newman, 2002). Biomolecules belonging to the same module interact with each other to carry out a specific biological function.

The rapid accumulation of genome-wide gene expression data allows the creation of gene co-expression networks by examining the co-expression patterns of genes over a large number of experimental conditions. In the gene co-expression network, a node is a gene, and an edge is drawn between gene A and B if the correlation coefficient between these two genes is above a threshold. Gene co-expression networks have proven useful in analyzing microarray data in model organisms including yeast, mouse, and human (Bergmann et al., 2004; Carlson et al., 2006; Freeman et al., 2004, 2007; Horvath and Dong, 2008; Lee et al., 2004; Stuart et al., 2003). In plants, since the complete sequencing of the *Arabidopsis thaliana* genome in 2,000, thousands of microarray experiments under diverse conditions have been conducted, and the array data have been deposited in public databases. Accordingly, genome-wide AGCNs have also been constructed by calculating the pairwise gene expression correlations over a large number of microarray experiments, ranging from over 300 arrays to more than 2,000 arrays (Aoki et al., 2007; Ma et al., 2007; Manfield et al., 2006; Mentzen and Wurtele, 2008; Obayashi et al., 2007; Wei et al., 2006).

In detecting gene functional modules (or clusters) from gene co-expression networks, a guide-gene approach is commonly used. In this approach, a set of genes with known functions, termed as guide genes (or *bait* genes), were used to query the gene co-expression network. A subnetwork comprising of the guide genes and the genes that were connected to the guide genes within a user-defined distance was retrieved. A gene module was then considered to be equivalent of the retrieved subnetwork itself (Ma et al., 2007), or it was extracted from the subnetwork using visualization tools (Aoki et al., 2007; Obayashi et al., 2007) or maximal-clique-finding method (Manfield et al., 2006). Using the guide-gene approach, one can find gene modules that are associated with a specific biological function or metabolic process (Aoki et al., 2007; Ma et al., 2007). However, the drawback of this approach is that a module found in this way might be incomplete and belong to a larger and more densely connected module (Aoki et al., 2007). In addition, using visualization to extract modules is subjective and affected by users' judgments. To avoid such drawbacks, an alternative approach, top–down approach (or non-targeted approach), is used to naturally partition the network into modules by applying graph clustering algorithms. Compared with the guide-gene approach that requires the prior knowledge about the seed genes, the top–down approach is relatively knowledge independent and novel hypotheses might be developed from the clustering result (Freeman et al., 2007; Lee et al., 2004; Mentzen and Wurtele, 2008).

In this chapter, we used a top–down approach to identify and evaluate gene functional modules from large *Arabidopsis* microarray data sets. First, we constructed

AGCN by using more than 1,000 high quality microarrays. Then, we analyzed the topological properties of the network and extracted modules from the network by using Markov Clustering (MCL) Algorithm (Van Dongen, 2000a). The functional coherence of the extracted modules was evaluated. In this chapter, we attempted to assess if there exists intrinsic modular structure in AGCN and find the relationship between the network topology and formation of modular structure by comparing the real biological network with random networks. We then focused our analysis on two gene functional modules, photosynthesis module and cell cycle module, that are central to plant growth and development. A close examination of the organization of these two modules reveals that both modules involve multiple biological processes coordinated at the transcriptional level. Although our findings are based on the analysis of *Arabidopsis* microarray data, the uncovered network properties and organization of gene functional modules may have implications in non-plant organisms and other types of biological networks such as protein interaction networks.

MATERIALS AND METHODS

Normalization and Pcc Calculation

The 1,094 arrays from AtGenExpress were normalized using the justMAS function in the simpleaffy package (version 2.8.0) (Wilson and Miller, 2005) downloaded from Bioconductor with the target value set to 500. After the normalization, genes satisfying the following two conditions were selected for the further analysis: (1) the ratio between standard deviation and mean of a gene's expression values over the 1,094 arrays is greater than 0.5; (2) the difference between a gene's maximal expression value and minimal value among the 1,094 arrays is greater than 32. After the filtration, the remained genes' expression values were $\log_{10}$ transformed. During the transformation, if a gene's expression value was less than one, its $\log_{10}$ transformed value was converted to 0 instead of a negative number. The Pearson correlation coefficient (Pcc) between two genes over the 1,094 arrays was calculated using their $\log_{10}$ transformed values.

We also applied our network-based approach to a small data set which consists of 14 samples. The data set profiles gene expression in response to cold stress which is part of AtGenExpress (Kilian et al., 2007). We used the above two criteria to select genes that showed significant changes across the 14 samples. After the filtration, the remained genes' expression values were $\log_{10}$ transformed, and the Pccs for gene pairs over the 14 samples were calculated.

Network Analysis

Node degree indicates the number of links connected by a node. Network density is defined as a ratio of the observed number of edges to all possible edges among the network nodes. The clustering coefficient (C_n) of a given node n was calculated as the following. Assuming the node n has k ($k \geq 2$) directly connected neighbors, then

$$C_n = \frac{e(k)*2}{k^*(k-1)}$$

where *e(k)* is the observed number of edges among the *k* neighbors. The $<C_n>$ represents the average clustering coefficient of the network over all nodes which have at least two neighbors. The clustering coefficient (C_k) with respect to the node degree k is the average over all nodes each of which has exact *k* neighbors.

Mass fraction, area fraction, and efficiency were used to quantify the overall quality of the network clustering. The mass fraction is defined as the following. Let *e* be an edge of the network. The clustering captures *e* if the two nodes connected by *e* belong to the same module (cluster). Now the mass fraction is the ratio between the joint weights (Pccs) of all captured edges over all modules and the joint weights of all edges in the network Van Dongen, (2000a). The area fraction, *AF*, is calculated as the following:

$$AF = \frac{\sum_{i=1}^{M} N_i * (N_i - 1)}{N^* (N - 1)}$$

where *M* is the number of modules extracted from the network, N_i is the number of nodes contained in the ith module, and *N* is the number of nodes in the network. A low area fraction indicates a fine-grained clustering whereas a high area fraction indicates a coarse clustering (Van Dongen, 2000a). The efficiency aims to balance between the objective to obtain a high mass fraction and the objective to keep the area fraction low. The formal definition of efficiency can be found in (Van Dongen, 2000b).

The random network, which assumed the same node degree distribution as the AGCN, was generated using the randomNodeGraph function in the R package Graph (version 1.15.6).

GO/Pathway Term Enrichment Analysis

The Gene ontology (GO) terms for *Arabidopsis* loci were downloaded from http://www.geneontology.org. The GO terms were then assigned to array probe sets based on the correspondence between the probe sets and loci obtained from the *Arabidopsis* Information Resource (TAIR) (Rhee et al., 2003). The GO term enrichment analysis was carried out by using BiNGO 2.0 (Maere et al., 2005). Bonferroni Family-Wise Error Rate (FWER) correction was used to control the false positive rate. If a GO term in a module showed a FWER corrected p-value less than 0.05 in comparing with the AGCN, which comprised of 6,206 probe sets, under a hypergeometric distribution, then the GO term was determined to be significantly enriched in this module.

For some module, more than 50 biological process GO terms were significantly over-represented. To simplify our functional annotation of the module, these GO terms were consolidated to obtain a small set of representative major GO terms. Firstly, the GO terms that were too general (e.g., macromolecule metabolism) were manually discarded. Secondly, based on parent-child relationships depicted in a hierarchical graph of over-represented GO terms (e.g., Figure 10A), the GO terms that were at the top level were manually retrieved from the graph to represent the GO terms that were at lower levels. Thirdly, genes annotated to the GO terms that were retrieved in Step 2 were inspected so that each gene is only associated with one GO term. Thus, the retrieved major GO terms are associated with non-overlapping gene sets.

The pathway information for *Arabidopsis* genes was obtained from AraCyc 4.0 (Mueller et al., 2003). The criteria to detect the significantly enriched pathway terms in a module are same as those to detect GO terms.

DISCUSSION

In this study, we used a top–down approach (or non-targeted approach) to naturally partition the genome-wide AGCN into gene modules based on the topological property of the network. We used an efficient graph clustering algorithm to identify modules from the AGCN. Compared with the traditional clustering analysis such as hierarchical clustering and k-means clustering, the network approach provides additional structural information regarding the connectivity of genes (Schadt and Lum, 2006). Genes belonging to the same module are not only highly correlated at the expression level but also densely connected to each other. Thus, compared with a cluster obtained from traditional clustering methods, a module detected from the network is a more tightly controlled structure which would be more biologically meaningful and resilient to data noise (Freeman et al., 2007; Ruan and Zhang, 20007; Schadt and Lum, 2006).

The constructed AGCN and its extracted modules showed the following properties: (1) The distribution of the node degree fits to a power law distribution (Figure 4A). Such distribution was also observed in the human gene co-expression network and conserved gene co-expression network derived from the human, fly, worm, and yeast comparisons (Freeman et al., 2004; Stuart et al., 2003). (2) The hub genes in the AGCN were densely connected to each other as shown by the clique structure formed by the 382 hub genes. Although it contradicts with a commonly held view that hub nodes tend not to link to each other, it has been recently reported that hub–hub interactions were not suppressed in a multi-validated high-confidence protein interaction network for yeast (Batada et al., 2006). In addition it was shown that nodes in gene co-expression networks tended to connect with the ones with similar degrees while the connections between highly and lowly connected nodes were suppressed (Bergmann et al., 2004). Here, we further argue that the preferential hub–hub connections would be necessary for the formation of modular structure in gene co-expression networks. The hub genes which are densely connected to each other would be self-contained in a single module. In our case, the 382 hub genes were embedded in Module 1. In contrast, for the random networks, which had the same distribution of node degrees as AGCN but in which the connection between a hub and a low degree gene would not be expected to be suppressed, our results clearly indicate the lack of modular structure. (3) The average clustering coefficient of AGCN was increased by more than one fold compared with the random networks, supporting the modular structure in AGCN. (4) Similar to the node degree distribution, the distribution of the module size also follows a power law distribution (Figure 4C). Interestingly, the distribution of the sizes of 400 complexes extracted from the yeast protein interaction network also displayed the power law distribution (Pu et al., 2007). The biological implication of the power law distribution of the module size remains to be further investigated.

The approach used in this study, constructing a gene co-expression network and naturally partitioning the network into modules, provided a systems-level understanding of the gene modules that coordinate multiple biological processes to carry out specific biological functions. Plants convert light energy to chemical energy through photosynthesis. Our results suggested that photosynthesis module in *Arabidopsis* involves a very large number (>1,000) of genes which participate in photosynthesis and related biological processes. The related biological processes encompassed protein biosynthesis, electron transport, cofactor metabolism, chloroplast organization and biogenesis, pigment metabolism, and vitamin metabolism. The GO terms for these biological processes were all significantly over-represented, suggesting the important roles of these biological processes in photosynthesis. Nevertheless, other biological processes which were not significantly over-represented might also play a role in photosynthesis. Different from most animals, plants develop continuously with new organs being developed throughout the lifetime of the plant (Schmid et al., 2005). The cell cycle regulation is one of the keys to the control of plant development. The cell cycle module detected from AGCN orchestrated the coordinated expression of hundreds of genes participating in cell cycle, DNA metabolism, and cytoskeleton organization and biogenesis. Interestingly, a human cell cycle module, which was obtained from an integrated analysis of ~2,000 cancer arrays, GO, and pathways, contained approximately the same number of genes (263) and similar gene function compositions (Segal et al., 2004). In this report, we studied the gene co-expression network and functional modules for *Arabidopsis*. The same approach should be applicable to other model organisms.

Genes in the same module are co-expressed across diverse conditions, suggesting the potential underlying co-regulation mechanism. The list of gene modules obtained in this study would provide a useful tool for the regulation investigation. One approach to linking co-expression to co-regulation is to examine the putative transcription factor binding sites (TFBS) in the promoters of the co-expressed genes (Obayashi et al., 2007).

RESULTS

We used 1,094 non-redundant Affymetrix ATH1 arrays from the AtGenExpress consortium to calculate the pairwise correlations between genes. These arrays were normalized to the same scale by employing MAS algorithm (see Materials and Methods). The AtGenExpress array data has been shown to be highly reliable and reproducible (Kilian et al., 2007; Schmid et al., 2005). Figure 1 shows the experimental conditions of these 1,094 ATH1 arrays. The high quality of the microarray data and diverse experimental conditions allow us to capture the true co-expression relationship between two genes.

To circumvent the correlation computed from noise, we used only the genes that showed significant changes across the 1,094 conditions (see Materials and Methods). Of the 22,746 *Arabidopsis* probe sets on the ATH1 chip, 16,293 (72%) were selected. Next, the genes' expression values were log transformed and Pcc was computed between each pair of the 16,293 genes (see Materials and Methods).

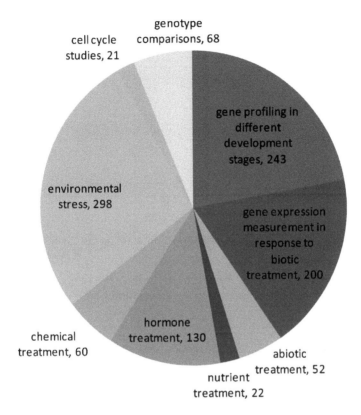

Figure 1. Composition of the 1,094 ATH1 arrays according to the experimental conditions they represent. For a detail description of the arrays, refer to the TAIR website http://www.arabidopsis.org/portals/expression/microarray/ATGenExpress.jsp.

To choose an appropriate Pcc cutoff value, we examined the changes in the node number, edge number, and network density as a function of Pcc cutoff values. As the cutoff value increased, both the node number and edge number decreased (Figure 2A); however, as the cutoff reached a relatively high value, the decreasing rate of edges became slower than that of nodes, which might lead to an increase in the network density. Indeed, as shown in Figure 2B, the network density showed minima around 0.70 Pcc cutoff value and increased thereafter. A Pcc cutoff value greater than 0.70 would be appropriate so that edges with high Pcc values would densely connect a decreasing number of nodes, which would facilitate the following detection of biologically meaningful modules (Aoki et al., 2007). In this study, Pcc cutoff value was set to 0.75 so that a relatively large number of nodes could be retained in the network. At this relatively stringent cutoff value, only the top 0.39% of all possible edges among the 16,293 genes with respect to their Pcc values were retained. The resulting AGCN contains 6,206 nodes, 512,936 edges, and a network density of 0.0266. For comparison, three random networks that each preserved the node numbers and node degrees of AGCN were also created (see Materials and Methods).

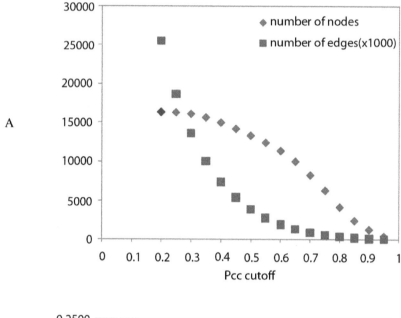

A

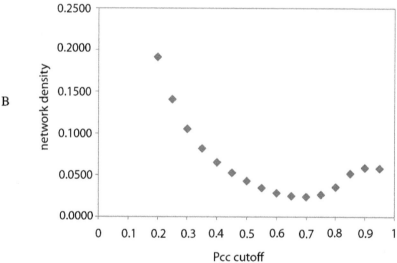

B

Figure 2. Choosing Pcc cutoff values. (A) The number of nodes and number of edges as a function of Pcc cutoff value. Only edges with Pcc greater than the cutoff value were used to construct the co-expression network. Only nodes connected by these edges were used in our network analysis. (B) Network densities at different Pcc cutoff values.

Network Topology

Figure 3A displays a layout of AGCN using the Cytoscape software package (Shannon et al., 2003). The network comprises 100 disconnected components. Within each component, each pair of nodes was directly or indirectly connected. The major component in the network has 5,743 (92.5%) nodes. The smallest component contains only two nodes, and 68 such components were found. The qualitative global topology of the AGCN is similar to that of the yeast protein interaction network which comprises a major component covering 93% of the network nodes and many small components (Batada et al., 2006).

A

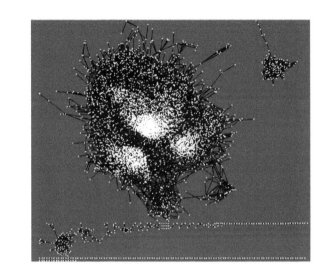

B

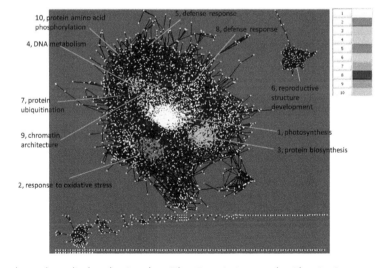

Figure 3. Network topology displayed using the yFiles Organic Layout algorithm in Cytoscape (Shannon et al., 2003). (A) Layout of the *Arabidopsis* gene co-expression network. A white rectangle represents a node (i.e., probe set). A black edge connecting two nodes indicates the co-expression relationship between these two nodes. (B) Mapping the 10 largest modules onto the network. The most over-represented biological process GO term was also shown with each module.

On average, each node in AGCN has 165 co-expression links, but the distribution of the node degrees is highly skewed. The distribution fits to a power law distribution with a tail (Figure 4A), indicating that the network is scale free. Interestingly, we found that the top 382 nodes (genes) in terms of their degrees connected to each other and formed a 382 member clique. Each of these 382 genes has at least 889 co-expression links. We then examined the immediate neighbors of these 382 genes. Surprisingly, these 382 genes were connected to only 1,099 other genes in AGCN, whereas the same set of genes were linked to 4,913, 4,964, and 4,970 other genes in the three random networks, respectively. Thus, unlike the random network, these hub genes did not reach out to the entire AGCN. They were rather densely connected only to a fraction of the network. Since a module is a subnetwork which is densely connected within itself but sparsely connected with rest of the network, these 382 genes and many of their densely connected neighbors will form a large module with the clique structure serving as the module's core. It was later confirmed by network clustering results (see below).

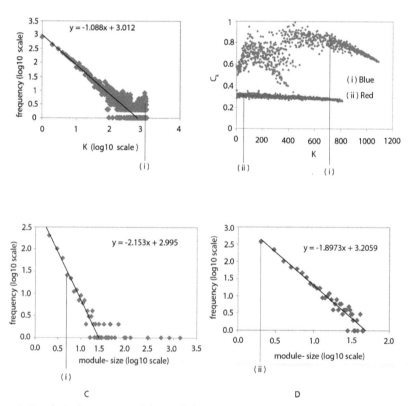

Figure 4. Topological properties of the *Arabidopsis* gene co-expresion network (AGCN) and the GGM network (Ma et al., 2007). (A) Distribution of the node degree (K) for AGCN. (B) Comparing the distribution of the clustering coefficient (Ck) with respect to the node degree between AGCN and random networks. The three random networks exhibited almost identical distributions. For clarity, only one random network's distribution was shown. (C) The distribution of the module size for AGCN. (D) The distribution of the module size for GGM network.

We analyzed the functions of the 382 hub genes using GO (see Materials and Methods). Of the 382 genes, 335 were annotated with cellular component GO terms. Interestingly, among the 335 genes, the products of 265 genes (79%) were located in chloroplast with a p-value as low as 1.87E-132. Table 1 lists the biological process GO terms that were significantly over-represented in the hub genes. The most dominant term was photosynthesis with respect to p-values. The 382 genes forming a co-expression clique exhibited a maximal degree of coordination at the transcriptional level, suggesting that these genes might be involved in a common biological function. Based on the GO analysis and the observation that several over-represented biological processes (e.g., electron transport, pigment metabolism, glucose metabolism) are coupled with photosynthesis, the 382 genes are very likely to function in photosynthesis. Additional evidence towards this conclusion is provided below.

Table 1. Comparing significantly enriched biological process GO terms in the 382 hub genes with Module 1.

GO Term	Hub-genes p-value[1]	Module 1 p-value
cell redox homeostasis	4.99E-05	NA
chloroplast organization and biogenesis	NA	7.01E-08
cofactor metabolism	1.86E-06	1.93E-18
electron transport	5.49E-06	1.80E-06
glucose metabolism	7.61E-05	NA
glycine catabolism	7.92E-05	NA
photosynthesis	1.38E-24	1.38E-52
pigment metabolism	3.48E-07	3.03E-12
protein biosynthesis	1.15E-07	3.05E-07
protein folding	5.70E-05	NA
vitamin metabolism	NA	8.41E-06

1. 'NA' indicates the GO term was not significantly over-represented.

To further evaluate the network topology, we analyzed the property of clustering coefficients. The clustering coefficient of a given node, C_n, measures how close the node n and its directly connected neighbors resemble a clique (see Materials and Methods). The AGCN exhibited an average clustering coefficient, $<C_n>$, of 0.640, whereas the three random networks exhibited an average clustering coefficient of 0.304, 0.305, and 0.306, respectively. That $<C_n>$ of AGCN is more than twice of the random network indicated the potential modularity in the co-expression network (Barabasi and Oltvai, 2004). The distribution of the clustering coefficient (C_k) with respect to the node degree (k) further distinguished the co-expression network from the random networks (Figure 4B). For each random networks, C_k was approximately a constant with respect to k, and the variation of C_k was small (stdev = {0.0156, 0.0162, 0.0164}). For the AGCN, C_k exhibited a complex relationship with k, and the variation of C_k was much larger (stdev = 0.107). The complex relationship between C_k and the node degree may affect how modules are organized in AGCN (Barabasi and Oltvai, 2004).

Network Clustering Analysis

We used the MCL algorithm to partition AGCN into gene modules. MCL is an efficient graph clustering algorithm based on the simulation of random walks within a graph. MCL has been applied to detect modules in yeast protein interaction networks (Pu et al., 2007) and protein family networks (Enright et al., 2002). A recent study, which evaluated four clustering algorithms for protein interaction networks, showed the superior performance of MCL in the identification of protein complexes (Brohee and Helden, 2006). The algorithm is very efficient and took only two minutes to perform clustering on the AGCN (the Linux command: mcl AT-cor-net-0d75 -I 1.8 -- abc -scheme 7) on a 3.6 GHz Intel Xeon CPU with 3 GB memory.

The MCL algorithm has an important parameter, the Inflation parameter (I). A higher value for I tends to produce a larger number of modules with a smaller module size. We tested different inflation values on the AGCN, as well as the three random networks. We used area fraction, mass fraction, and efficiency (see Materials and Methods) to assess the overall quality of the network clustering. Since a module is a densely connected subnetwork and the connections between modules are sparse, clustering on a network with the intrinsic modular structure should produce a small area fraction but a large mass fraction close to one. This is indeed the case for the AGCN (Figure 5A). For example, when I was set to 1.5, clustering on AGCN captured 97.9% of the entire edge masses by using only 9.3% of the network area, reflecting the presence of modular structure in AGCN. In contrast, with the same inflation value, all three random networks had to use 93% of the network area to capture a similar mass fraction (Figure 5A), suggesting the absence of modular structure. An appropriate value for I is between 1.5 and 3.0. Within this range, clustering on the co-expression network used 510% of the area to capture more than 85% of the entire edge masses. The above analysis is purely mathematical. Recently Brohee and van Helden evaluated the MCL algorithm in identifying protein complexes from yeast protein interaction networks, and they chose 1.8 as the optimal value for I based on the analysis of 42 artificial biological networks that simulated the data sets obtained from high-throughput experiments (Brohee and Helden, 2006). They also found that when I was set at 1.8, MCL was resilient to network noise.

With the inflation value set to 1.8, MCL detected 527 modules from the AGCN. The 10 largest modules were mapped to the network (Figure 3B). Similar to the node degree distribution, the module size distribution is also highly skewed. The largest module had 1,381 nodes whereas 86% of the modules had fewer than 10 nodes. The average size of the modules is 11.8 and the median is 3. The log–log plot of frequency versus module size demonstrates that the distribution of the module size followed a power law distribution with tails (Figure 4C). With the same parameter setting, the largest module extracted from the three random networks contained 5,398, 5,403, and 5,528 nodes, respectively. And the size of the second largest module only ranged from 9 to 13 nodes. The module size distribution further indicates the lack of modular structure in the random network.

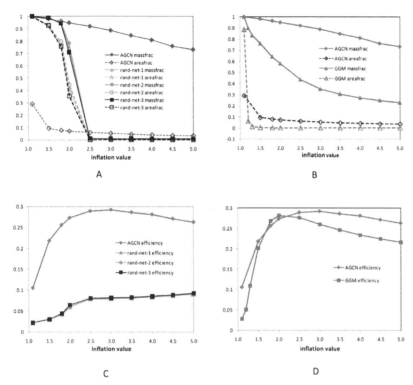

Figure 5. Assessment of the quality of network clustering. (A) Comparing the effects of inflation values on area fraction and mass fraction between AGCN and three random networks. Clustering on a network with the intrinsic modular structure should produce a small area fraction but a large mass fraction close to one. This is indeed the case for the AGCN. In contrast, all three random networks had to use a large area fraction to capture a large mass fraction, suggesting the absence of modular structure. (B) Comparing the effects of inflation values on area fraction and mass fraction between AGCN and GGM network. (C) Comparing the effects of inflation values on efficiency between AGCN and three random networks. The efficiency aims to balance between the objective to obtain a high mass fraction and the objective to keep the area fraction low. A higher efficiency indicates a better performance on network clustering by using some mathematical criteria. A formal definition of efficiency can be found in (Van Dongen, 2000b). (D) Comparing the effects of inflation values on efficiency between AGCN and GGM network.

Module Annotation

Since genes belonging to the same module are co-expressed across diverse conditions, functional coherence among the module members is expected. We carried the enrichment analysis of biological process GO terms in 317 modules containing three or more members. The 127 of the 317 modules (40%) had GO terms that were significantly over-represented (i.e., FWER-adjusted pvalue < 0.05, see Materials and Methods). We categorized these 127 modules by manual annotations (Figure 6A). Not surprisingly, the largest group is associated with "response to stimulus" (Figure 6B), the major theme of microarray experiments in AtGenExpress (Figure 1).

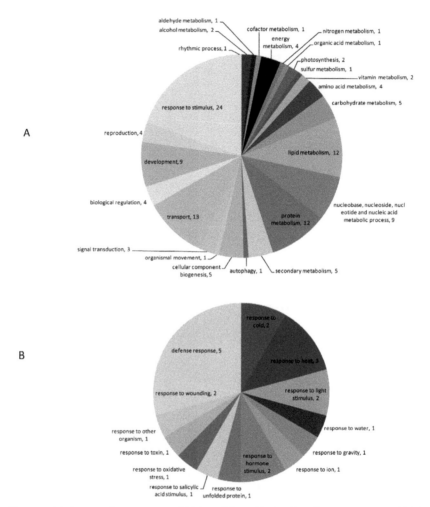

Figure 6. (A) Functional annotations of 127 modules with significantly over-represented biological process GO terms. The number associated with each annotation indicates the number of modules annotated to that category. See our website for the list of 127 modules. (B) Composition of the 24 modules that were annotated to response to stimulus.

Furthermore, we observed that 46 of the 317 modules (14.5%) in AGCN had over-represented GO terms with FWER-adjusted p-values below 5E-4; whereas all three random networks had less than 1% of the modules with over-represented GO terms at such significance level (Figure 7A). The annotations for these 46 modules include both central metabolic processes and specific cellular functions (Table 2). In addition, we compared the clustering result for AGCN using 1.8 inflation value with that using 3.0. Although MCL clustering showed maximal efficiency at 3.0 (Figure 5C), the clustering result at 1.8 inflation value produced a higher percentage of functionally coherent modules (Figure 7A).

Table 2. Significantly enriched GO terms in AGCN modules (I = 1.8).

Module	# Annotated Genes[1]	GO Terms[2]	Genes in GO Term[3]	p value[4]
1	1208	photoshynthesis	97/107	1.38E-52
2	684	response to oxidative stress	30/88	1.43E-07
3	438	protein biosynthesis	95/204	5.68E-52
4	256	DNA metabolism	59/107	9.07E-52
5	104	defense response	18/205	4.79E-08
6	87	reproductive structure development	15/100	3.16E-11
7	78	protein ubiquitination	5/14	1.00E-06
8	59	defense response	19/205	1.65E-13
9	48	establishment and/or maintenance of chromatin architecture	6/43	1.71E-06
11	39	cell wall modification	5/32 2	64E-06
15	36	response to wounding	7/54	4.50E-08
18	33	cuticle biosynthesis	5/8	3.45E-10
21	29	toxin metabolism	5/25	1.54E-07
27	23	glucosinolate biosynthesis	4/9	3.12E-08
28	18	RNA processing	5/65	1.65E-06
31	20	monosaccharide metabolism	5/47	5.67E-07
33	17	response to heat	10/47	1.69E-17
34	18	secondary cell wall biosynthesis (sensu Magnoliophyta)	6/8	1.51E-14
35	17	protein biosynthesis	11/204	1.75E-12
41	16	enzyme linked receptor protein signaling pathway	5/51	2.49E-07
42	15	organic acid metabolism	9/186	2.38E-10
43	13	response to auxin stimulus	7/69	6.62E-11
46	14	lipid metabolism	7/194	1.93E-07
50	12	cellular respiration	6/15	1.34E-13
56	11	leaf development	4/28	1.87E-07
67	10	starch metabolism	8/20	3.18E-19
79	9	indoleacetic acid metabolism	3/4	1.28E-08
80	8	phenylpropanoid metabolism	6/44	5.71E-12
119	6	response to heat	6/47	3.14E-13
122	6	lipid transport	4/47	7.49E-08
140	5	nitrogen compound metabolism	5/129	7.23E-09
154	4	wax biosynthesis	2/7	8.65E-06
165	4	glutamate biosynthesis	2/2 4	12E-07
170	4	electron transport	4/219	2.64E-06
173	4	fatty acid beta-oxidation	2/9	1.48E-05
180	4	purine transport	2/3	1.24E-06
190	4	response to water deprivation	4/41	2.87E-09
201	2	Glycolysis	2/16	8.24E-06
205	3	RNA splicing, via transesterification reactions with bulged adenosine as nucleophile	2/5	2.06E-06

Table 2. (Continued)

Module	# Annotated Genes[1]	GO Terms[2]	Genes in GO Term[3]	p value[4]
213	4	response to heat	4/47	5.05E-09
224	3	fatty acid beta-oxidation	2/9	7.41E-06
238	3	sulfolipid biosynthesis	2/2	2.06E-07
266	3	ovule development	2/12	1.36E-05
273	3	phenylpropanoid metabolism	3/44	5.06E-07
311	3	Proteolysis	3/130	1.37E-05
315	3	response to iron ion	2/3	6.18E-07

1. The number of genes which were assigned with biological process GO terms in a module.
2. Only the most over-represented GO term was listed for each module.
3. The two values listed in this column refer to the number of genes associated with the over-represented GO term in the module and the number of genes associated with the same GO term in the network.
4. The p value indicated the probability that a module contains equal or larger number of genes associated with the GO term under a hypergeometric distribution.

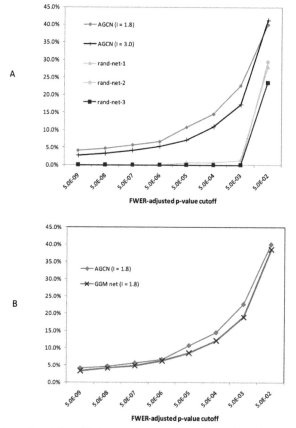

Figure 7. Percentage of modules with three or more members that had significantly over-represented biological process GO terms using different p-value cutoffs. For example, the data points at p-value cutoff of 5.0E-02 indicate the percentage of modules that had enriched GO terms with Bonferroni Family-Wise Error Rate (FWER) adjusted p-values less than 5.0E-02. (A) Comparing AGCN with three random networks. (B) Comparing AGCN with GGM network.

In addition to analyzing the enrichment of GO terms in modules, we also analyzed the over-representation of pathway terms (see Materials and Methods). However, only 10% of the 6,206 genes in AGCN have been annotated as metabolic pathway genes. Among the 39 modules, each of which contained at least three annotated pathway genes, 26 were detected with significantly enriched pathway terms. Ten of them (FWER-adjusted p-value < 5E-4) were listed in Table 3. The table also indicates the correspondence between the enriched pathway terms and biological process GO terms.

Table 3. Significantly enriched pathway terms in AGCN modules (I = 1.8).

Module	# Pathway Genes[1]	Pathway Term[2]	GO Term[3]	Genes in Pathway Term[4]	p value
1	192	photosynthesis, light reaction	Photosynthesis	27/27	4.69E-15
3	22	*de novo* biosynthesis of purine nucleotides	purine nucleoside monophosphate biosynthesis	7/10	2.84E-09
4	5	de novo biosynthesis of pyrimidine deoxyribonucleotides	DNA metabolism	3/6	5.02E-06
11	11	homogalacturonan degradation	cell wall modification	10/34	6.47E-13
45	6	acetyl-CoA biosynthesis (from citrate)	acetyl-CoA biosynthesis	2/2	7.82E-05
67	6	starch degradation	starch metabolism	5/11	3.66E-09
79	8	glucosinolate biosynthesis from tryptophan	indoleacetic acid metabolism	5/5	7.45E-11
80	6	flavonoid biosynthesis	phenylpropanoid metabolism	4/6	3.67E-08
140	3	tryptophan biosynthesis	nitrogen compound metabolism	3/5	2.53E-07
273	3	salicylic acid biosynthesis	aromatic compound biosynthesis	2/3	4.69E-05

1. The number of genes which were annotated as pathways genes in a module.
2. Only the most over-represented pathway term was listed for each module.
3. The column lists the over-represented GO term that matches or relates to the pathway term for each module.
4. The two values listed in this column refer to the number of genes annotated to the over-represented pathway in the module and the number of genes annotated to the same pathway in the network.

The effectiveness of our approach is best illustrated by the correspondence of these computational modules with actual biological entities. Three of these modules are examined in detail from this perspective and are presented below.

Module 1—Photosynthesis

Module 1, the largest module in AGCN, had 1,381 nodes, 399,922 edges, and a density of 0.42. As shown previously, the 382 hub genes in AGCN formed a clique and altogether connected to 1,099 neighbors. All of the 382 hub genes and 927 of their neighbors were included in Module 1 and constituted 95% of the module members.

The significantly over-represented biological process GO terms detected in Module 1 are depicted in Figure 8A. They were consolidated into seven major GO terms based on their hierarchical relations (Figure 8B, see Materials and Methods). Five of the seven major GO terms are also significantly over-represented in the 382 hub genes (Table 1). The p-values for protein biosynthesis and electron transport GO terms are similar between Module 1 and the hub genes, respectively. However, the p-values for the other three GO terms (photosynthesis, cofactor metabolism, and pigment metabolism) in Module 1 are much more significant than those in the hub genes (Table 1), indicating that Module 1 was formed by recruiting more functionally related genes to the 382 member clique.

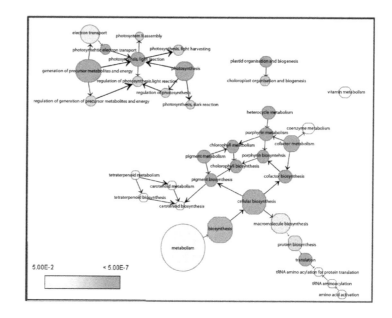

A

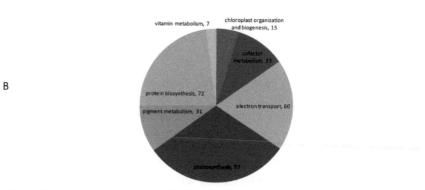

B

Figure 8. Functional analysis of Module 1. (A) Significantly over-represented biological process GO terms detected in Module 1. Each colored circle represents an over-represented GO term. The color scale indicates the p value of the over-represented GO term. An arrow from GO term A to Go term B indicates that A is the parent of B. (B). Seven major biological process GO terms retrieved from (A). The number following each major GO term refers to the number of genes that were annotated to that category. See our website for the gene lists.

Among the seven major GO terms, photosynthesis is the most over-represented biological process in Module 1 (Table 1), and three other processes (electron transport, pigment metabolism, and chloroplast organization and biogenesis) bear direct physiological connections/associations with photosynthesis. We show below that the remaining three major processes (cofactor metabolism, protein biosynthesis, and vitamin metabolism) are also strongly integrated with photosynthesis, respectively. Thus, photosynthesis becomes the uncontested umbrella process for Module 1.

The 33 cofactor metabolism genes in Module 1 included six genes involved in ATP biosynthesis, seven genes in NADPH regeneration, and 12 genes in the biosynthesis of photosynthetic electron carriers such as Fe-S cluster, quinones, and hemes.

Among the 72 protein biosynthesis genes, the potential locations of the products of 63 genes could be assigned to chloroplast based on their cellular component GO terms, genes' annotations (Rhee et al., 2003), and literatures. These 63 genes included 41 genes which might function as the structural constituents of the chloroplast ribosome, eight genes involved in translation initiation/elongation/release, and seven genes involved in tRNA aminoacylation. These protein biosynthesis genes are probably involved in the synthesis of photosystem proteins inside chloroplast. For example, *RPS17* and *RPL9* encode two chloroplast ribosomal proteins, and the transcripts of these two genes were much more abundant in leaves and stems than they were in roots (Thompson et al., 1992). In a mutation of *Arabidopsis* where the RPS17 expression was dramatically reduced, the activity of the photosystem I (PSI) was significantly reduced (Woo et al., 2002). The *HCF107* gene encodes a protein localized to the chloroplast membrane (Sane et al., 2005). The experimental results demonstrated the critical role of *HCF107* in the 5'-end processing/stability and/or translation of the *psbH* (*photosystem II protein H*) gene as well as in the translation of the *psbB* gene (Sane et al., 2005). The *HCF109* encodes a peptide chain release factor 2, which is involved in the process of translational termination in chloroplasts (Meurer et al., 2002). In the *HCF109* mutant, the protein abundances for two ATP synthase subunits, the photosystem *I PsaC*, the photosystem *II PsbB*, and *PsbZ* were substantially reduced (Meurer et al., 2002).

The seven vitamin metabolism genes are involved in vitamin B1 (two genes), B2 (one gene), B6 (one gene), C (two genes), and E (one gene) biosynthesis. Among the seven genes, *AT5G28840* and *VTC2* encode enzymes involved in the Ascorbic Acid (AsA, vitamin C) biosynthesis pathway. It was shown that the light regulation of AsA biosynthesis in *Arabidopsis* leaves is dependent on the photosynthetic electron transport chain (Yabuta et al., 2007). On the other hand, AsA is a potent antioxidant which could detoxify the reactive oxygen generated by photosynthesis and adverse environmental conditions (Conklin et al., 2000). Ascorbate-deficient mutant of *Arabidopsis* exhibited the symptoms of chronic photooxidative stress when grown in high light (Muller-Moule et al., 2004). Other vitamins such as vitamin B6 and E could also function as potent antioxidants and protect plants from the photooxidative stress (Havaux et al., 2005; Titiz et al., 2006).

With respect to over-represented pathway terms in Module 1, both light reaction and dark reaction of photosynthesis were significantly over. The pathways to synthesize two photosynthetic pigments, chlorophyll, and carotenoid, were also over-represented. When we looked at cellular component GO terms, the products of 59% of 1,148 annotated genes in Module 1 could be assigned to the chloroplast (p-value = 4.9E265).

Following the functional analysis of genes in Module 1, we examined their transcriptional activities under different conditions. Here, we focused our examination on the 382 hub genes which formed a clique and served as the core of the module. The gene expression behavior of the hub genes should be a typical representation of Module 1. Overall, the hub genes exhibited a tightly controlled co-expression pattern across the 1,094 conditions profiled in AtGenExpress (Figure 9). Particularly, these hub genes showed an oscillatory expression pattern over the 272 conditions which consisted of nine environmental stresses (Kilian et al., 2007). The average expression level for each hub gene over the 136 conditions sampled from the shoot tissue is higher than that over the remaining 136 conditions sampled from the root tissue. The average fold increase over all 382 hub genes is 45. In another experimental setting, gene expressions in response to light stimulus were compared with those under darkness conditions (contributed by Thomas Kretsch to AtGenExpress). We found that 97% of the 382 hub genes showed higher expression levels with the treatment of 4 hr continuous white light compared with the treatment of 4 hr continuous darkness; whereas only 54% of the 6,206 genes in AGCN showed higher expression levels upon the light treatment. Thus, the 382 hub genes are highly expressed in shoots and up-regulated by light.

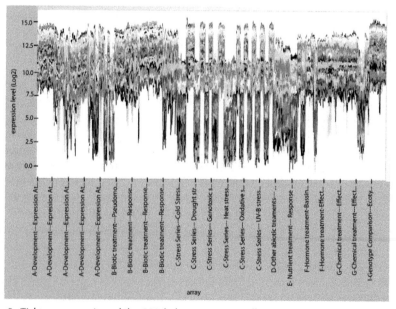

Figure 9. Tight co-expression of the 382 hub genes across all 1,094 arrays in AtGenExpree. The figure was generated using MetaOmGraph, a component of the MetNet bioinformatics platform (Wurtele et al., 2003).

Based on the enrichment analysis of biological process and cellular component GO terms, pathway information, and gene expression data, Module 1 is likely to carry on the biological function of photosynthesis by coordinating more than 1,000 genes' transcriptional activities. Based on this module-level annotation, many genes in Module 1 with unknown functions would be hypothesized to be linked to photosynthesis (Horan et al., 2008).

Module 4Cell Cycle

Module 4 has 280 nodes, 5,685 edges with a density of 0.15. Through consolidation, three major biological process GO terms were retrieved from the hierarchical relations of the GO terms that were over-represented in Module 4 (Figure 10A): DNA metabolism (52 genes), cell cycle (33 genes), and cytoskeleton organization and biogenesis (26 genes). Altogether, these 111 genes account for 63% of the genes in the module with known biological processes.

Among the 33 cell cycle genes, 25 encode the cell cycle regulators that included 14 cyclins, 5 cyclin-dependent protein kinases (CDK), two members of the E2F transcription factors (E2F3 and DEL3), WEE1, MAD2, TSO2, and PCNA1. The E2F transcription factors play important roles in pathways related to cell division, DNA repair, and differentiation (Sozzani et al., 2006). WEE1, a protein kinase, controls cell cycle arrest by functioning as a DNA replication checkpoint (De Schutter et al., 2007). The MAD2 is a mitotic spindle checkpoint protein. The TSO2 is a small subunit of ribonucleotide reductase (RNR) which is critical for cell cycle progression, DNA damage repair, and plant development (Wang and Liu, 2006). The TSO2 mutants resulted in developmental defects, including callus-like floral organs and fasciated shoot apical meristems (Wang and Liu, 2006). The PCNA1 is a proliferating cell nuclear antigen, which is involved in DNA replication, DNA repair, chromatin remodeling, cell cycle regulation, and other functions (Raynaud et al., 2006). Interestingly, PCNA1 is transcriptionally regulated by E2F (Egelkrout et al., 2002).

The 52 DNA metabolism genes in Module 4 included 23 genes involved in DNA replication, and 23 genes involved in chromatin assembly (mainly *histone H2A/H2B/H3/H4*) and modification (histone phosphorylation/methylation). The DNA metabolism process is apparently integrated with the cell cycle. Since cells orchestrated the coordinated progression through the cell cycle, the genes involved in DNA metabolism in Module 4 are likely subject to the cell-cycle regulation. This is indeed the case. For example, *ORC1A, ORC1B, ORC3*, and *ORC4* encode the subunits of Origin Recognition Complex (ORC) which is involved in the initiation of DNA replication. The expressions of these four *ORC* genes are all regulated by E2F (Diaz-Trivino et al., 2005). Another target of the E2F transcriptional factor, FAS1, encodes the chromatin assembly factor-1 (CAF-1) large subunit. Loss of FAS1 caused the inhibition of mitotic progression and triggered the endocycle program (Ramirez-Parra and Gutierrez, 2007). The regulation of *H4* genes by another cell cycle regulator, TSO2, was demonstrated by the result that in a tso2 mutant, H4-expressing cells in flowers were dramatically increased compared with wild type, suggesting a prolonged S-phase in the mutant (Wang and Liu, 2006).

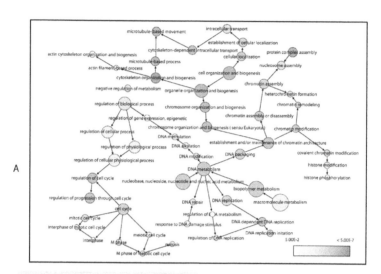

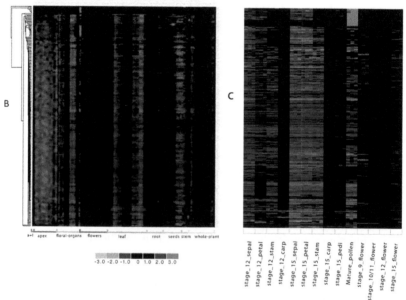

Figure 10. Functional analysis of Module 4. (A) Significantly over-represented GO terms detected in Module 4. (B) Co-expression patterns of 280 module genes over the 237 arrays which made up a gene expression map of *Arabidopsis* development (Schmid et al., 2005). In the heat map, each row represents a gene, and each column represents an array. Prior to hierarchical clustering, a gene's expression values over the 237 arrays were processed so that they had a zero mean and unit standard deviation. Arrays sampled from the same tissue were grouped together. "a + l" represents the tissue that includes both shoot apex (vegetative) and young leaves. The heat map was generated using dChip software (Li and Wong, 2001). (C) A closer examination of the expression pattern of 280 module genes in different floral organs and whole flower tissues at different development stages. To generate the heat map, genes' expression values were extracted from the 280 x 237 data matrix, which were used to produce the heat map depicted in (B). Stage_XX represents a flower development stage, "stam" represents stamen, "carp" represents carpel, "pedi" represents pedicel. For each experimental condition (e.g., stage_12_sepal), three replicates were measured.

Among the 26 cytoskeleton organization and biogenesis genes, 19 encode kinesin motor proteins, one encodes γ-tubulin, and four encode actin-binding proteins. The kinesin motor proteins move along microtubules and play a role in mitosis by functioning in spindle formation, chromosome movement and cytokinesis (Vanstraelen et al., 2006). Since multiple kinesin genes were detected in the co-expression module, their products might act cooperatively and play a role in the formation of mitotic microtubule arrays such as phragmoplast (Lee et al., 2007; Vanstraelen et al., 2006). It was shown that several members of the kinesin protein family were probably regulated by CDK phosphorylation (Vanstraelen et al., 2006). The γ-Tubulin, is required for centrosomal and noncentrosomal microtubule nucleation and coordinates late mitotic events in *Arabidopsis* (Binarova et al., 2006). The functional role of actin cytoskeleton in the progression of cell cycle was well described in Yu et al. (2006).

Since DNA metabolism, microtubule, and actin organization and biogenesis carried by the genes in Module 4 are all integrated with the cell cycle, the module is likely to carry on the function of cell cycle and cell proliferation. Since the cell cycle regulation is one of the keys to the control of plant development (Beemster et al., 2003), we selected a data set from AtGenExpress which made up a gene expression map of *Arabidopsis* development (Schmid et al., 2005). The data set has 237 arrays which profiled many development stages, various tissues/organs, wild type and different mutants.

Figure 10B shows the expression profiles of the 280 genes in Module 4 across the 237 arrays. Among different tissues, most of the 280 genes exhibited the highest expression levels in the shoot apex regardless of development stages, indicating that the module is active in the tissue enriched for cell dividing cells. The expression map of *Arabidopsis* development profiled the transcript abundance from the whole flower tissues from development Stages 9–15. The map also included profiles from the four major floral organs (sepal, petal, stamen, carpel) at two development Stages (12 and 15), pedicels at Stage 15 and mature pollen. As shown in Figure 10B, most of the genes in Module 4 were also highly expressed in flowers, another tissue enriched for cell dividing cells. However, they declined from relatively high expression to low expression (with respect to the average expression value of a gene over the 237 arrays) as the flower evolved from the development Stages 9–15 (Figure 10C). With respect to the flower organs, most of the genes showed relatively high expression in carpels at Stage 12 (Figure 10C). Such a distinct expression pattern implies that the cell-cycle specific module may play a role in the flower development. Indeed, three module genes, *ICU2*, *HTA8*, and *MET1*, function in the regulation of flower development. The *ICU2* encodes a catalytic subunit of the DNA polymerase which is essential for the cell cycle by initiating DNA replication (Barrero et al., 2007). The ICU2-1 mutant derepressed the expression of a number of regulatory genes including the ones involved in flowering time, floral meristem, and floral organ identity (Barrero et al., 2007). In the upstream, the expression of *ICU2* is likely to be cell-cycle regulated based on the observation that its promoter contains an E2F binding site motif (Barrero et al., 2007). The second gene, *HTA8*, encodes a histone H2A protein. The gene knockdown experiment suggested the potential role of *HTA8* in the regulation of flower development through the activation of FLC, a central floral repressor (Choi et al., 2007). In human, a replication-dependent *H2A* gene was regulated by E2F in the early S phase of the cell cycle (Oswald et al., 1996). The third gene, *MET1*, encodes a cytosine methyl-transferase. Demethylation of DNA brought about by a *MET1* antisense caused early flowering in vernalization-responsive *Arabidopsis* (Finnegan et al., 1998). The *MET1* was recently designated as a proliferation gene, and its expression is likely to be cell cycle dependent (Beemster et al., 2005). Thus, studying these three genes and prob-ably other module members may help elucidate the mechanism underlying the linkage between the cell cycle regulation and the control of flower development.

In contrast to the shoot apex and flowers, nearly all of the genes in Module 4 showed relatively low expression in leaves across all development stages (Figure 10B).

Module 67Starch Metabolism

Compared with Module 1 and 4, Module 67 is a relatively small module with 10 nodes, 23 edges and a density of 0.51. Eight of 10 genes in Module 67 are involved in starch metabolism (p-value = 3.2E-19) based on their GO annotations. The other two genes are *AT3G46970* and *AT2G28900*. The *AT3G46970* encodes a cytosolic α-glucan phosphorylase which was suggested to function as an enzyme of starch degradation (Smith et al., 2004). The *AT2G28900* encodes an outer plastid envelope protein which was involved in the import of protochlorophyllide oxidoreductase A (Reinbothe et al., 2004). The expression of *AT2G28900* gene was shown to be modulated by sucrose and responsive to a starchless mutant (Gonzali et al., 2006). How this sugar sensing gene communicates with other starch metabolism genes remains to be elucidated.

Comparison with GGM Network

In this study, we used standard Pcc to measure the degree of co-expression between two genes and connected them in the network if their Pcc is above a certain cutoff value. One concern with this approach is due to the transitive property of the standard correlation coefficient, which implies that if gene A and gene B are correlated with gene C, respectively, then A and B should be expected to correlate as well (Free-man et al., 2004; Ma et al., 2007). If this were true, then every gene with at least two neighbors in the AGCN would have a clustering coefficient close to 1. The plot displaying the relationship between the clustering coefficient (C_k) and node degree (k) would approximately be a straight line of $C_k = 1$. However, as shown in Figure 4B, C_k showed a complex relationship with the node degree. Some sets of genes had relatively high clustering coefficients, suggesting the tight co-expression; while others had low clustering coefficients, indicating loose co-expression. As an example, among the 19 disconnected components each of which had exactly three nodes, 17 compo-nents had two edges connecting three genes whereas only two components had three edges, indicating only limited co-expression transitivity in AGCN. Nevertheless, we compared AGCN constructed in this study with the *Arabidopsis* gene network derived from a modified GGM, which used partial correlation as well as the standard Pearson correlation to select significantly correlated gene pairs (Ma et al., 2007).

In their study, a network of 18,625 edges connecting 6,760 genes was obtained by using GGM (Ma et al., 2007). The number of genes contained in the GGM-based network is close to that in AGCN constructed in this study. On the other hand, the aver-age connectivity of a node in the GGM-based network was 5.5, which is much smaller than that (165 links/node) in AGCN. Such a sparsely connected network, together with the number of highly connected genes being less than what would be expected according to the power-law distribution (Figure 3 in (Ma et al., 2007)), would be a challenge for computational methods to reliably detect a large number of modules (Zhang Shihua et al., 2007). Using the guide-gene approach, the authors retrieved subnetworks from the GGM-based network. Each subnetwork included a seed gene and the genes that were within certain connecting steps from the seed gene. A functional module was then considered to be equivalent of the retrieved subnetwork itself. The disadvantage of this approach is that the retrieved subnetwork might be embedded within a larger module or it may include extra noisy genes. Nevertheless, the GO terms for many biochemical pathways (e.g., sulfate assimilation, cellular response to phosphate starvation, glyco-lipid metabolism, leucine catabolism, tryptophan metabolism, starch catabolism), cell wall metabolism, and cold response were significantly enriched in the subnetworks retrieved from the *GGM*-based gene network (Ma et al., 2007). These GO terms were also significantly over-represented in the modules extracted from the AGCN (see the data files in our website). However, with the guide-gene approach, the distinct major modules corresponding to photosynthesis (Module 1 in AGCN), ribosome assembly and protein biosynthesis (Module 3 in AGCN), and DNA metabolism and cell cycle (Module 4 in AGCN), which are central to the plant growth and development, seemed to be absent from their report (Ma et al., 2007) probably because the correlations between many genes involved in these functions were determined to be insignificant by their partial correlation standard.

Since the gene modules of AGCN were detected by using the top–down approach, for a direct comparison, we also applied the same approach to the GGM network. Again, MCL algorithm was used to naturally partition the GGM network into modules. The functional coherence of obtained gene modules was assessed by using biological process GO terms (see our website for the results). In general, clustering on the GGM network produced a larger number of modules but smaller module size compared with the clustering result of AGCN. For example, at 1.8 inflation value, MCL detected 1,132 modules from the GGM network with the largest module only containing 45 nodes. With the same inflation value, MCL detected 527 modules from the AGCN with the largest modules containing 1,381 nodes. Nine additional modules in AGCN also have more than 50 nodes. Although a large number of relatively small modules were detected in GGM network, interestingly, the distribution of module size fits very well to a power law distribution (Figure 4D). A power law distribution was also observed for module size of AGCN (Figure 4C).

Since a module is a densely connected subnetwork and the connections between modules are sparse, clustering on a network with internal modular structure should produce a large mass fraction close to one. However, it is not the case for GGM network (Figure 5B). For example, at 1.8 inflation value, clustering on GGM network captured 64% of the entire edge masses whereas AGCN captured 96% of edge masses. We also compared biological process GO term enrichment results between GGM and AGCN. Clustering on AGCN produced a slightly higher percentage of functionally coherent modules (Figure 7B). In the GGM network, the module with the most over-represented photosynthesis GO term (module 200, p-value = 2.60E-16) has nine genes, the cell-cycle module (Module 46, p-value = 6.77E-11) has 23 genes, and protein biosynthesis module (Module 119, p-value = 2.08E-17) has 13 genes. In comparison, the corresponding gene functional modules in AGCN had much larger module sizes, and they were detected with the over-represented GO terms that were much more statistically significant (Module 1, 3, 4 in Table 2).

To use GGM approach to construct genome-wide *Arabidopsis* gene network, a large number of samples are required. Since there are more than 20,000 genes in *Arabidopsis*, a sample size comparable to the gene number is required to assess a full partial correlation for every gene pair (Ma et al., 2007). In contrast, the standard Pcc approach does not require a large sample size. It can be used to analyze small data sets that are typically seen in focused microarray experiments. As a proof of concept, we applied our network-based approach to a relatively small data set. The data set profiles global gene expression in shoot in response to continuous cold stress (4°C). The gene expression was measured at 7 time points (0, 0.5, 1, 3, 6, 12, and 24 hr) (Kilian et al., 2007). For each time point two samples serving as replicates were analyzed. Thus, the data set consists of a total of 14 samples. Of the 22,746 *Arabidopsis* probe sets on the ATH1 chip, 4,915 (22%) were selected as the genes that showed significant changes over the 14 samples (see Materials and Methods). The Pcc value for each pair of the 4,915 genes was calculated and sorted. The top 0.39% of gene pairs with respect to their Pcc values were used to construct the cold induced gene co-expression network. The same percentage of gene pairs were retained in the 1,094-array AGCN. The resulting cold induced gene network has 1,700 nodes, 46,671 edges, and a network density of 0.0323.

Following the construction of cold induced gene network, MCL algorithm was used to partition the network into modules. Eighty-four modules with at least three genes were obtained at 1.8 inflation value. Of these modules, 18 (21.4%) had biological process GO terms that were significantly over-represented (see our website for the results). Among the enriched GO terms, "response to auxin stimulus" is the most over-represented (Module 31, p-value = 4.13E-8). It has been reported that auxin responsive genes were regulated by cold stress which may contribute to the alteration of plant growth to coordinate with cold (Lee et al., 2005). "Cellular carbohydrate metabolism" is the second most over-represented biological process GO term, and it was identified in Module 1 (p-value = 9.31E-08). The module includes genes involved in starch, sucrose, trehalose, and glucose metabolism. Starch and sugars play an important role in the biochemical adaption of plant to cold (Cook et al., 2004; Fatma et al., 2007; Stitt and Hurry, 2002). Other GO terms for biological processes involved in transcription regulation, cellular defense, stress response, and signal transduction are also significantly over-represented. Taken together, these results can help us better understand the molecular mechanisms of plant cold responses. It demonstrates that our network-based approach can be applied to analyze small data sets and identify gene modules important to specific biological questions.

CONCLUSION

In this study, we used a network-based approach to identify gene functional modules from large microarray data sets of *Arabidopsis thaliana*. The study reveals new insight into the topological properties of biological networks. The preferential hub–hub connections might be necessary for the formation of modular structure in gene co-expression networks. The study also reveals new insight into the organization of gene functional modules.

KEYWORDS

- **AGCN**
- ***Arabidopsis* gene**
- **AtGenExpress**
- **Gene functional modules**
- **Gene ontology**

AUTHORS' CONTRIBUTIONS

Linyong Mao designed the project, conducted the project, and wrote the manuscript. John L Van Hemert developed some data analysis tools. Sudhansu Dash modified the manuscript. Julie A Dickerson oversaw the project and modified the manuscript. All authors read and approved the final manuscript.

ACKNOWLEDGMENT

Funding: National Science Foundation (grant DBI-0604755).

Chapter 5

Gene Expression and Physiological Responses in Mexican Maize Landraces under Drought Stress and Recovery Irrigation

Corina Hayano-Kanashiro, Carlos Calderón-Vázquez,
Enrique Ibarra-Laclette, Luis Herrera-Estrella, and June Simpson

INTRODUCTION

Drought is one of the major constraints for plant productivity worldwide. Different mechanisms of drought-tolerance have been reported for several plant species including maize. However, the differences in global gene expression between drought-tolerant and susceptible genotypes and their relationship to physiological adaptations to drought are largely unknown. The study of the differences in global gene expression between tolerant and susceptible genotypes could provide important information to design more efficient breeding programs to produce maize varieties better adapted to water limiting conditions.

Changes in physiological responses and gene expression patterns were studied under drought stress and recovery in three Mexican maize landraces which included two drought tolerant (Cajete Criollo (CC) and Michoacán 21 (M21)) and one susceptible (85-2) genotypes. Photosynthesis, stomatal conductance, soil and leaf water potentials were monitored throughout the experiment and microarray analysis was carried out on transcripts obtained at 10 and 17 days following application of stress and after recovery irrigation. The two tolerant genotypes show more drastic changes in global gene expression which correlate with different physiological mechanisms of adaptation to drought. Differences in the kinetics and number of up- and down-regulated genes were observed between the tolerant and susceptible maize genotypes, as well as differences between the two tolerant genotypes. Interestingly, the most dramatic differences between the tolerant and susceptible genotypes were observed during recovery irrigation, suggesting that the tolerant genotypes activate mechanisms that allow more efficient recovery after a severe drought.

A correlation between levels of photosynthesis and transcription under stress was observed and differences in the number, type and expression levels of transcription factor families were also identified under drought and recovery between the three maize landraces. Gene expression analysis suggests that the drought tolerant landraces have a greater capacity to rapidly modulate more genes under drought and recovery in comparison to the susceptible landrace. Modulation of a greater number of differentially expressed genes of different TF gene families is an important characteristic of the tolerant genotypes. Finally, important differences were also noted between the tolerant landraces

Abiotic stress is a major limiting factor for plant growth and food production in many regions of the world and its effects will become more severe as desertification claims more of the world's arable land. Among environmental stresses, drought has the greatest effect on agriculture worldwide (Vinocur and Altman, 2005), affecting more than one-fifth of the tropical and subtropical areas used for maize production (Ribaut and Ragot, 2007). As an example, in Mexico around 80% of all maize cultivated is grown under rain-fed conditions (Aquino et al., 2001), where the possibilities for alleviating water stress are limited (Ribaut and Ragot, 2007). Therefore, an urgent need exists to develop drought-tolerant varieties either by conventional breeding or by genetic engineering in order to cope with the rising demand for maize to feed both humans and animals.

Due to a unique genome structure and continuous human selection for over 7,000 years, maize is one of the most plastic plant species in terms of its adaptation to different environmental conditions, capable of growing at high and low altitudes and in tropical, subtropical, and temperate climates. This genetic variability has been exploited to produce locally adapted drought tolerant maize cultivars for the dry tropical areas of Indonesia, Kenya, Mexico, and Colombia (Pingali and Pandey, 2001). Currently marker-assisted selection (MAS) is used in the development of maize germplasm with improved stress tolerance (Bruce et al., 2002) based on quantitative trait loci (QTL's) affecting root architecture, leaf abscisic acid (ABA) concentration and other drought-related traits (Tuberosa et al., 2007). Despite these efforts, improvement programs for drought stress in maize have advanced slowly and substantial research is needed to adapt the improved genetic materials to particular environmental conditions (Pingali and Pandey, 2001), where they should not only withstand greater levels of drought but also perform well under optimal conditions (Ribaut and Ragot, 2007). Moreover, local landrace accessions could provide novel alleles that will complement strategies based on existing stress-adaptation mechanisms (Reynolds et al., 2007).

During evolution, plants have acquired a myriad of developmental and metabolic strategies to optimize water uptake and efficiently balance this with water utilization during vegetative growth and reproduction (Parry et al., 2005), making drought tolerance a complex multigenic trait. In the past decade, research to unravel the molecular processes involved in drought tolerance has received special attention (Chaves et al., 2003), for reviews see (Bray, 1997; Ingram and Bartels, 1996; Shinozaki and Yamaguchi-Shinozaki, 1997). Physiological studies have shown that sugars, sugar alcohols, amino acids, and amines function as osmolytes, protecting cellular functions from the effects of dehydration and are known to accumulate under drought stress conditions in different plant species (Seki et al., 2007). Reduction in vegetative growth, stomatal closure and a decrease in the rate of photosynthesis (Mahajan and Tuteja, 2005) are among the earliest responses to drought, protecting the plant from extensive water loss (Chaves et al., 2003).

More recently, genomic technologies have provided high-throughput integrated approaches (Bartels and Sunkar, 2005) to investigate global gene expression responses not only to drought but also to other abiotic stresses (Chaves et al., 2003). Microarray profiling under drought stress has been carried out in different plant species such as *Arabidopsis* (Kawaguchi et al., 2004; Oono et al., 2003; Seki et al., 2002), rice (Rabbani et al., 2003), barley (Ozturk et al., 2002; Talamé et al., 2007), and wheat (Mohammadi et al., 2007). These studies identified differentially expressed transcripts of genes involved in photosynthesis, ABA synthesis and signaling, biosynthesis of osmoprotectants, protein stability and protection, reactive oxygen detoxification, water uptake and a myriad of transcription factors including several members of the zinc finger, WRKY, and bZIP families. To date gene expression studies in maize in response to water stress have investigated different organs such as roots (Poroyko et al., 2007) and developing kernels (Yu and Setter, 2003) or particular developmental stages (Zheng et al., 2004). However, no reports have addressed comparisons between the drought stress responses of susceptible and tolerant maize genotypes or genotypes that have been reported to possess different tolerance mechanisms.

Mexican maize genotypes with apparently different mechanisms for achieving drought tolerance have been reported. For instance, CC, cultivated mainly in Oaxaca State, Mexico, has a high tolerance to low water content in the soil and a long vegetative cycle with slow growth until the rains arrive when a rapid response in terms of growth and recovery occurs (Pérez, 1979). The M21 from the Purépecha highlands in Michoacán State (Mexico) was described as a landrace with a clear response to drought and cold stress (Fischer et al., 1983). The mechanism of tolerance of M21 was termed "latency" and consists of prolonging the vegetative stage under drought stress without flowering and a rapid return to normal growth and completion of the reproductive cycle even when the rains begin. The M21 is more resistant to permanent wilting in seedlings in comparison to other maize genotypes and has a higher transpiration rate under well irrigated conditions as compared to conditions of limiting water resources (Fischer et al., 1983).

The aim of this study is to analyze the differences in physiological responses and gene expression of one susceptible (85-2) and two drought-tolerant (CC and M21) maize landraces. The three genotypes were subjected to intermediate (10 days without water) and severe (17 days without water) drought stress treatments followed by recovery irrigation and global gene expression were evaluated at the different time points using a 56K oligonucleotide maize microarray. The results confirm that different physiological responses and different gene expression patterns occur under drought stress and recovery in the two tolerant genotypes, and provide insights as to how changes in gene expression could lead to drought tolerance and recovery in maize. Expression patterns of genes involved in photosynthesis and carbohydrate and proline metabolism, those encoding transcription factors and those known to be involved in other abiotic stress responses were studied in more detail, providing information on the correlation between the physiological and gene expression responses of the three genotypes, and allowing the identification of specific genes and expression patterns associated with particular metabolic pathways in each of the three landraces.

MATERIALS AND METHODS

Plant Material

Three Mexican maize genotypes were used for this work. The CC and M21 are considered to be drought tolerant landraces and were supplied by the International Maize and Wheat Improvement Center (CIMMYT), whereas 85-2 is considered to be susceptible from field observations, and was supplied by the Instituto Nacional de Investigaciones Forestales, Agricolas y Pecuarias (INIFAP)-Mexico.

Growth Conditions

Seeds were treated with NaOCl (10%) for 30 min then washed with distilled water before sowing. Plants were grown in 15 L plastic pots in a substrate of 92.46% sand and 7.44% clay under greenhouse conditions in the months of July to September in 2005 and 2006 at CINVESTAV, Irapuato, Mexico. Temperatures were between 19°C and 32°C and relative humidity was 60 ± 5%. Maize plants were watered daily to soil capacity until application of stress, and fertilization was applied using a slow release fertilizer (Triple 17, Profer Mix, 4 g for each pot-17:17:17 NPK). Long Ashton Solution (Phillips and Jennings, 1976) was added once a week until application of drought stress.

Drought Stress Treatments

Thirty day old plants were subjected to a progressive water deficit by leaving them unwatered for 17 days (severe stress) and then given recovery irrigation. Control plants were watered daily to maintain soil water content close to field capacity. Soil and leaf water potentials were measured daily for 17 days of the drought treatment. At day 0, 10, and day 17 of drought stress and following the recovery irrigation samples for RNA extraction were collected.

Leaf (ψ_l) and Soil (ψ_s) Water Potential and Gas Exchange

Leaf water potential (ψ_l) was measured both pre-dawn (6 am) and at midday (12 pm) in control and stressed plants with a psychrometer model C-52 sample chamber (WESCOR, Inc., Logan, Utah) and a dew point microvoltimeter (model HR-33T, WESCOR, Inc., Logan Utah), for each individual plant on the most recent fully expanded leaf. Soil water potential (ψ_s) was measured using model PST-55 psychrometers (WESCOR, Inc., Logan, Utah) placed in the center of each pot at a depth of 15.5 cm.

Photosynthesis and stomatal conductance were analyzed using a portable Li-6200 photosynthesis system (Li-Cor, Lincoln, NE, USA) every 2 hr between 7 am and 7 pm on the most fully expanded leaf for both control and drought stressed plants at 0, 10, and 17 days drought stress and on recovery irrigation. The values were normalized with a foliar area of 9 cm².

Microarray Design

The Maize Oligonucleotide Array (MOA) from http://www.maizearray.org was used in this study. The MOA contains about 57,000 individual spots on two slides (A and B) and putatively contains all maize genes identified when slides were obtained. Array annotation and composition is available at www.maizearray.org. A loop design was used in order to contrast the gene expression differences between genotypes under each treatment. Single samples analyzed included two independent biological replicates and two technical replicates. The biological replicates were obtained by pooling the leaves of five representative plants from each cultivar under a particular treatment (0, 10, and, 17 days of drought stress and recovery irrigation). A total of 24 sets (48 slides) of microarray hybridizations were carried out, including direct and dye swap comparisons.

RNA Isolation and Labeling, Microarray Hybridization, and Image Processing

Total RNA from pooled leaves of five control and five stressed plants for each cultivar and each time point was isolated using the TRIZOL reagent (*Invitrogen*) and then re-purified with the Concert Plant RNA Purification reagent (*Invitrogen*). To ensure the purification of high quality RNA samples, the RNeasy MinElute Cleanup kit (Qiagen) was used following the manufacturer's instructions. Purified total RNA was then labeled according to the protocols recommended at http://www.maizearray.org. Probe concentrations were determined in a NanoDrop spectrophotometer ND-100 (Nano-Drop Technologies Inc., Wilmington-DE, USA). Three micrograms of cRNA of each probe was used per slide. Hybridization, washing, and scanning were performed as described in (Calderón-Vázquez et al., 2008).

Microarray Normalization and Data Analysis

Raw data from the 48 slides was imported into the R 2.2.1 software (http://www.R-project.org) and background correction was carried out. Normalization of the corrected signal intensities within slides was carried out using the "printtiploess" method (Yang et al., 2002) and between slides using the Aquantile method. Both methods were implemented using the LIMMA package (Smyth et al., 2003). All microarray data reported in this chapter is described in accordance with MIAME guidelines and have been deposited in NCBI's Gene Expression Omnibus.

The analysis was basically performed as described in (Calderón-Vázquez et al., 2008). A mixed linear model analysis (Gibson and Wolfinger, 2004; Wolfinger et al., 2001) was conducted for each printed oligonucleotide by using the SAS mixed procedure (SAS 9.0 software, SAS Institute Inc., Cary, NC, USA). Direct comparisons between genotypes under a particular treatment were done on each slide. The design permitted the evaluation of the differences in gene expression between the three genotypes under a specific drought stress or recovery irrigation treatment but also whether differences were treatment dependent by including the data from different treatments in the mixed model and looking for gene specific effects. Normalized data were log2 transformed and then fitted into mixed model ANOVAs using the Mixed procedure with two sequenced linear models considering as fixed effects the dye, cultivar, treatment, and cultivar treatment. Array and array dye were considered as random effects. The Type 3 F-tests and p-values of the genotype*treatment and genotype model terms were explored and significance levels for those terms were adjusted for by the False Discovery Rate (FDR) method (Benjamini and Hochberg, 1995). Estimates of differences in expression were calculated using the mixed model. Based on these statistical analyses, the spots with an FDR less or equal to 5% (FDR ≤ 0.05) and with changes in signal intensity between stressed and control leaves of two fold or higher were considered as differentially expressed.

Accession Numbers

The microarray data have been deposited in NCBI's Gene Expression Omnibus (Edgar et al., 2002) and are accessible through GEO Series accession number: GSE14728 (http://www.ncbi.nlm.nih.gov/geo/query/acc.cgi?acc=GSE14728).

Functional Annotation and Metabolic Pathway Analysis using MapMan Software and BioMaps

Functional annotation and metabolic pathway analysis were performed as described by Calderón-Vázquez et al. (2008). Genes differentially expressed according to the selected parameters (FDR < 0.05 and Fold ± 2) were visualized and clustered with the standard correlation method using GeneSpring 7.0 software (Silicon Genetics, Redwood City, CA). The FiRe 2.2 macro Excel® (Microsoft) (Garcion et al., 2006) was used to facilitate the handling of the microarray information. Due to the limited functional annotation in maize, the functional classification in the mapping files that structure the *Arabidopsis* genes from the Affymetrix ATH1 array into distinct metabolic and cellular processes from the MapMan program (Thimm et al., 2004) was used. Differentially expressed maize genes were functionally annotated by performing a BLAST alignment against the TAIR *Arabidopsis* database release 6.0 (www.arabidopsis.org) and to PLANTA database (TIGR).

The annotations of the mapping files for the best match to the TAIR protein database (with at least an Expected Value of 1.0E-10) were assigned to the corresponding maize ortholog. MapMan software (Thimm et al., 2004) was employed to show the differences in gene expression in different cellular and metabolic processes. Ratios were expressed in a log2 scale for importing into the software and changes in expression were displayed via a false color code (Thimm et al., 2004).

In addition to the MapMan software, the microarray data was analyzed using a tool called BioMaps (Gutiérrez et al., 2007) at the Virtual-Plant site (www.virtual plant.org). This tool helps relate differential expression data with functional categories based on the functional classification by the Munich Information Center for Protein Sequences (MIPS) annotation, taking into account the best match to the TAIR protein database and was utilized to identify the common functional categories related to drought stress among the three landraces and/or the tolerant landraces.

Application of Pearson's Chi-Squared Test

In order to verify the statistically significant difference among the three landraces of the differentially expressed genes along the microarray analysis a Pearson's Chi-squared test was performed using R version 2.7.1 (2008-06-23) software (http://www.R-project.org) for some functional category and for each treatment of drought stress and recovery irrigation.

DISCUSSION

The aim of this study was to compare changes at the physiological and global gene expression levels of two drought tolerant and one susceptible maize genotype in response to the gradual application of drought stress, in order to identify the general responses of maize to drought and possible differences in the mechanisms employed to achieve tolerance. Although some genetic variation exists within each landrace, the repetition of the drought experiment in two different years and the use of replicates of each landrace to obtain physiological and gene expression data produced consistent landrace specific data. Monitoring of soil water potentials throughout the application of drought stress and on recovery ensured that levels of stress were adequate and equivalent for all three landraces.

Physiological Responses: Photosynthesis, Stomatal Conductance, and Water Potential

It is well documented that upon water deficit, most plants respond rapidly by stomatal closure to avoid excessive water loss and by establishing physiological and molecular responses to prevent irreversible damage to the photosynthetic machinery (for a review see (Chaves et al., 2003)). These two processes are closely linked since stomatal closure results in a decline in the rate of photosynthesis (Foyer et al., 1998; Pelleschi et al., 1997). Therefore, leaf water potential, stomatal conductance and rate of photosynthesis were monitored at different stages throughout the experiment and revealed different responses in the three landraces for these physiological parameters. The M21 showed a more rapid and drastic reduction in stomatal conductance and rate of photosynthesis than CC and 85-2. The CC showed a more gradual and less pronounced decline in leaf water potential, rate of photosynthesis and stomatal conductance at 10 days stress, whereas at 17 days stress the rate of photosynthesis and stomatal conductance dropped sharply. The susceptible landrace, 85-2 had the highest drop in leaf water potential which correlated with a lower decrease in stomatal conductance. In previous work the characteristic of drought resistance called "latency" observed in M21, was associated with early stomatal closure in comparison to the susceptible controls and that stomatal hypersensitivity was a trait common to several drought resistant maize lines (Muñoz et al., 1983). The M21 strategy to sharply drop photosynthesis rate may be advantageous when short periods of severe drought stress are experienced, however it could have a negative effect under prolonged drought stress even though the overall stress is less severe. Under the latter conditions the CC strategy of gradual reduction in photosynthesis rate and higher leaf water potential may allow the plant to survive for longer periods of low water availability. At recovery irrigation, both CC and M21 show a rapid and strong increase in photosynthesis as compared to the response in 85-2, suggesting that the drought-tolerant genotypes may share a mechanism of rapid recovery after drought not present in the susceptible landrace.

Molecular Drought Stress Responses Common to the Three Maize Genotypes

Analysis of the general overview of the global changes in gene expression in response to drought for the tolerant and susceptible landraces shows that there are several common alterations, albeit to a significantly different degree in terms of the number of transcripts differentially expressed and in the expression levels of genes involved in different metabolic and cellular processes. The earliest response to water deficit is stomatal closure to protect the plants from extensive water loss (Chaves et al., 2003; Mahajan and Tuteja, 2005) and consequently the inhibition of photosynthesis (Chaves et al., 2003). In this respect, the first obvious common response of the three genotypes is the decrease in transcript level of photosynthesis-associated genes during drought stress. In particular, genes encoding components of photosystem II (PSII), and to a lesser extent of photosystem I (PSI), are repressed during drought stress. A reduction in the components of PSI and II would prevent the photo-oxidation of the photosynthetic apparatus and the formation of free radicals that are harmful for the cell, although as discussed below there were significant differences in the expression of photosynthesis-associated genes that were observed only in the drought tolerant genotypes or specific to either of them. Another feature in the general maize gene response to drought was the induction of genes encoding heat shock protein (HSP) and late embryogenesis abundant protein (LEA). Genes encoding HSP17, HSP22, HSP70, and HSP90 were induced in the three genotypes under drought stress. These proteins prevent detrimental effects of stress by preventing protein aggregation, protecting non-native enzymes from degradation and assisting in protein refolding (Wang et al., 2003). Induction of these genes under drought stress was observed in previous studies on drought stress in barley (Talamé et al., 2007) and rice (Rabbani et al., 2003) dehydration in *Arabidopsis* (Seki et al., 2002) and polyethylene glycol (PEG) treatment in maize (Jia et al., 2006). Interestingly, only two LEA genes, one belonging to Group 1 and the other to Group 3, were identified as induced in all three genotypes, suggesting that as a whole the LEA protein family might not play a major role, at least under our experimental conditions, in the general drought stress response in maize.

Plant growth and response to stress conditions is largely under the control of hormones (Mahajan and Tuteja, 2005). In particular, ABA has been associated with the promotion of stomatal closure and plays an important role in the tolerance response of plants to drought and high salinity (Yamaguchi-Shinozaki and Shinozaki, 2006). In the present study, genes encoding enzymes related to ABA synthesis (ZEP and NCED) were induced at 10 days stress in the three landraces. These genes were shown to be up-regulated by dehydration in *Arabidopsis* (Yamaguchi-Shinozaki and Shinozaki, 2006). As mentioned-above, NCED is a key enzyme of ABA biosynthesis; *At-NCED3* was strongly induced by dehydration and high salinity and its overexpression improved dehydration stress tolerance in transgenic plants, indicating the important role in ABA accumulation during dehydration (Yamaguchi-Shinozaki and Shinozaki, 2006). The induction of *HVA22* under environmental stresses such as ABA, cold and drought has been reported in barley (Shen et al., 2001). It was also reported that the ectopic expression of ABI3 conferred a freezing tolerance in transgenic ABI3 *Arabidopsis* plants (Tamminen et al., 2001). The up-regulation of transcripts for *ABI3* and *HVA22* exclusively in M21 under stress suggests the existence of specific ABA signaling stress response in this landrace that could be important for the drought tolerance.

Reduction in carbon fixation and the inhibition of photosynthetic activity by drought also alters the carbohydrate metabolic equilibrium (Xue et al., 2008). For plants, carbohydrate-based regulation represents an especially valuable mechanism for adjusting to environmental changes (Koch, 1996). An increase in β-amylase transcripts in all three landraces under drought suggests that when levels of photosynthesis drop, carbohydrates stored as starch may be mobilized from the chloroplasts. This could lead to the increase in glucose levels observed for M21 and 85-2.

Glucose in addition to a structural role, functions as a signal molecule in both hexokinase dependant and independent pathways (Xiao et al., 2000). Although hexokinase transcripts were up-regulated under drought stress the genes encoding the other enzymes needed to produce MI from glucose were down-regulated suggesting that the increase in MI content observed at 10 days in CC and 85-2 and at 10 and 17 days in M21 could be the result of changes regulated at the translational or post-translational level. The fact that M21 has a high MI content could indicate another drought tolerance strategy, since MI is implicated in many aspects of metabolism including: osmoregulation, auxin physiology, cell wall and membrane metabolism, and signaling among others (Hegeman et al., 2001). Increases in transcript levels in response to drought, both at the transcriptional and/or post-transcriptional level, requires the participation of components of signaling pathways that activate transcription and/or mRNA stabilization. In this study components of signal transduction pathways related to Ca^{+2} signaling and G-proteins (CDPK, Ca^{+2}-binding EF, Rho, and Rab GDP dissociation inhibitor and calmodulin) were the only ones identified as differentially regulated common in the three genotypes under stress. The involvement of Ca^{+2} signaling in response to osmotic and ionic stress has been well documented (Bartels and Sunkar, 2005). Signal transduction networks usually include TFs and their cognate cis-acting elements (Yamaguchi-Shinozaki and Shinozaki, 2006) that activate a cascade of genes encoding proteins and enzymes that may act together to enhance tolerance to multiple stresses (Bhatnagar-Mathur et al., 2008). Previous studies have revealed that plant responses are complex requiring the participation of several TFs, some of which are transcriptionally activated during drought stress. Most of these TFs fall into several large families, such as *AP2/ERF, bZIP, NAC, MYB, CysHis2 zinc finger*, and *WRKY* gene families (Umezawa et al., 2006). The three maize genotypes analyzed in this study showed a common up-regulation of several genes encoding TF belonging to the *C2H2, G2-like, HB, MADS*, and *MYB* gene families and a homeodomain leucine zipper protein Hox7, which could be considered to be induced in the general response of maize to drought. Some of the genes encoding TFs common to the three maize landraces, such as MADS and homeodomain leucine zipper TFs, were also up-regulated in wheat (Mohammadi et al., 2007), MYB in *Arabidopsis* (Seki et al., 2002), and C2H2 under PEG stress in maize (Jia et al., 2006). This suggests that some of the responses in differential gene expression to drought stress are probably modulated by the same types of TFs and involve similar signal transduction pathways.

Differential Responses Among the Three Landraces Under Drought Stress

In order to elucidate differences between the tolerant and the susceptible landraces we identified gene families that are over-represented or have differences in their expression level in the tolerant genotypes with respect to the susceptible one (Figure 1). In fact, one might expect more changes in the tolerant cultivars, which should carry alleles of genes that contribute to increased tolerance (Degenkolbe et al., 2009).

Although it is quite possible that some genes responsible for drought tolerance might not be inducible or repressible during stress, the identification of differentially expressed genes in drought tolerant genotypes, could provide important information about the metabolic and cellular processes that are ultimately responsible for stress tolerance. In this respect, TFs are important in regulating the expression of downstream stress-regulated genes (Li et al., 2008). It was found 18 *TF* genes are differentially expressed under drought stress in both tolerant genotypes but not in 85-2 such as some members of the *AP2, bHLH, C2C2, C2H2, C3H, zinc finger, CCAAT binding factor (HAP2),* and *WRKY* gene families. Of these, one AP2/EREPB, one C2C2 (H-protein promoter binding factor 2b), one C2 domain containing C2H2 zinc finger family, one zinc finger CCCH-type and the CCAAT binding protein NF-YA were induced under stress only in the two tolerant genotypes. The C2C2 and CCAAT families have been proposed to play an important role in drought tolerance (Spollen et al., 2008). It was also reported that the expression of *NFYA5*, a member of NF-YA (HAP2) was strongly induced by drought stress or ABA treatments in *Arabidopsis* and suggested to play an important role in controlling stomatal aperture and drought resistance (Li et al., 2008). On the other hand, AP2/EREPB domain proteins include DREB or CBF proteins which bind to dehydration response elements (DRE) or C-repeats (Bartels and Sunkar, 2005), also shown to be involved in improved stress tolerance to drought, high salinity and freeze in *Arabidopsis* (Yamaguchi-Shinozaki and Shinozaki, 2006) and rice (Shinozaki and Yamaguchi-Shinozaki, 2007). Additionally, 37 and 7 TF genes were specifically up-regulated under stress for M21 and CC respectively. In M21 some of these genes showed a 3- to 8-fold higher change in expression to 85-2 and CC. These TFs could be important for the regulation of drought stress responsive genes and might be involved in the tolerance mechanism. The higher number of up-regulated TF genes observed for M21 could explain the more rapid and wide-ranging responses observed in this landrace and therefore its tolerance mechanism.

In CC a remarkable difference was the induction of a *NAC TF* gene. NAC proteins are known to function as transcription activators in cooperation with the ZFHD (zinc finger homeodomain) proteins and the overexpression of these genes significantly increased drought tolerance (Yamaguchi-Shinozaki and Shinozaki, 2006). Different alleles of the same TF genes may also produce different responses between the tolerant landraces and between the tolerant and susceptible landraces, as well as a careful orchestration of gene expression leading to tolerance or susceptibility to a greater or lesser extent. Taken together these results suggest that some tolerance mechanisms are similar whereas others are specific to each landrace.

Drought stress tolerance requires changes in water transport to allow cells and tissues to adapt to the stress situation. Aquaporins facilitate osmosis by forming water-specific pores as an alternative to water diffusion through the lipid bilayer thus increasing the permeability of the membrane (Bartels and Sunkar, 2005). Although several aquaporin genes were differentially expressed in the three genotypes, CC and particularly M21 had more up-regulated genes encoding NIP, TIP, and PIP aquaporin as compared to 85-2. These results suggest that in the tolerant genotypes the activation of greater number of aquaporin genes would facilitate water flux to maintain cellular homeostasis.

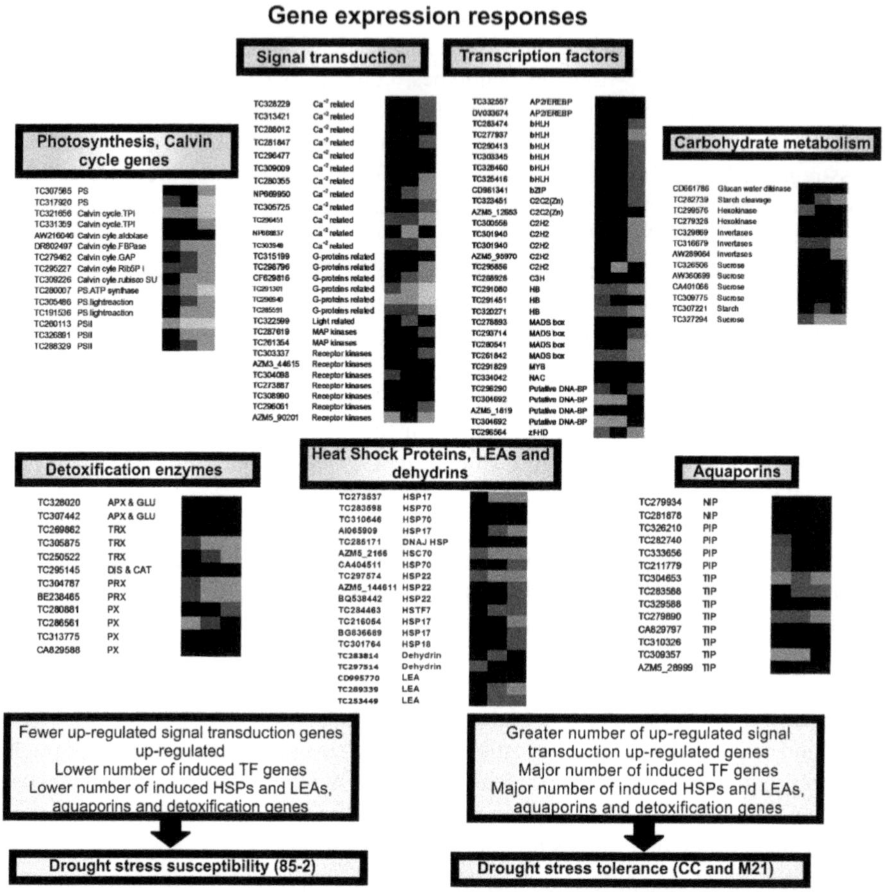

Figure 1. General view of gene expression responses in the three maize landraces under drought stress. Gene expression was monitored at 10 and 17 days stress and differences in gene expression levels were observed between the landraces. Transcripts encoding signal transduction, transcription factors, HSP, detoxification enzymes, and aquaporins are shown. Gene identifiers correspond to the accession numbers from the corresponding databases as reported in Maize Oligonucleotide Array Annotation GAL files version 1.13 (http://www.maizearray.org/maize_annotation.shtml). Microarray data were visualized using the FiRe 2.2 Excel macro (Garcion et al., 2006). A ≥ 2 fold change is shown in red, a fold change ≤ 0.5 in green and no change in black (FDR ≤ 0.05). Left column: 85-2, middle column: CC and the right column: M21. The PS: photosystem, TPI: triosephosphate isomerase, FBPase: fructose-1,6-bisphosphatase, GAP: glyceraldehyde 3-phosphate dehydrogenase, Rib5PI: ribose 5-phosphate isomerase-related, Rubisco SU: ribulose bisphosphate carboxylase small subunit, PSII: photosystem II, MAP kinase: mitogen-activated protein kinase, AP2/EREBP: AP2/Ethylene-responsive element binding protein family, bHLH: Basic Helix-Loop-Helix family, C2C2: C2 domain-containing protein, similar to zinc finger and C2 domain protein, C2H2: C2 domain-containing protein, similar to zinc finger and C2 domain protein, C3H: zinc finger (CCCH-type) family protein, HB: Homeobox transcription factor, Putative DNA BP: putative DNA binding protein, zf-HD: zinc finger homeobox, APX & GLU: ascorbate peroxidase and glutathione related, TRX: thioredoxin, PRX: peroxiredoxin, PX: peroxidase, HSP17: 17 kDa class I heat shock protein, HSP70:heat shock protein 70, HSC70: heat shock cognate 70 kDa protein, HSP22: 22.0 kDa ER small heat shock protein, DNAJ HSP: DNAJ heat shock protein, HSP18: 18.1 kDa class I heat shock protein, LEA: late embryogenesis abundant protein, NIP: NOD26-like membrane integral protein, PIP: Plasma membrane intrinsic protein, and TIP: tonoplast intrinsic protein.

A secondary effect of dehydration is the increase in reactive oxygen species (ROS) (Bartels and Sunkar, 2005) and consequently the requirement of an enhanced activity of antioxidant enzymes (Chaves et al., 2003) to protect cells from oxidative injury under drought stress (Bhatnagar-Mathur et al., 2008). In this case, increased transcript levels for thioredoxin and peroxidases were observed for both tolerant genotypes but mainly in M21, whereas in the susceptible line the up-regulation of these genes was rare. These results suggest that the tolerant genotypes and in particular M21 activate the expression of genes that could allow them to better cope with ROS.

Although it was observed that the induction of HSPs is a common response in maize, a greater number of genes encoding HSPs were induced under drought stress in the tolerant landraces (24 and 31 for CC and M21) as compared to the susceptible one. In particular, a higher number of members of the HSP17 family were induced in the tolerant landraces. A positive correlation between the levels of several HSPs and stress tolerance has been previously reported (Bartels and Sunkar, 2005; Sun et al., 2002; Vinocur and Altman, 2005). Since small HSPs have a long half life, it has been proposed that they might play an important role during stress recovery (Sun et al., 2002) therefore, signaling pathways that lead to an increased expression of small HSP might be preferentially activated in drought tolerant maize genotypes in comparison to susceptible ones.

Decrease photosynthetic activity under drought is due to reductions in stomatal conductance and Rubisco activities leading to lower carbon fixation (Xue et al., 2008) that consequently results in the over-reduction of components within the electron transport chain, generating ROS (Mahajan and Tuteja, 2005). In our study, it was found that genes related to PSI and PSII were down-regulated mainly in the tolerant landraces, suggesting that these genotypes reduce the activity of the components of the PSs to avoid the generation of large amounts of ROS under drought stress. A similar situation was also observed in rice drought tolerant cultivars under drought stress, where a higher number of members of gene families related to PSI and PSII were down-regulated in tolerant rather than sensitive cultivars (Degenkolbe et al., 2009). We also found that a larger number of Calvin cycle-related genes such as transcripts for Rubisco, phosphoglycerate kinase, GADPH, TPI, and FBPase were repressed under stress in both tolerant genotypes, and to a greater extent in M21, suggesting that a general repression of photosynthesis-related genes occurs in maize under drought. Ten photosynthesis-related genes that were repressed at 17 days stress were induced at recovery most of them were shared between the two tolerant landraces. Together these results suggest that the maize drought-tolerant genotypes analyzed in this study (more evident in M21) more broadly reduce electron transport in the PSI and PSII systems, probably to reduce the effect of photoxidation and the synthesis of enzymes involved in carbon fixation to avoid spending energy and resources under conditions of low CO_2 availability.

Osmotic Adjustment by Proline Accumulation in the Three Maize Landraces Under Stress and at Recovery

Proline is probably the most widely distributed osmolyte in plants (Bartels and Sunkar, 2005) and is implicated in responses to various environmental stresses (Mahajan and

Tuteja, 2005). Besides osmotic adjustment other roles such as protection of plasma membrane integrity, as an energy sink or reducing power, a source for carbon and nitrogen, or a hydroxyl radical scavenger (Bartels and Sunkar, 2005) as well as preservation of enzyme structure and activity (Chaves et al., 2003) have been proposed for this molecule in osmotically stressed plant tissues. It was found that M21 had the highest accumulation of proline under 10 and 17 days stress, followed by CC, whereas 85-2 showed the lowest level of proline accumulation. The up-regulation of genes encoding P5CS in CC and a putative ornithine aminotransferase, involved in an alternative proline biosynthetic pathway (via ornithine aminotransferase) in M21 at 17 days stress could explain the higher accumulation of proline in the tolerant genotypes. The P5CS is a rate-limiting enzyme for the biosynthetic pathway via glutamate in higher plants (Kishor et al., 2005; Vendruscolo et al., 2005) and it has been reported that concomitant to the accumulation of proline, an increase in the expression of P5CS is observed in a salt-tolerant genotypes but not in sensitive genotypes exposed to salt stress (Kishor et al., 2005). Our results suggest that both proline biosynthetic pathways are active in maize and that depending upon the genotype only one of them is activated during drought stress.

At recovery irrigation, a strong correlation between the decrease in proline content and the up-regulation of genes involved in proline degradation: pyrroline-5-carboxylate dehydrogenase (P5CDH) and the down-regulation of P5CS were observed in the three maize landraces. Although the proline content at recovery irrigation decreased in the three landraces, M21 had the highest content of proline after recovery irrigation. High proline content has been associated with increased recovery capacity (Schafleitner et al., 2007).

Differential Molecular Responses Under Recovery in the Three Maize Landraces

On recovery a total of 2,567 and 2,765 up- and down-regulated transcripts were identified in the three genotypes, of which 1,466 genes were found to be inversely regulated between stress and recovery (induced during stress and repressed during recovery and vice versa). Figure 2 shows a general scheme of the responses observed in the three landraces during the recovery process. The observation that the greatest number of differentially expressed genes was found at recovery, suggests a rapid and global re-activation of general plant metabolism following severe stress. As expected, the number of down-regulated genes encoding HSPs was greater in the tolerant than in the susceptible landraces, particularly in M21. Further, the two tolerant landraces shared differentially expressed genes related to signaling including receptor kinases, G-proteins, Ca^{+2} signaling, and phosphoinositide metabolism during recovery irrigation. The fact that CC and M21 have a higher number of induced signaling genes at recovery (48 and 47 genes respectively) than 85-2 (32 genes) indicates the possibility that the tolerant genotypes adjust their metabolism more efficiently during the recovery process. Genes related to carbon metabolism were up-regulated in tolerant landraces, however in 85-2 most of these genes were constitutive and this could be one of the key differences between tolerant and susceptible responses. Genes encoding Calvin cycle enzymes, PSI, PSII, and photosynthesis-related enzymes were also up-regulated mainly in the tolerant genotypes on recovery. The up-regulation of genes involved in photosynthesis and Calvin cycle is in accordance with the increase in the rate of photosynthesis and stomatal conductance observed after recovery in CC and M21. In this context, the up-regulation of peroxiredoxins genes shown to protect DNA, membranes, and certain enzymes against damage by removing H_2O_2 and hydroxyl radicals (Bartels and Sunkar, 2005), during recovery irrigation in the two tolerant landraces, could represent a protective mechanisms against the production of ROS during a rapid re-activation of photosynthetic activity. The down-regulation during recovery of genes encoding aquaporins, supports their importance in stress tolerance. The expression patterns of genes encoding TFs were distinct for the three landraces, suggesting different responses during recovery. A greater number of *TF* genes including bHLH, MADS, and MYB were found to be both up- and down-regulated in the tolerant landraces in comparison to 85-2.

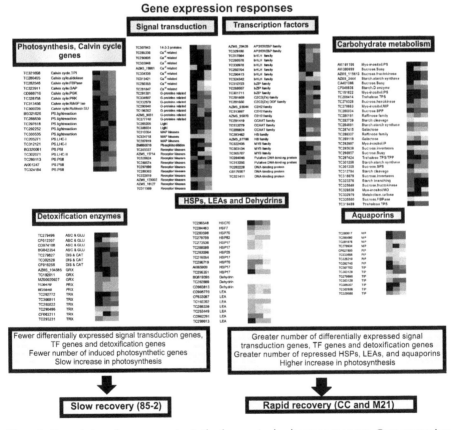

Figure 2. General view of gene expression in the three maize landraces on recovery. Gene expression was monitored at 10 and 17 days stress and differences in gene expression levels were observed between the landraces. Transcripts encoding signal transduction, transcription factors, HSP, detoxification enzymes, and aquaporins are shown. Gene identifiers correspond to the accession numbers from the corresponding databases as reported in Maize Oligonucleotide Array Annotation

Figure 2. (*Caption Continued*)

GAL files version 1.13 (http://www.maizearray.org/maize_annotation.shtml). Microarray data were visualized using the FiRe 2.2 Excel macro (Garcion et al., 2006). A ≥ 2 fold change is shown in red, a fold change ≤ 0.5 in green and no change in black (FDR ≤ 0.05). Left column:85-2, middle column: CC, and the right column: M21. The TPI: triosephosphate isomerase, FBPase: fructose-1,6-bisphosphatase, GAP: glyceraldehyde 3-phosphate dehydrogenase, PGK: phosphoglycerate kinase, PRK: phosphoribulokinase, Rib5P Iso: ribose 5-phosphate isomerase-related, Rubisco SU: ribulose bisphosphate carboxylase small subunit, LHC-I: Light harvesting chlorophyll a/b binding protein of PSI, PSI: photosystem I, LHC-II: Light harvesting chlorophyll a/b binding protein of PSII, PSII: photosystem II, MAP kinase: mitogen-activated protein kinase, AP2/EREBP: AP2/Ethylene-responsive element binding protein family, bHLH: Basic Helix-Loop-Helix family, C2C2: C2 domain-containing protein, similar to zinc finger and C2 domain protein, C2H2: C2 domain-containing protein, similar to zinc finger and C2 domain protein, CCAAT: CCAAT-binding transcription factor, HB: Homeobox transcription factor, MIP:myo-inositol-1-phosphate synthase, Susy: sucrose-phosphate synthase, TPS: trehalose-6-phosphate synthase, SPP: sucrose-phosphatase, IP: nositol monophosphatase, TPS/TPP: trehalose-6-phosphate synthase/phosphatase, MO: myo-inositol monophosphatase, TPS: trehalose-6-phosphate synthase, ASC & GLU: ascorbate and glutathione related, GRX: glutaredoxins, PRX: peroxiredoxin, TRX: thioredoxin, HSC70: heat shock cognate 70 kDa protein, HSF7: heat shock factor protein 7, HSP70:heat shock protein 70, HSP83: heat shock protein 83, HSP17: 17 kDa class I heat shock protein, HSP25: 25.3 kDa small heat shock protein, LEA: late embryogenesis abundant protein, NIP: NOD26-like membrane integral protein, PIP: Plasma membrane intrinsic protein, and TIP: tonoplast intrinsic protein.

A significant difference was found in the recovery response of M21 which showed the greatest changes in gene expression in almost all of the functional categories of the three genotypes (except for signaling), suggesting that this genotype possess a recovery mechanism that responds rapidly by activating metabolic processes on recovery. Expression changes in CC also reflect these responses albeit at a lower level and 85-2 is the least responsive even though it shares some common genetic background with M21.

RESULTS

Physiological Effects of Drought Stress

Changes in Leaf and Soil Water Potentials During Drought Stress Treatments

To ensure that plants were grown under the required drought stress conditions, soil and leaf water potentials were monitored throughout the experiment. As shown in Figure 3A, soil water potentials (ψ_s) were similar for all genotypes throughout the experiment. To determine whether plant water status differed between the three genotypes, leaf water potentials (ψ_l) were monitored at 10 and 17 days of drought stress and after recovery irrigation. Irrigated plants of all three genotypes maintained relatively constant levels of leaf water potential (ψ_l) of between −0.52 and −0.53 MPa (Figure 3B) throughout the experiment. Leaf water potentials (ψ_l) in stressed plants became more negative as the level of stress increased, −0.98 to −1.17 MPa after 10 day of stress and −1.06 to −1.23 MPa for 17 days of stress. At 10 days stress, the two drought tolerant genotypes showed slightly higher levels of leaf water potential as compared to the susceptible genotype. At 17 days stress, 85-2 and M21 showed a similar decrease in water potential, whereas CC still maintained higher leaf water potential. Ten hours after the recovery irrigation, all three genotypes showed a similar increase in water potential, to levels only slightly lower than before the drought stress treatment.

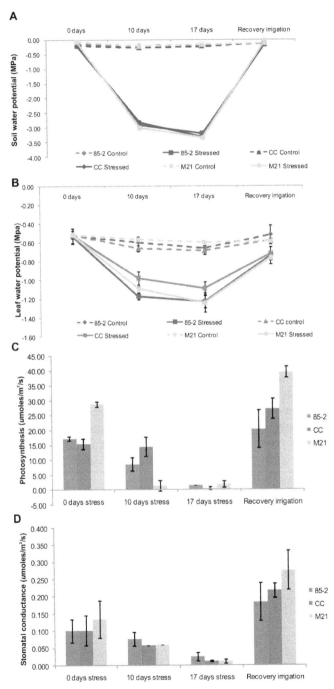

Figure 3. Physiological parameters of maize plants under drought stress and recovery irrigation. (A) Soil water potential, (B) Leaf water potential, (C) Photosynthetic rate, and (D) stomatal conductance. Data are from three measurements from different samples with standard error.

Rates of Photosynthesis

To determine the effect of drought on photosynthetic activity in the three genotypes, rates of photosynthesis were determined at 10 and 17 days of drought stress and 12 hr after the recovery irrigation. Under well-watered conditions, M21 showed the highest photosynthetic rate followed by CC and 85-2. Drought stress treatments caused reductions in rates of photosynthesis in all three landraces. The M21 showed the most rapid reduction of photosynthetic rate, dropping by 77.5% after 10 days and 86.92% after 17 days stress in comparison to the levels in irrigated plants. The CC showed the slowest decrease in photosynthetic rate, decreasing by only 30.55% after 10 days of drought and by 46% after 17 days of drought, whereas 85-2 showed a reduction of 52.2% and 88.4% after 10 and 17 days of drought stress, respectively (Figure 3C). Whereas M21 showed the fastest reduction in photosynthesis rate during drought treatment and a more rapid recovery of photosynthetic activity upon recovery irrigation, CC appeared to maintain a higher rate of photosynthesis during drought and showed a slower recovery on irrigation. The 85-2 showed an increase in rate of photosynthesis following the recovery irrigation but to a lesser extent than M21 (Figure 3C).

Stomatal Conductance

Drought stress also caused a gradual decrease in stomatal conductance in all three landraces (Figure 3D) as stress became more severe. Values for irrigated plants before stress were similar for all three landraces, although M21 showed a slightly higher value. At 10 days of drought stress, M21 showed a sharp drop (76.92%) in stomatal conductance compared to the irrigated plants, whereas for CC and 85-2 stomatal conductance decreased only 20% and 14.3% respectively. At day 17 under drought stress all three landraces showed a significant decrease in stomatal conductance (85.71% for 85-2, 90% for CC, and 84.61% for M21) as compared to the corresponding value prior to the stress treatment. Upon recovery irrigation all three landraces showed a rapid increase in stomatal conductance of 94.12%, 93.33%, and 88.89% (for 85-2, CC and M21 respectively), greater than the corresponding value prior to the stress treatment (Figure 3D).

Variation in Sugar Concentration

Analysis of glucose and myo-inositol content (Figures 4A and 4B) indicated a rise in both these sugars to a peak at 17 days stress in M21, whereas CC showed a drop in glucose levels as drought stress progressed and a peak in myo-inositol levels at 10 days stress. Landrace 85-2 showed the highest levels for both sugars at 10 days stress. Sucrose levels were maintained relatively constant in all three landraces throughout the drought experiment with a slight rise to a maximum at 17 days stress in all cases (data not shown).

Proline Content

Drought stress causes changes in amino acid metabolism in general and in particular accumulation of proline has been correlated with osmoprotection in several plant species (Bartels and Sunkar, 2005; Chaves et al., 2003). Determination of proline levels in the three landraces under drought stress and recovery showed a 2.5 fold increase at 10 days stress in 85-2 and a 7–9-fold increase in CC and M21. At 17 days stress, proline content in 85-2 had increased by four fold and by approximately 14 fold in CC and M21. However, M21 was the landrace with greatest proline accumulation. On recovery irrigation proline levels fell in all landraces to two fold higher than prior to drought stress in 85-2 and around five fold greater in CC and M21 (Figure 4C). These results suggest that osmoprotection by proline accumulation may be an important factor in achieving tolerance which is shared by both tolerant landraces.

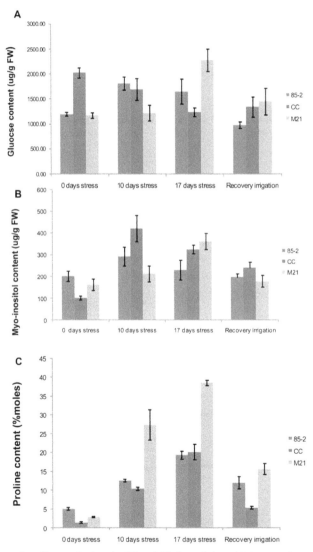

Figure 4. Sugar and proline content under 10 and 17 days of drought stress and recovery irrigation. (A) Glucose content, (B) Myo-inositol content, and (C) Proline content. Data are the means of three different samples with standard error.

Transcription Profiling

The MOA was used to analyze the differences in gene expression under drought stress between the three maize landraces. A numerical comparison of differentially expressed transcripts between the three genotypes under different drought stress treatments is shown in Figures 5A and 5B. Differences in number, level of expression and type of responsive genes can be seen between the different landraces and under the different drought stress treatments. In general, changes between stress treatments and untreated plants (both up- and down-regulated) were greatest for M21. The CC showed an intermediate response and 85-2 the lowest number of differentially expressed genes. Throughout the text all the differences in the numbers of up- or down-regulated genes between the three landraces are statistically significant as determined by chi-square analysis unless otherwise stated. Although certain genes were differentially expressed in all three landraces many were specific to each landrace.

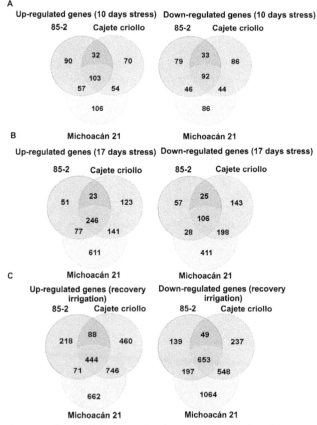

Figure 5. Venn diagrams of up- and down-regulated transcripts under drought stress and on recovery irrigation. (A) Differentially expressed genes at 10 days stress, (B) Differentially expressed genes at 17 days stress, and (C) Differentially expressed genes at recovery irrigation. Number of genes with at least two fold change and FDR ≤ 0.5 are shown for each landrace identified by the name above the circle.

At 10 days drought stress, 103 up-regulated and 92 down-regulated genes were common to the three landraces. M21 showed a higher number of specific, differentially expressed transcripts (106 up- and 86-down regulated) compared to 85-2 (90 up- and 79 down-regulated) and CC (70 up- and 86 down-regulated genes) as seen in Figure 5A. As the level of drought stress increased, the number of differentially expressed genes also increased. At 17 days stress, a total of 246 up-regulated and 106 down-regulated genes were common to all three landraces. At the same time point, the susceptible landrace (85-2) showed the lowest number (51 up- and 57 down-regulated), CC showed an intermediate number (123 up- and 143 down-regulated genes) and M21 the highest number (611 up- and 411 down-regulated genes) of specific differentially expressed genes (Figure 5B). At 17 days of stress, the number of differential transcripts shared by the tolerant landraces (141 up- and 198 down-regulated) was also greater than those shared between either of the tolerant landraces and the susceptible landrace.

In total, a greater number of differentially expressed transcripts were observed upon recovery irrigation in comparison to those found at 10 and 17 days of stress (Figure 5C). Under these conditions 444 transcripts were up- and 653 down-regulated common in all three landraces. A similar pattern of expression to that observed at 17 days stress was also observed at recovery irrigation, with 85-2 showing the lowest level (218 up- and 139 down-regulated), CC intermediate (460 up- and 237 down-regulated) and M21 the highest level (662 up- and 1064 down-regulated) of genotype specific differentially expressed genes. The number of differentially expressed transcripts shared by the tolerant landraces (746 up- and 548 down-regulated) was also higher than those shared between the tolerant landraces and 85-2. Moreover, we found that 65.49% of up-regulated genes at 17 days stress were repressed on recovery; whereas 55.68% of the genes repressed at 17 days stress were induced at recovery. We also found that 8.13% of the genes were induced both at 10 and 17 days stress.

The differential expression pattern observed in the microarray experiment was evaluated for 16 genes using qRT-PCR. Most of the genes showed the same expression pattern in both the microarray experiment and the qRT-PCR analysis. Differences were mainly observed at the quantitative level, with the qRT-PCR analysis showing in general a higher fold of induction or repression than the microarray analysis. Similar, quantitative differences between qRT-PCR and microarray data have been reported previously (Morcuende et al., 2007) and (Calderón-Vázquez et al., 2008).

Functional Classification of Differentially Expressed Transcripts

Due to the limited functional annotation currently available for maize transcripts, functional classification of differentially expressed transcripts was carried out using the MapMan hierarchical ontology software (Thimm et al., 2004) and BioMaps at the Virtual-Plant site (www.virtual plant.org) as described in Materials and Methods with similar results. However, the analyses presented here were based on MapMan software. A general overview of the metabolic and cellular processes, for which differentially expressed genes were identified, are shown in Figures 6A, 6B, and 6C for 85-2, CC, and M21 respectively for 17 days stress and Figures 6D, 6E and 6F for 85-2, CC, and M21 respectively for recovery irrigation. These global maps of differentially expressed genes illustrate that the tolerant genotypes showed more wide-ranging metabolic and cellular responses during drought stress and recovery irrigation than 85-2.

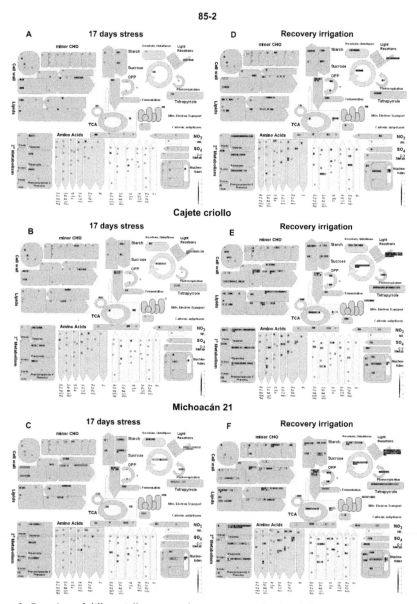

Figure 6. Overview of differentially expressed transcripts involved in different metabolic processes under stress and recovery irrigation. (A) Genes at 17 days stress in 85-2, (B) Genes at 17 days stress in CC, (C) Genes at 17 days stress in M21, (D) Genes at RI in 85-2, (E) genes at RI in CC, and (F) Genes at RI in M21. Gene transcripts that are induced or repressed are shown in red or green coloring respectively as shown in the color bar in each panel. The MapMan sotware was used to show the different functional categories involved. (CC: Cajete criollo, M21: Michoac6n 21).

Photosynthesis and Carbohydrate Metabolism

To determine whether the observed changes in rates of photosynthesis described above correlated with changes in gene expression, the effect of drought stress and recovery irrigation on the expression of genes encoding components of the photosynthetic machinery was analyzed in detail. At 10 days stress 4, 1, and 2 up-regulated (differences not statistically significant) and 13, 17, and 22 down-regulated genes related to photosynthesis were found in 85-2, CC, and M21 respectively. At 17 days stress, all differentially expressed photosynthesis-related genes were down-regulated in all three genotypes (85-2: 11, CC: 28, and M21: 45). The only notable exception was a transcript for a putative fructose-bisphosphate aldolase which was up-regulated in M21. A general view of this data is shown in Figures 7A, 7B, and 7C for one representative member of each photosynthesis-related gene family. Interestingly, CC and M21 showed more down-regulated gene families than 85-2. For instance, three Calvin cycle-related genes were down-regulated in 85-2 as compared to seven in CC and 12 in M21. Down-regulated transcripts of Calvin cycle genes such as triosephosphate isomerase (TPI), fructose-1,6-bisphosphatase (FBPase), Rbcs (RuBisCO small subunit), and Rubisco activase were identified.

On recovery irrigation the pattern of expression was reversed, with an increase in differential expression of photosynthesis-related genes in all three genotypes. The responses of the landraces were low (17), intermediate (61), and high (81) in terms of numbers and levels of up-regulation of specific genes for 85-2, CC, and M21 respectively (Figures 7D, 7E, and 7F). With respect to Calvin cycle-related genes 6, 19, and 21 were up-regulated in 85-2, CC, and M21 respectively.

Sugar metabolism is closely linked to photosynthesis and differences in accumulation of glucose and myo-inositol between the genotypes were observed as described above. In this context, 8, 15, and 28 genes related to carbohydrate metabolism were found to be up-regulated on recovery irrigation in 85-2, CC, and M21 respectively. A transcript for β-amylase is induced in all three genotypes at 10 and 17 days stress and to a much greater extent in M21 at 17 days stress whereas on recovery irrigation three transcripts for β-amylase were repressed in M21. Two transcripts for hexokinases (HXK) were found to be up-regulated at 17 days stress in M21 and one in 85-2 whereas in CC they did not reach the two fold level (1.84 and 1.4 fold change). On recovery three hexokinase transcripts were repressed in M21, but in 85-2 constitutive expression was observed whereas in CC a 0.64 fold change was observed; suggesting that changes occur in glucose metabolism under stress that are then reversed or repressed on recovery. Perhaps surprisingly, the genes encoding enzymes involved in synthesis of myo-inositol (Ins (3) P synthase and MI monophosphatase) are repressed or remained constant under stress in all three genotypes in spite of the fluctuations in myo-inositol levels described above. On recovery, however, M21 shows a slight increase in the expression of these genes.

Induction and repression of genes associated with proteins involved in HXK dependent and independent signaling pathways as proposed for *A. thaliana* was also observed. For example genes associated with HXK dependant glucose signaling such as *Cab* and *RbcS* were repressed under stress but induced on recovery, whereas PLD was induced under drought and repressed on recovery for all three landraces. For the HXK independent pathway, AGPase and PAL were all down-regulated or constitutive in CC and 85-2 but up-regulated in M21 at 17 days stress, while all were down-regulated or constitutive in all three landraces on recovery.

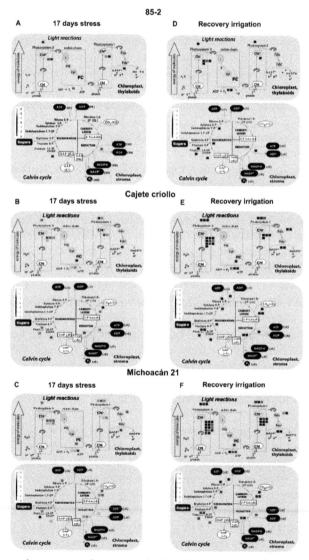

Figure 7. Differential expression of genes involved in photosynthesis under drought stress and at recovery irrigation. (A) Genes differentially expressed at 17 days stress in 85-2, (B) Genes differentially expressed at 17 days stress in CC, (C) Genes differentially expressed at 17 days stress in M21, (D) Genes differentially expressed at RI in 85-2, (E) Genes differentially expressed at RI in CC, and (F) Genes differentially expressed at RI in M21. Gene transcripts that are induced or repressed are shown in red or green coloring respectively as shown in the color bar in each panel. (CC: Cajete criollo, M21: Michoac6n 21).

Genes associated with sucrose metabolism were mainly constitutive or repressed at 17 days stress; however, we found one and four transcripts up-regulated for sucrose synthase (Susy) in CC and M21 respectively at 17 days stress. On recovery, most of the genes related to sucrose metabolism were induced in CC and M21, whereas levels of expression of these genes in 85-2 remained constant.

Aminoacid Metabolism

Proline Metabolism

The expression patterns of genes encoding enzymes involved in proline synthesis and degradation agreed well with the levels of proline determined. A pyrroline-5-carboxylate synthase gene was up-regulated under stress in CC and constitutive in M21 and 85-2 under 17 days stress. In contrast an ornithine aminotransferase involved in a different proline biosynthetic pathway was up-regulated specifically in M21 under stress. This may indicate preferential use of one or the other proline biosynthetic pathways in the tolerant landraces under drought stress. A transcript for proline oxidase involved in proline degradation showed a slight decrease (0.95 and 0.58 fold) in CC and M21 respectively under drought but was up-regulated in 85-2. On recovery irrigation, a transcript for pyrroline-5-carboxylate dehydrogenase (P5CDH) involved in proline degradation was up-regulated in the tolerant landraces but constitutive in 85-2. Two transcripts for proline oxidase were up-regulated in CC and one in 85-2 during recovery irrigation, no changes were detected in M21. On the other hand, a gene encoding pyrroline-5-carboxylate synthetase (P5CS), involved in proline synthesis, was down-regulated under recovery irrigation in M21, although 85-2 and CC showed 0.79 and 0.54 fold changes respectively.

Signaling and Abiotic Stress-Related Genes

Metabolic responses to different abiotic stresses are often shared; therefore in order to compare changes in expression of previously characterized stress-related genes between the three landraces under drought stress, functionally annotated transcripts were grouped into different categories: hormone metabolism, signaling, transport, detoxification, heat-shock proteins (including dehydrins and LEA), and abiotic stress genes. In general, fewer stress-related genes were differentially expressed at 10 days as compared to 17 days stress. Although similar patterns of up- and down-regulation were observed for all categories in all three landraces, M21 showed the highest changes in transcript abundance of differentially expressed genes in all categories and many differentially expressed genes were unique to M21. On recovery irrigation, in general the number of induced genes in each category was lower than those that were repressed. The CC and M21 showed very similar patterns of up-regulation although the number of genes observed in each category was slightly higher for M21, with the exception of "abiotic stress genes" where more up-regulated transcripts were observed for CC. Landrace 85-2 showed a poor response in the number of up-regulated transcripts in comparison to the other landraces. The greatest difference between the three landraces was observed for down-regulated transcripts on recovery irrigation. The CC and 85-2 showed very similar patterns of down regulation where the numbers of transcripts in each category did not exceed 40. In contrast for M21 most categories showed down-regulation of between 40 to >70 transcripts representing a two-fold difference in comparison to the other two landraces (Figure 8).

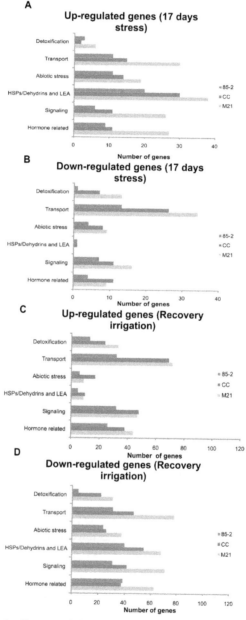

Figure 8. Functional classification of abiotic stress genes under 17 days of drought stress and recovery irrigation. (A): Up-regulated genes under 17 days of drought stress, (B): Down-regulated genes under 17 days of drought stress, (C): Up-regulated genes under recovery irrigation, and (D): Down-regulated genes under recovery irrigation.

Genes associated with signal transduction such as calcium dependent protein kinases (CDPKs), G-proteins, and receptor kinases were both up- and down-regulated under drought. M21 proved to be the landrace with most differentially expressed signaling genes at 17 days stress in comparison to the other two maize landraces. From 42 differentially expressed genes, seven were common to the two tolerant landraces and 27 genes were unique to M21 including genes encoding to: calcium binding protein, CDPK, calmodulin, GTP binding protein, and phosphoinositides. For the recovery process, 170 differentially expressed genes related to signaling were identified, of which 32 were common to the two tolerant landraces including genes encoding receptor kinases, G-proteins, Ca + 2 signaling, and phosphoinositides.

In the heat shock category, a greater number of up-regulated genes encoding HSPs under 17 days stress were observed, especially in the tolerant maize landraces (24 and 31 for CC and M21 respectively) in comparison to 85-2 (16). Perhaps surprisingly only HSP17 and LEA transcripts were up-regulated under stress and in common with HPS18, HSP70, DNaJ, HSF, and dehydrins; these transcripts were strongly repressed on recovery. Other abiotic stress-related genes, such as those for cold and drought/salt stress were in general induced under stress and repressed on recovery as would be expected. Several transport associated transcripts showed little change under drought but showed both up- and down-regulation on recovery irrigation including those associated with amino acids or metals, ABC, metabolite, and Pi transporters. A pattern of up-regulation under stress and down-regulation on recovery was observed for the aquaporin genes. At 10 days stress, two up-regulated transcripts for tonoplast intrinsic proteins (TIPs) were identified only in the two tolerant landraces. One of these TIPs was also up-regulated at 17 days stress only in the two tolerant maize landraces. Two transcripts for nodulin-like intrinsic proteins (NIPs) were up-regulated at 17 days stress and at recovery irrigation were down-regulated. Interestingly the sugar transport associated genes were up- and down-regulated in both stages, reflecting the changes in sugar metabolism and transport which occur in relation to photosynthesis levels under drought and recovery. In relation to genes associated with detoxification, only peroxidases and thioredoxins genes were up-regulated. These genes showed both up and down regulation on recovery as did ascorbate and glutathione metabolism and dismutases and catalases genes. The only other transcripts which clearly showed repression under drought and induction on recovery were those associated with the peroxiredoxins. Only one transcript for a peroxiredoxin was found to be induced at 10 days stress in M21 and this cultivar and CC showed the highest number of induced peroxiredoxin genes at recovery in relation to 85-2.

Hormone-Related Responses

Many hormone-related genes were found to be differentially expressed under stress and on recovery such as those related to ABA, auxins, cytokinins, and ethylene metabolism. The plant hormone ABA plays a central role in many aspects of response to various stress signals (Bartels and Sunkar, 2005; Mahajan and Tuteja, 2005) and has been shown to participate in drought and high salinity-tolerance mechanisms (Wasilewska et al., 2008). In this study, differential expression of genes involved in ABA metabolism was observed under drought stress and recovery irrigation. The 9-*cis* epoxycarotenoid dioxygenase (*NCED*) gene was up-regulated in 85-2 and M21 at 10 days stress. This enzyme is thought to be involved in the rate-limiting step in ABA biosynthetic pathway (Chaves et al., 2003; Xiong et al., 2002). At 17 days stress, a gene named ABA insensitive 3 (ABI3) was up-regulated in M21. This gene encodes a TF which may be involved in multiple hormonal signaling pathways related to ABA and auxins (Brady et al., 2003). The *HVA22* gene known to be highly induced under drought, ABA, cold and salt stresses was also up-regulated only in M21. In addition at recovery irrigation, 11 down-regulated transcripts related to ABA metabolism were identified. During recovery irrigation, transcripts for AREB2 (ABA-responsive element binding protein), ABRE (ABA-responsive element binding protein) and a protein phosphatase 2C involved in ABA signal transduction were down-regulated in the three landraces and five transcripts for *HVA22* were down-regulated in at least one of the landraces.

Cytokinins are hormones that play an essential role in plant growth and development (Brugiére et al., 2003). In this study, cytokinin-related genes showed few changes under drought but were very strongly up-regulated on recovery with few genes down-regulated. Under 10 days stress a gene for cytokinin oxidase was up-regulated only in 85-2. Cytokinin oxidase regulates the levels of cytokinin in plants by degrading it irreversibly (Massonneau et al., 2004). At 17 days stress differentially expressed genes related to cytokinin signal transduction were identified. However, at recovery many up-regulated genes related to cytokinin signal transduction were observed: 14, 17, and 18 genes in 85-2, CC, and M21 respectively (differences not statistically significant). Most were found to be response regulator genes (*ARR*). Only three genes related to cytokinin metabolism were down-regulated in at least one of the three landraces.

Auxin-related genes showing induction of 2, 1, and 5 transcripts at 10 days stress in 85-2, CC, and M21 respectively were also identified (differences not statistically significant). At 17 days stress 7 up-regulated and 6 down-regulated genes for auxin metabolism were identified and a transcript for GH3-like protein enzyme that conjugates amino acids to indole 3-acetic acid (Park et al., 2007) was induced higher in M21. At recovery the number of differentially expressed genes increased 18 and 28 for up- and down-regulated genes respectively in at least one of the three landraces. Notably, the GH3 gene showed repression only in M21.

Ethylene is another hormone that is related to abiotic stress responses. However, it was found that ethylene related genes showed no strong response under stress although a transcript for a putative Fe/ascorbate- dependent oxidoreductase was induced in CC and M21 at 10 days stress and at 17 days stress in M21. One ACC oxidase gene was repressed only in M21 whereas in CC it was found to be constitutive and in 85-2 the value was higher than 0.5 fold change. An ethylene response factor was also down-regulated in CC and M21 at 17 days stress. At recovery, many genes related to ethylene metabolism were found to be down-regulated. The Fe/ascorbate- dependent oxidoreductase gene was repressed under this treatment in CC and M21, but not in 85-2. Transcripts of ACC oxidases were constitutively expressed in M21 but repressed in CC. A down-regulated transcript for the ethylene response factor (ERF) was identified in 85-2 and another in 85-2 and M21. However, a different ERF gene was also induced in CC and M21 at recovery.

Transcription Factors

Table 1 compares the number of differentially expressed transcription factor (*TF*) genes for each of the principal *TF* gene families and for each landrace. All *TF* gene families analyzed showed differential expression in the three landraces with differences in the patterns of induction/repression. In general the response of each landrace followed the pattern described previously for numbers of differentially expressed transcripts: 85-2 low, CC intermediate, and M21 high. A total of 121 differentially expressed genes encoding TF were identified in this study under drought stress. Among these, tolerant landraces (CC and M21) showed more genes induced and repressed for bHLH, WRKY, MYB, C2C2 and C2H2, HB, and CCAAT (HAP2) TF families (24 and 78 respectively) compared to 85-2. Further, more members of the AP2/EREBP, HB, and MADS families were up-regulated specifically in M21 under stress. Both CC and M21 had greater numbers of up- and down-regulated TFs in the unclassified group as com-

Table 1. Principal transcription factor gene families differentially regulated under drought stress and recovery irrigation.

Treatment	Drought stress						Recovery irrigation					
Cultivar	85-2	95-2	CC	CC	M21	M21	85-2	85-2	CC	CC	M21	M21
Regulation	Up	Down	Up	Down	Up	Down	Up	Down	Up	Down	Up	Down
AP2/EREBP	1	0	2	0	4	1	1	2	3	1	5	6
bHLH	1	0	2	2	6	3	1	2	6	5	10	8
bZIP/Putative BZIP	2	0	0	0	2	1	3	1	5	2	5	6
C2C2	0	1	2	1	3	0	0	2	5	5	6	5
C2H2	0	2	0	3	0	1	1	5	7	5	7	5
CCAAT:	0	0	1	0	1	1	0	0	2	1	2	3
-HAP2	0	0	1	0	1	1	0	0	1	0	1	0
-HAP3	0	0	0	0	0	0	0	0	0	0	0	3
-HAP5	0	0	0	0	0	0	0	0	1	1	1	0
DNA binding	1	1	0	2	3	0	3	3	6	5	9	5
Finger TF	0	2	0	2	0	2	0	2	1	2	1	2
G2 like	1	3	1	4	1	2	1	4	3	5	3	3
HB	3	0	3	2	5	1	4	5	5	7	8	6
HD leucine zipper	1	0	1	0	1	0	1	0	1	1	1	1
MADS	3	0	2	0	6	0	3	4	2	4	6	7
MYB	3	3	2	4	7	5	5	6	7	8	11	12
NAC	0	0	1	1	0	0	1	2	1	3	2	3
Unclassified	2	2	5	3	6	5	7	7	18	11	24	16
WRKY	0	2	0	3	0	1	2	2	1	4	1	3
Zinc finger/HD	1	0	0	2	1	2	3	2	9	2	9	6
Total	19	16	22	29	46	24	36	49	83	71	111	98

pared to 85-2. On recovery irrigation, 202 differentially expressed genes encoding TF were identified. The tolerant landraces showed significantly different responses with more induced and repressed genes (*CC 104* and *M21 139 TFs* respectively) compared to the susceptible landrace 85-2 for *AP2, NAC, MADS, CCAAT, HB, bHLH,* and *bZIP TF* families which most of them were induced under stress but repressed on recovery irrigation. One CCAAT family member, HAP3 previously shown to confer drought tolerance by overexpression of NF-YB (HAP3) in *Arabidopsis* and maize (Nelson et al., 2007) is specifically down-regulated on recovery in M21 whereas transcripts for HAP2 and HAP5 were up-regulated in CC and M21 also after recovery irrigation. The AP2/EREBP and MADS families also showed patterns specific to M21 on recovery irrigation. The M21 showed the greatest number of responsive TF gene families, suggesting that this is the most responsive landrace at the level of differential gene expression under drought stress as well as under recovery irrigation.

CONCLUSION

Differences in rates of photosynthesis, stomatal conductance, sugar and proline accumulation, and gene expression patterns were identified between the three landraces. In many cases, in comparison to the tolerant landraces, 85-2 failed to respond or responded more weakly. Important differences were also noted between the tolerant landraces that probably underlie different mechanisms of achieving tolerance: CC may have an advantage under prolonged drought periods due to a gradual reduction of photosynthesis and stomatal conductance, whereas M21 with the capacity for latency, with a rapid reduction of photosynthesis and efficient recovery responses may perform better under short periods of severe drought stress. Although necessarily the most outstanding differences have been emphasized in this study, subtle differences between the landraces should not be overlooked. Differences in response mechanisms were also supported by the detailed changes in gene expression patterns under drought conditions. Modulation of a greater number of differentially expressed genes from different TF gene families could be an important characteristic of the tolerant landraces, many belonging to families previously implicated in stress responses such as members of the *AP2/EREBP, bHLH, HB, CCAAT,* and *MYB TF* gene families. Furthermore, the genes encoding hormones, aquaporins, HSPs, and LEAs and detoxification enzymes were induced to a greater extent in the tolerant landraces again suggesting more efficient responses in these genotypes. In the case of recovery from the drought stress, the most important feature was the speed and scope of changes in gene expression, which differed between the three genotypes. This report emphasizes the most outstanding differences between drought-tolerant and susceptible genotypes and identified potential regulators of the drought and recovery processes in maize. The task in hand is now to characterize the expression patterns and responses unique to each landrace and the genes or specific alleles involved in order to compare with other commercial maize germplasm and suggest possible breeding or transgenic strategies to improve drought tolerance. Modulation of expression of specific transcription factor genes has already proved successful in improving drought stress (Wang et al., 2003). In this chapter several additional TF families were identified and the regulatory effects of these genes in particular should be studied in more detail.

KEYWORDS

- **Abscisic acid**
- **Cajete Criollo**
- **Maize oligonucleotide array**
- **Michoacón Hexokinases**
- **Transcription factor**
- **Trios phosphate isomerase**

ACKNOWLEDGMENTS

The authors thank Pedro Cervantes and Fernando Hernández for their assistance in the greenhouse experiments, Dr. Víctor Olalde and Iván Gasca for their valuable help in the photosynthesis measurements and data analyses; Dr. Josué Altamirano and Dr. Juan Jose Peña-Cabriales for equipment and assistance in measurements of soluble sugars and Dr. Octavio Martinez de la Vega and Araceli Fernández for their help with the statistical analysis. We also would like to thank Susana M. L. Fuentes-Guerra and Flor M. X. Zamudio-Hernandez for qRT-PCR analysis. We are in debt to Dr. J. L. Pons Hernández for providing the 85-2 maize line and to Dr. Suketoshi Taba (CIMMYT) for providing the Cajete Criollo and Michoacán 21 maize seeds and Marcial Rivas (CIMMYT) for the information provided about these maize collections.

Chapter 6

Regulators in Transgenic *Arabidopsis* Plants

Wei Wei, Jian Huang, Yu-Jun Hao, Hong-Feng Zou, Hui-Wen Wang, Jing-Yun Zhao, Xue-Yi Liu, Wan-Ke Zhang1, Biao Ma, Jin-Song Zhang, and Shou-Yi Chen

INTRODUCTION

Soybean (*Glycine max* (L.) Merr.) is one of the most important crops for oil and protein resource. Improvement of stress tolerance will be beneficial for soybean seed production.

Six *GmPHD* genes encoding Alfin1-type PHD finger protein were identified and their expressions differentially responded to drought, salt, cold, and ABA treatments. The six *GmPHDs* were nuclear proteins and showed ability to bind the *cis*-element "GTGGAG." The N-terminal domain of *GmPHD* played a major role in DNA binding. Using a protoplast assay system, we find that *GmPHD1* to *GmPHD5* had transcriptional suppression activity whereas *GmPHD6* did not have. In yeast assay, the *GmPHD6* can form homodimer and heterodimer with the other *GmPHDs* except *GmPHD2*. The N-terminal plus the variable regions but not the PHD-finger is required for the dimerization. Transgenic *Arabidopsis* plants overexpressing the *GmPHD2* showed salt tolerance when compared with the wild type plants. This tolerance was likely achieved by diminishing the oxidative stress through regulation of downstream genes.

These results provide important clues for soybean stress tolerance through manipulation.

Drought and high salinity are the major factors to affect plant growth and productivity. These environmental stresses cause the changes of physiological and biochemical processes through alteration of gene expressions. Genes induced by various abiotic stresses are classified into two groups. The products of the first group are effector proteins that protect cell membrane system, hold water, control ion homeostasis, and scavenge ROS. These proteins include the key enzymes required for osmoprotectants, LEA proteins, aquaporin proteins, chaperones, and detoxification enzymes. The products of the second group are regulatory proteins that control perception of signal, signal transduction, and transcriptional regulation of gene expression, including protein kinases, enzymes involved in phoshoinositide metabolism and transcription factors. Several transcription factor families have been found to be induced by drought and salt stresses, such as DREB, ERF, WRKY, MYB, bZIP, and NAC families (Hasegawa et al., 2000; He et al., 2005; Liao at al., 2008; Seki et al., 2003; Zhou et al., 2008; Zhu, 2002). DREB1A and AtMYB2 improved the drought and salt tolerance of transgenic plants when transferred into *Arabidopsis* (Abe et al., 2003; Kasuga et al., 1999). Alfin1, a PHD finger protein, was identified as a salt-induced transcriptional factor and enhanced the stress tolerance by ectopic expression in transgenic plants (Bastola et al., 1998).

The PHD finger was first named from the product of the *Arabidopsis HAT3.1* gene in 1993 (Schindler et al., 1993). After that a number of PHD finger proteins have been identified throughout eukaryotic kingdom. The PHD finger is a conserved Cys4-HisCys3 type zinc finger domain similar to RING finger and LIM domain (Aasland et al., 1995; Borden and Freemont, 1996; Dawid et al., 1998). Plenty of evidences suggest that the PHD finger proteins are most likely to be chromatin-mediated transcriptional regulators. PHD finger proteins such as transcriptional cofactor P300 and CBP are histone acetyltransferases (HATs) that covalently modify the N-terminal tails of histones (Kalkhoven et al., 2002). As subunits of histone aceyltransferase or histone deacetylase complexes, PHD finger proteins are required for transcriptional activation or transcriptional repression, such as ING1 (Skowyra et al., 2001), Pf1 (Yochum and Ayer, 2001), TIF1 (Venturini et al., 1999), and KAP1 (Schultz et al., 2001). A PHD finger protein, Alfin1, is characterized as a transcriptional factor that can bind to the promoter of MsPR2 and enhance the expression of MsPR2 at the transcriptional level (Bastola et al., 1998; Winicov, 2000; Winicov and Bastola, 1999; Winicov et al., 2004). Moreover, PHD fingers often occur with SET, Bromo, and chromodomains that provide additional evidence for correlation with chromatin (Aasland et al., 1995; Anderson et al., 2002; Tripoulas et al., 1996). Taken together, there are three possible functions related to chromatin for PHD finger: (1) like other zinc fingers, it might be a DNA or RNA-binding domain; (2) similar to the RING and the LIM domain, it may be a protein-protein interaction domain; (3) it may interact with the flexible histone tails or the central part of the histones (Pena et al., 2006; Shi et al., 2006; Wysocka et al., 2006). Recent studies have suggested two other functions of PHD finger. The PHD finger of MEKK1 and MIR is suggested to act as E3 ubiquitin ligase (Coscoy and Ganem, 2003; Lu et al., 2002). However, two groups argued that the PHD finger of MEKK1 and MIR is more similar to RING domain (Aravind et al., 2003; Scheel and Hofmann, 2003). Furthermore, the PHD fingers of ING2 and AIRE1 are proposed to be phosphoinositide receptors (Kuzmichev et al., 2002). But other group did not observe the PIPs binding activity of AIRE1 and there is no more evidences supporting the PHD finger as PIPs receptor (Bottomley et al., 2005). To define the role of PHD finger, researchers tend to believe that PHD fingers in diverse proteins might share the common function. However, some PHD fingers may be different from other PHD fingers in both the sequence similarity and function. In addition, many PHD fingers were identified to interact with specific proteins.

Although the PHD-domain-containing proteins have been extensively studied, the protein functions may not solely depend on the PHD finger, and their roles in plant abiotic stress responses were largely obscure. In this study, we identified six *GmPHDs* from soybean as a specific set of PHD finger proteins. They are responsive to various abiotic stresses at transcription level. Five of the six proteins had transcriptional suppression activity in plant cells. The N-terminal region of *GmPHD* was mainly responsible for DNA binding. Overexpression of *GmPHD2* enhanced the salt tolerance of transgenic *Aarabidopsis*, and this may be achieved by scavenging of reactive oxygen species (ROS).

MATERIALS AND METHODS

Plant Materials and Treatments

The soybean population of recombinant inbred lines derived from Jindou23 (JD23, salt- and drought-tolerant variety) and Huibuzhi (HBZ, salt- and drought-sensitive variety) were used. Two-week-old seedlings from 24 salt and drought-tolerant lines and 24 salt and drought-sensitive lines were placed on Whatman filter paper at 23°C and with 60% humidity for dehydration. After 1 hr and 3 hr, one leaf from each seedling was harvested and combined for RNA isolation to construct the stress-tolerant and stress-sensitive RNA pool for cDNA-AFLP. The seedlings were immersed with the roots in 200 mM NaCl or 100 µM ABA and maintained for the indicated times. For cold treatment, seedlings were placed at 0°C. For drought treatment, seedlings were placed on Whatman filter paper at 23°C and with 60% humidity. Roots, cotyledons, stems, leaves, and 10-day-old developing seeds from soybean were collected for RNA analysis.

Gene Cloning and RT-PCR Analysis

The cDNA-AFLP was conducted as described (Wang et al., 2005). Based on the obtained sequence encoding a PHD finger domain, six *GmPHD* genes were identified and cloned by RT-PCR or RACE. Homologous genes from *Medicago truncatula*, rice and *Arabidopsis* were also identified. Cluster analysis was conducted using the MEGA 4.0 program.

Stress-responsive genes were examined by RT-PCR with primers as follows: for At1g21230, 5'-gtaggtagaaacatatgtgg-3' and 5'-GTGTTCCCATGTAAGCGAAG-3'; for AT5G39110, 5'-GATCCAAGTCCACTTCAAGAC-3' and 5'-CAACATT-GACGTCTAACTG-3'; For AT1G76430, 5'-GCTCCTTTGGTTGTGGCTTCT-3' and 5'-CTAGGAACCAATTGGCTGAGGC-3'; For AT2G36270, 5'-CAACAAGCAG-CAGCAGCTGCAG-3' and 5'-GGATTAGGTTTAGGATTAGTGGG-3'; For AT3G09940, 5'-GTTTGTGCTGGAACTGGAG-3' and 5'-CAGTACAGATTCTC-CAACG-3'; For AT5G19890, 5'-CTTGTGCTGATATCCTCACTTT-3' and 5'-GT-GATCATTCTGATACACACGA-3'; For AT1G49570, 5'-GTTGGAGAATATAA-CAGCCAAG-3' and 5'-CCATTACACACAAACGTAACAC-3'; For AT4G08770, 5'-GGAAACCAGAGTGTATTGGTAG-3' and 5'-GTGATCATTCTGATACACAC-GA-3'; For At1g73660, 5'-AGAATTTGGGAGATGGAGTGG3' and 5'-CCTTAC-CAATTCACTATTCAC-3'; For CBF2, 5'-ATGTTTGGCTCCGATTACG-3' and 5'-ATAGCTCCATAAGGACACGT-3';ForSTRS1,5'-ATGGCTGGACAAAAGCAA-GA-3' and 5'-CATATCAAGCATTCGATCTGC-3'; For STRS2, 5'-ATGAATTC-CGATGGACCCAA-3' and 5'-GACCTCATCAGATACTGTGG-3'; For At1g68875, 5'-ATGACAGAACTCAAATGGAT-3' and 5'-CTAGTTAGACTGTGGTGCCA-3'; For At1g02200, 5'-ATGGCCACAAAACCAGGAGT-3' and 5'-GAATATCATGGA-GAGAGAGG-3'; For At5g07550, 5'-ATGTTTGAGATTATTCAGGC-3' and 5'-TTA-GACGCCGGAACCTGCTG-3'.

Real-Time Quantitative PCR

The *GmPHDs* were amplified with the following primers: for *GmPHD1*, 5'-AT-GGACTCTCGCACGTATAA-3' and 5'-GTGGTACTTCTTCAGCAGGT-3'; for

GmPHD2, 5'-ATGGACGGTGGTGGAGTGAA-3' and 5'-CCTTCCGCAGGTAAAT-
TAAC-3'; for *GmPHD3*, 5'-ATGGAGGCGCTAAGTCGCTC-3' and 5'-AAGCTCT-
GGAGGAACTTCTT-3'; for *GmPHD4*, 5'-ATGGAGGCAGGTTACAATCC-3' and
5'-CAGGGGGCACCTCCTCAGCT-3'; for *GmPHD5*, 5'-ATGGAAGGAGTACCG-
CACCC-3' and 5'-GCACTTCCTCAACAGGCAAA -3'; for *GmPHD6*, 5'-ATGGA-
CAGTGGAGGACACTA-3' and 5'- GGAACTTCTTCAGCAGGCAA -3'. Real-time
PCR were performed on MJ PTC-200 Peltier Thermal Cycler based on previous pro-
tocol (Liao et al., 2008b). The results were analyzed using Opticon Monitor™ analysis
software 3.1 (Bio-Rad). Each experiment had four replicates and was repeated twice.

Localization of the GmPHD-GFP Fusion Proteins
The GmPHD coding sequences were amplified with the primers containing BamHI
and SalI sites. The products were fused to the 5' end of GFP to generate the pUC-
GmPHDs-GFP constructs (Xie et al., 2003). The pUC-GFP vector was used as control. These constructs were introduced into *Arabidopsis* protoplasts by PEG-mediated
transfection. After culturing for 20 hr, the fluorescence of GFP was visualized under
fluorescence microscope.

Dimerization of GmPHDs
Interaction of GmPHDs was investigated by co-transforming plasmids into the yeast
strain YRG2 according to the manual (Stratagene). The pBD-GmPHDs and pAD-
GmPHDs were made by insertion of the GmPHD coding region into pBD vector or
pAD vector. Each of the six pBD-GmPHDs, together with each of the pAD-GmPHDs,
was co-transformed into YRG2. The PHD finger sequences of each GmPHDs were
amplified by PCR and then introduced into pBD vector to generate pBD-G1PHD,
pBD-G2PHD, pBD-G3PHD, pBD-G4PHD, pBD-G5PHD, and pBD-G6PHD. These
plasmids, together with pAD-GmPHD6, were co-transformed into YRG2. The pBD-
GmPHD5, pBD-G5N (amino acids 1 to 115), pBD-G5V (amino acids 116 to 197),
pBD-G5PHD (amino acids 198 to 252), pBD-G5NV (amino acids 1 to197), and pBD-
G5PV (amino acids 116 to 252), together with pAD-GmPHD6, was also co-transformed into YRG2. The pBD vector or pAD vector, together with the corresponding
recombinant plasmids, was co-transformed into yeast cells as negative controls. The
yeast transformants were plated onto SD-His/Trp/Leu plus 3-AT and the growth was
examined. X-gal staining was performed to examine the LacZ reporter gene expression (Wang et al., 2003).

Transient Assay for Transcriptional Activation/Inhibition Activity of GmPHDs in *Arabidopsis* Protoplast System
Reporter plasmid 5XGAL4-LUC and internal control pPTRL (*Renilla reniformis*
Luciferase driven by 35S promotor) were kindly provided by Dr. Masaru Ohme-Takagi.
5XGAL4-LUC contains five copies of GAL4 binding element and minimal TATA
region of 35S promoter of Cauliflower Mosaic Virus (CaMV), located upstream of
the firefly gene for luciferase (Ohta et al., 2001). Expression vector pRT-BD was constructed by insertion of the GAL4DBD coding region into pRT107 vector by Sac I/
Xba I digestion. And the positive control (35S-BD-VP16) was constructed by insertion of

VP16, a herpes simplex virus (HSV)-encoded transcriptional activator protein, into pRT-BD vector.

For effector plasmids used in Figure 4A, the coding regions of GmPHDs were digested by BamHI/Sal I, and cloned into pRT-BD vector to generate 35S-BD-GmPH-Ds. For effector plasmids in Figures 4B and C, the coding regions of GmPHDs were digested by BamHI/Sal I, and cloned into pRT107 vector to generate 35S-GmPHDs, which will not compete for the GAL4 binding elements in reporter plasmid 5XGAL4-LUC when incubated with the 35S-BD-VP16. The truncated coding sequences of GmPHD2 were also cloned into pRT107 to obtain 35S-G2N (amino acids 1 to 117), 35S-G2V (amino acids 118 to 196), 35S-G2PHD (amino acids 197 to 253), 35S-G2NV (amino acids 1 to 196), and 35S-G2VP (amino acids 197 to 253). The *Arabidopsis* Dof23 (At4g21030) was used as a non-interactive control when incubated with 35S-BD-VP16.

The ratios in Figure 4 indicate μg of each plasmid. The effectors, reporter, and internal control were co-transfected into *Arabidopsis* protoplasts. After culturing for 16 hr, Luciferase assays were performed with the Promega Dual-luciferase reporter assay system and the GloMaxTM20-20 luminometer (Liao et al., 2008b).

Gel Shift Assay

The genes for GST-GmPHDs fusions and various domains of the GmPHD4 and GmPHD2 were cloned into pGEX-4T-1, and the proteins were expressed in *E. coli* (BL21) and purified according to the manual. A pair of oligonucleotides 5'-AATTC GGATCCGTGGAGGTGGAGGTGGAGGTGGAGGT'GGAGGGTACCGAGCT-3' and 5'-CGGTACCCTCCACCTCCACCTCCACCTCCACCTCCAC'GGATCCG-3' was synthesized. The two sequences contained five tandem repeats of "GTGGAG." The double-stranded DNA was obtained by heating oligonucleotides at 70°C for 5 min and annealing at room temperature in 50 mM NaCl solution. Gel shift assay was performed as described (Wang et al., 2005).

Generation of GmPHD2-Transgenic Plants and Performance of the Transgenic Plants Under Salt-Stress

The coding sequence of the GmPHD2 was amplified by RT-PCR using primers 5'-GGAGGATCCATGGACTCTCGCACGTATAATCC-3' and 5'-TGTGGTAC-CGGGCCGAGCTCTCTTGTTAC-3', and cloned into the BamHI/KpnI sites of the pBIN438 under the control of CaMV 35S promoter. The homozygous T3 seeds were analyzed.

Seeds were plated on NaCl medium for germination tests. Plates were placed at 4°C for 3 days and then incubated in a growth chamber under continuous light at 23°C. Each value represents the average germination rate of 80–100 seeds with at least three replicates. For salt-stress tolerance tests, 5-day-old seedlings on MS agar medium were transferred on MS agar medium supplemented with 0, 50, 100, 150, and 200 mM NaCl respectively. The phenotypes were observed two weeks later. The seedlings under 150 mM NaCl treatment were further transferred into soil and grown for two weeks under normal conditions. Then the root length and height of plants were measured.

Oxidative Stress Tolerance Test and Physiological Parameters

Seeds were plated on the MS containing different concentrations of paraquat for germination tests. Each value represents the average germination rate of 80–100 seeds with at least three replicates.

Ten-day-old seedlings were transferred onto the MS plates containing 150 mM NaCl and maintained for 5 days. Plant leaves were cut and submerged in 1 mg/ml 3′,3′-diaminobenzidine (DAB) solution for 6–8 hr and then fixed with solution of ethanol/lactic acid/glycerol (3:1:1, V/V/V). Brown color indicates presence of the hydrogen peroxide.

Measurement of peroxidase (POD) activity and relative electrolyte leakage were performed according to previous descriptions (Cao et al., 2007; Maehly and Chance, 1954).

DISCUSSION

The present study identified six GmPHD proteins from soybean plants. These proteins shared high identity and belonged to a small family with the PHD finger in the C-terminal end. The transcriptional regulatory activity, DNA binding ability and nuclear localization were revealed for these proteins, indicating that the GmPHD proteins represent novel transcription regulators. The roles of one of these GmPHD proteins, GmPHD2, were investigated through transgenic approach and we find that the GmPHD2 improved stress tolerance in plants.

Among the six proteins, the GmPHD1 to GmPHD5 had transcriptional suppression activity in plant protoplast assay whereas the GmPHD6 did not have such ability, suggesting their different roles in transcriptional regulation. The suppression activity of the five GmPHDs may depend on the presence of the V domain as judged from the GmPHD2 analysis (Figure 5C). The V domain had similar suppression ability as the whole GmPHD2 did. Removal of the V domain disclosed the strong inhibitory effects of the N or PHD domain, suggesting that the V domain may have a regulatory role for the function of the N and/or PHD domain during transcription.

Three domains can be defined for the six GmPHDs as exemplified in GmPHD2. These included the N domain in the N-terminal conserved region (amino acids 1 to 117), the PHD domain in the C-terminal conserved region with a PHD finger (amino acids 197 to 253), and the V domain in the variable region between the N and the PHD domains (amino acids 118 to 196). In addition to transcriptional regulation, roles of these domains in protein dimerization, localization, and DNA-binding were also studied. The PHD finger has been regarded as a protein-protein interaction domain (Linder et al., 2000; Townsley et al., 2004). The present GmPHD6 can form homodimer. It can also form heterodimers with GmPHD1, GmPHD3, GmPHD4, and GmPHD5. However, these interactions were not mediated by the PHD finger, but rather by the NV region as in the case of GmPHD5 (amino acids 1 to 197). This fact indicates that the PHD fingers in different proteins may have different roles. The possibility that the PHD finger of GmPHDs may interact with other unknown proteins cannot be excluded. Recently, the PHD fingers of the GmPHD/Alfin-like proteins from *Arabidopsis* have been found to bind to histone post-translational modifications H3K4me3/2 (Lee et al., 2009). Another PHD-containing protein ORC1, the large subunit of the origin recognition complex involved in defining origins of DNA replication, can bind to H3K4me3 with its PHD domain and regulate transcription (Sanchez and Gutierrez, 2009). Therefore the GmPHD proteins may interact with H3K4me3/2 via the PHD domain and form dimers through the NV region. In addition to the roles in interactions, the PHD domains also play some roles in nuclear localization or retention because removal of this domain in GmPHD2, GmPHD5, and GmPHD6 led to cytoplasmic distribution of the protein (Figure 4).

The six GmPHD proteins showed high identity to the Alfin1 from alfalfa, which also has a PHD finger at the C-terminal end (Bastola et al., 1998). Further comparison revealed that the GmPHD5 had highest identity (89%) with the Alfin1, suggesting that the GmPHD5 may be an orthologue of Alfin1 in soybean. The other five GmPHDs may be paralogues of the Alfin1. Alfin1 has been found to enhance the *MsPRP2* gene expression by binding to the element GTGGNG (Bastola et al., 1998; Winicov and Bastola, 1999). However, five out of the six GmPHDs had transcriptional suppression activity in protoplast assay system. This difference may reflect the divergence of the transcriptional regulatory mechanism between the Alfin1 and the GmPHDs. It is possible that the GmPHD proteins may first suppress gene expression and then indirectly affect expressions of other genes.

The PHD finger has been proposed to bind to DNA or RNA as many other zinc fingers do (Aasland et al., 1995; Schindler et al., 1993). However, from the solution structure of the PHD finger from KAP-1, no structural features typical of DNA binding proteins are observed (Capili et al., 2001). These studies imply that the DNA-binding ability is equivocal for the PHD fingers in different proteins. Despite the discrepancy, the Alfin1 and the present six GmPHD proteins all showed specific DNA binding ability (Winicov and Bastola, 1999). However, further domain analysis of the GmPHD4 and GmPHD2 disclosed that the N domain had strong DNA binding ability whereas the PHD domain showed no or only slight DNA-binding activity (Figure 7D). These results suggest that the N domain but not the PHD domain was mainly responsible for DNA binding. It is interesting to find that presence of the V domain has some effects on DNA-binding. In the case of GmPHD4, the V domain plays an inhibitory role on DNA-binding activity of the N domain, whereas in GmPHD2, the V domain promotes DNA-binding in the presence of PHD domain (Figure 7D). This phenomenon, together with the roles of the V domain in regulation of the N and/or PHD-mediated transcriptional suppression (Figure 5C), suggests that a specific regulatory mechanism existed for GmPHD/Alfin1-type transcription regulators. It is possible that the PHD domain of this type of proteins interacts with histone for chromatin regulation whereas N domain binds to DNA. The two coordinate reactions may thus lead to transcriptional suppression, with the regulation from V domain and the NV-mediated dimerization. It should be mentioned that although all the six GmPHD proteins are highly conserved, each one may also have specificities in terms of gene expression (Figures 3, 4), transcriptional regulation (Figure 5), dimerization (Figure 6A), and DNA binding ability (Figure 7). The V domain may determine the specificity of each protein in regulation of these processes. However, how these are realized requires further investigation.

Eighty-three canonical PHD finger proteins have been identified in *Arabidopsis* (Lee et al., 2009). Only several proteins containing the PHD finger domain have been studied in plants. However, except the conserved C4HC3 residues, other sequences in the PHD domains are divergent. The functions of these proteins are also different, ranging from regulation of anther development and male meiosis (Reddy et al., 2003; Wilson et al., 2001; Yang et al., 2003) to regulation of vernalization and flowering (Greb et al., 2007; Piñeiro et al., 2003; Sung and Amasino, 2004; Sung et al., 2006), disease resistance (Eulgem et al., 2007), apical meristem maintenance (Saiga et al., 2008), specification of vasculature and primary root meristem (Thomas et al., 2009), and embryogenesis and sister-chromatid cohesion (Sebastian et al., 2009). Unlike the PHD-containing proteins above, the present GmPHDs shared high homology only with Alfin1 from *Medicago sativa*. The Alfin1 can be induced by salt stress and enhance salt tolerance in the transgenic plants (Winicov and Winicov, 2000; Bastola, 1999). The present six *GmPHD* genes were differentially expressed in drought- and salt-tolerant JD23 and drought- and salt-sensitive HBZ in response to salt, drought, cold, and ABA treatment, indicating that this subset of genes may have specific roles in multiple stress responses. In most cases, these genes were induced to a higher intensity in the tolerant JD23 cultivar than that in the sensitive HBZ, suggesting that the genes may contribute to the stress tolerance of the JD23 cultivar. Different from the specific Alfin1 expression in roots, the six *GmPHD* genes were expressed in multiple organs. We selected the GmPHD2 for transgenic analysis because this protein had the least homology (67%) with the Alfin1. Overexpression of the GmPHD2 improved salt tolerance of the transgenic plants, indicating that proteins with transcriptional repression can also confer stress tolerance.

The GmPHD2 may confer salt tolerance through control of ROS signaling and ROS scavenging. The ROS scavengers have been reported to eliminate the cytotoxic effects of ROS under different stresses (Apel and Hirt, 2004; Mittler et al., 2004). Consistently, the transgenic plants overexpressing the GmPHD2 were more tolerant to oxidative stress, and had higher levels of POD activity and lower levels of hydrogen peroxide production under salt stress. It is likely that the GmPHD2 confers salt tolerance at least partially by diminishing oxidative stress. Other possibility may also exist due to the fact that many other genes, e.g., *ABA signaling* gene *ABI5*, were regulated by the GmPHD2 protein. The GmPHD2-regulated gene expression seemed to be different from that regulated by the Alfin1. The Alfin1 has been found to regulate the MsPRP gene expression by binding to the *cis*-element in the promoter region of this gene (Bastola et al., 1998). It is therefore possible that each PHD-type transcriptional regulator may contribute to the salt tolerance through upregulation of a specific subset of genes. It should be noted that although the GmPHD2 had transcriptional suppression activity, it still can enhance downstream gene expressions. This may be achieved through indirect regulation or via protein interactions. Several genes including *CBF2*, *STRS1*, *STRS2*, and *At1g73660* were also down-regulated in GmPHD2-transgenic plants (Figure 9A). These genes are negative regulators of stress tolerance (Gao and Xiang, 2008; Kant et al., 2007; Novillo et al., 2004) and may be the direct target of the GmPHD2.

In soybean plants, we have identified six GmPHD proteins. In other plants examined, similar number of genes was found. Because their differential expression patterns in response to various stresses and different mechanisms for transcriptional regulation, it is possible that each *GmPHD* gene has specificity in regulation of stress responses. However, these genes may also generate coordinate responses for stress tolerance through protein interaction or transcriptional regulation within this small gene family. Further investigation should reveal such possibilities and improve our understanding of the functions of this gene family in regulation for a variety of stress responses.

RESULTS

Cloning and Structural Analysis of the *GmPHD* Family Genes

A gene fragment (256 bp) encoding a PHD finger was identified during cDNA-AFLP analysis using stress-tolerant lines and stress-sensitive lines from the population of the recombined inbred lines derived from the soybean Jindou No. 23 (JD23, drought- and salt-tolerant) and Huibuzhi (HBZ, drought- and salt-sensitive). Expression of the corresponding gene was higher in the stress-tolerant pool than that in the stress-sensitive pool (data not shown). After EST assembly, a full-length gene *GmPHD1* (*DQ973812*), was obtained, which encoded a PHD finger protein of 253 amino acids. Further searching and assembly of soybean EST sequences revealed five other members of this gene family, namely *GmPHD2* to *GmPHD6* (*DQ973807, DG973808, DQ973809, DQ973810,* and *DQ973811*). The GmPHD3 and GmPHD6 were partial in 5′- and 3′-end respectively, and their full-length open reading frames were obtained using RACE method.

GmPHDs exhibited 70–88% identities with each other. Comparison of the amino acid sequences of these six members of GmPHD family revealed that the N-terminal regions and C-terminal regions were extremely conserved, indicating that these two parts may have significant function (Figure 1). The C-terminal region is identified as PHD finger, which is a conserved C_4HC_3 type zinc finger. However, the fourth cysteine was changed to arginine in GmPHD6. This variation was also found in the homologues of rice (data not shown).

The soybean GmPHDs showed 67–89% sequence identity to alfalfa Alfin1 (L07291) (Bastola et al., 1998). Homologues of GmPHDs were also present in many other plant species such as *Arabidopsis*, rice, *Medicago*, and *Solanum tuberosum*. In *Medicago truncatula*, seven homologues were found and five were full-length sequence termed MtPHD1 to 5 (*EF025125, EF025126, EF025127, EF025128,* and *EF025129*). MtPHD5 was almost identical to the Alfin1 (Bastola et al., 1998). From *Arabidopsis* and rice databases, seven homologues were also found respectively, including *AT1G14510, AT2G02470, AT3G11200, AT3G42790, AT5G05610, AT5G20510, AT5G26210, Os04g0444900, Os05g0163100, Os07g0233300, Os03g0818300, Os05g0419100, Os01g0887700,* and *Os07g0608400*. The GmPHDs has an overall identity of 68–72% compared to these homologues. The cluster analysis revealed that the GmPHD2, GmPHD4 and GmPHD6 were more closely related whereas GmPHD1 and GmPHD3 grouped with MtPHD1 and MtPHD3 respectively (Figure 2). The GmPHD5 was clustered with MtPHD5 and may be more divergent when compared with the other GmPHD proteins (Figure 2).

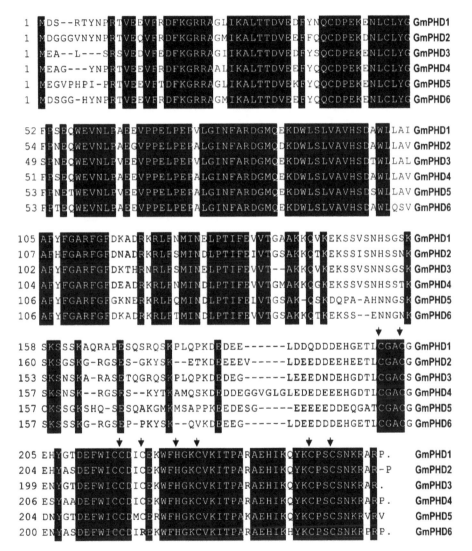

Figure 1. Alignment of the amino acid sequences of the six GmPHD proteins. Identical residues are shaded in black. The C-terminal region is PHD finger and arrows mark the most conserved residues C_4HC_3 in the finger.

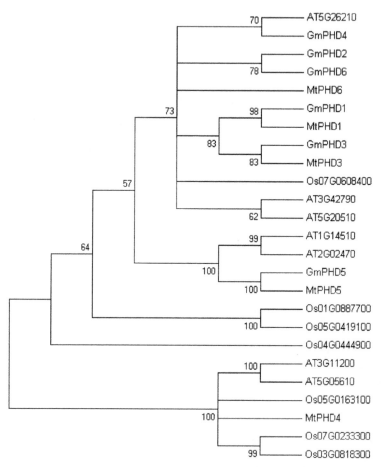

Figure 2. Cluster analysis of the PHD finger proteins from different plants. The analysis was performed by using the MEGA 4.0 program with neighbor joining method and with 1,000 replicates. Numbers on the figure are bootstrap values. The sequences are from soybean (GmPHDs), *Medicago* (MtPHDs), *Arabidopsis* and rice plants.

Expression Profiles of GmPHDs Under Various Stresses

Expressions of the six GmPHDs were investigated in JD23 and HBZ in response to different treatments, including high NaCl, drought, ABA, and cold (Figure 3). All of the genes were induced in response to drought, but showed differences in responses to the other stresses. One of the genes, *GmPHD4*, was induced in response to all four conditions while the other five genes were induced in response to two or three conditions. Interestingly, the *GmPHD4* and *GmPHD5* were the only genes induced in response to low temperature and in both cases, this was only observed in the more stress-tolerant line JD23. These results indicate that the six *GmPHD* genes were differentially regulated in response to various treatments, and in most cases, the inductions of the *GmPHD* genes were stronger in stress-tolerant JD23 than those in stress-sensitive HBZ.

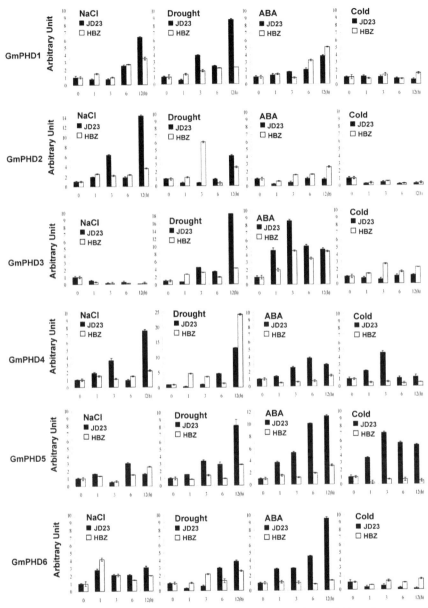

Figure 3. Expression of the six *GmPHD* genes in stress-tolerant cultivar JD23 and stress-sensitive cultivar HBZ under various treatments. Two-week-old soybean seedlings were subjected to treatments with 200 mM NaCl, 100 µM ABA, cold and drought, and total RNA was isolated for real-time quantitative PCR analysis.

The expressions of the six *GmPHDs* were examined in different organs of soybean plants. Figure 4A showed that all the six genes had relatively higher expression in cotyledons, stems and leaves, but low expression in roots and developing seeds.

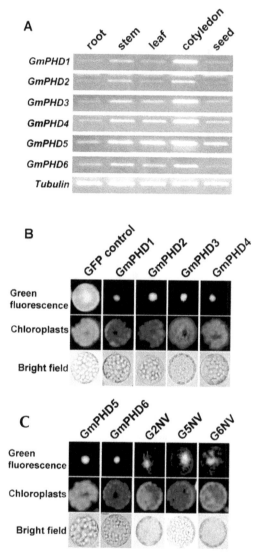

Figure 4. Organ-specific expression and subcellular localization of the *GmPHDs*. (A) Expression of *GmPHDs* in different organs of soybean plants revealed by RT-PCR. A Tubulin fragment was amplified as an internal control. (B) Subcellular localization of GmPHD proteins in *Arabidopsis* protoplasts as revealed by green fluorescence of GmPHD-GFP fusions or GFP control. For each panel, the photographs were taken in the dark field for green fluorescence (upper), for red fluorescence indicating chloroplasts (middle), and in the bright light for the morphology of the cells (lower). G2NV: the NV domain of GmPHD2; G5NV: the NV domain of GmPHD5; G6NV: the NV domain of *GmPHD6*.

Subcellular Localization of *GmPHDs*

Majority of PHD finger proteins are nuclear proteins but some of them are membrane proteins (Goto et al., 2003; Mansouri et al., 2003). Constructs containing the *GmPHDs-GFP* fusion genes in the plasmid pUC18 were generated. The fusion genes and *GFP* control in pUC18 driven by the cauliflower mosaic virus (CaMV) 35S promoter were transformed into *Arabidopsis* protoplasts, and the protein expression was revealed by the green fluorescence of the fused GFP protein under a fluorescence microscope. All the six GmPHD proteins were targeted to nucleus of the cells, while the control GFP protein was observed in the cytoplasm (Figure 4B). When the PHD domain (amino acids 198 to 252) was removed from the GmPHD5, the resulted G5NV truncated protein can be visualized in the cytoplasm although the protein was still abundant in the nuclear region (Figure 4B). Similarly, when the PHD domain was removed from the GmPHD2 or GmPHD6, the resulted G2NV or G6NV was localized in the cytoplasm and the nuclear region (Figure 4B). These results indicated that the six GmPHD proteins were nuclear proteins and the PHD domain may play a role in nuclear localization or nuclear retention of the GmPHD proteins.

Transcriptional Regulation Activity of *GmPHDs*

The PHD finger proteins have been reported to have the transcriptional activation activity (Halbach et al., 2000). We examined the transcriptional activation activity of GmPHDs in protoplast system. As shown in Figure 5A, among the six proteins compared, five (except the GmPHD6) was found to have inhibitory effect on reporter gene activity when compared to the negative BD control, possibly implying that the five proteins GmPHD1 to GmPHD5 can suppress the transcription of the reporter gene to different degrees. The GmPHD6 appeared not to have such inhibitory activity. To further investigate if the GmPHD proteins have any effect on VP16-mediated transcriptional activation, we included each of the six GmPHD proteins with the positive control VP16 transcription factor in the assay system. Figure 5B showed that the five proteins GmPHD1 to GmPHD5 had inhibitory effects on VP16-promoted gene expression, suggesting that the five proteins may mainly play roles in transcriptional suppression. On the contrary, the GmPHD6 did not show such ability. A Dof-type transcription factor Dof23 from *Arabidopsis* did not have significant effect on VP16 transactivation activity.

The GmPHD family members contained conserved N-terminal region, a variable middle part and a conserved C-terminal PHD finger domain. The three regions of the GmPHD2, namely N (N-terminal, amino acids 1 to 117), V (Variable, amino acids 118 to 196), and PHD (PHD domain, amino acids 197 to 253) were investigated for their effects on VP16-mediated transcriptional regulation. Figure 5C showed that the GmPHD2, V domain, NV (amino acids 1 to 196) and VP (amino acids 197 to 253) all had similar inhibitory effects on VP16 transcriptional activation. However, the single N or PHD domain appeared to have stronger roles in transcriptional suppression than the other versions examined, suggesting the importance of the N and PHD domain in transcriptional regulation. Addition of the V to the N domain (G2NV) abrogates the inhibition, suggesting that the regulatory effects may target different molecular aspects as determined by structure of the protein.

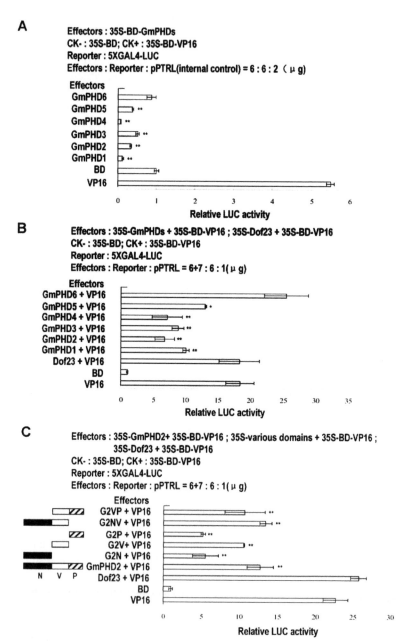

Figure 5. Transcriptional regulation activity of *GmPHDs* in protoplast assay. (A) Effects of the *GmPHDs* on reporter gene expression as revealed by relative LUC activity. The GAL4 DNA-binding domain (BD) and VP16 were used as negative and positive controls respectively. "**" indicate highly significant difference (P < 0.01) compared to BD value. (B) Effects of the GmPHDs on VP16-mediated *LUC* gene expression. The *Arabiodpsis* Dof23 was used as a non-interactive control. (C) Effects of various domains of the GmPHD2 on VP16-mediated *LUC* gene expression. For (B) and (C), "*" and "**" indicate significant difference (P<0.05 and P<0.01 respectively) compared to VP16 value.

Analysis of the GmPHDs Dimerization

Previous studies have shown that a few PHD finger proteins can form homo- or heterodimers by PHD finger (Fair et al., 2001). We then examined if the GmPHDs can dimerize by using the yeast two-hybrid assay. Figure 6A showed that cells transformed with pAD-GmPHD6 plus pBD-GmPHD1, pBD-GmPHD3, pBD-GmPHD4, pBD-GmPHD5, or pBD-GmPHD6 could grow on SD/His-/Trp-/Leu- medium with 10 mM 3-amino-1,2,4-triazole (3-AT). Also, the blue color was observed in the X-gal staining with these transformed cells (Figure 6A). These results indicate that the GmPHD6 can form homodimer and heterodimers with other GmPHDs except GmPHD2. However, other combinations of the GmPHD proteins did not generate any interactions (data not shown). We further examined if the PHD finger is involved in the interaction. Figure 6B showed that the cells harboring the PHD fingers and the pAD-GmPHD6 could not grow on SD/His-/Trp-/Leu- medium plus 3-AT and did not have positive X-gal staining, demonstrating that there was no interaction between GmPHD6 and PHD fingers of GmPHDs.

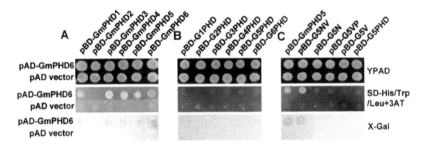

Figure 6. Dimerization ability of the six GmPHD proteins. (A) Dimerization between the GmPHD6 and other GmPHD proteins as revealed by transformant growth on YPAD and SD-His/Trp/Leu plus 3-AT, and by X-gal staining. (B) The PHD finger is not responsible for the dimerization. The yeast transformants containing pAD-GmPHD6 plus each of the PHD finger constructs or pBD vector, were examined for cell growth and X-gal staining. (C) The NV region of the GmPHD5 mediates the interaction between GmPHD5 and GmPHD6. The yeast transformants harboring the pAD-GmPHD6 plus different truncated versions of pBD-GmPHD5 were examined for cell growth and X-gal staining. Truncated proteins: GmPHD5(1–252), G5NV(1–197), G5N(1–115), G5PV(116–252), G5V(116–197), and G5PHD(198–252).

To further determine the interaction domain, we focused on the interaction between GmPHD6 and GmPHD5. Constructs harboring various domains of GmPHD5 in pBD vector were made and transformed into YRG-2 cells with pAD-GmPHD6 or pAD vector (negative control). Figure 6C showed that only the cells containing pBD-GmPHD5 or pBD-G5NV plus pAD-GmPHD6 grew well and exhibited blue color in the X-gal staining. Removal of the V region from the G5NV protein abolished growth of the corresponding transformants, suggesting that the extremely acidic V region has substantial influence on the interactions between GmPHD proteins. The cells from other combinations could not grow on selection medium and no positive X-gal staining was observed (Figure 6C). These results indicate that the NV region (amino acids 1 to 197 in GmPHD5) of GmPHDs may be the protein-protein interaction domain that functions in dimerization between GmPHD proteins.

DNA Binding Activity of the GmPHDs

Alfin1, a homologue of GmPHDs from *Medicago sativa*, showed DNA binding activity to the conserved core of GNGGTG or GTGGNG (Bastola et al., 1998). To identify if the present *GmPHDs* has any DNA binding activity, we performed gel-shift analysis. Bacterially expressed GST-proteins were isolated and purified (Figure 7A). Five tandem repeats of the sequence GTGGAG were annealed, labeled and incubated with the six purified GST-GmPHD fusion proteins. All six GmPHDs formed a complex with the labeled GTGGAG and the signal was dramatically decreased by addition of unlabeled DNA probe (Figure 7B). These results indicate that all the six GmPHDs specifically bind to the GTGGAG element *in vitro*.

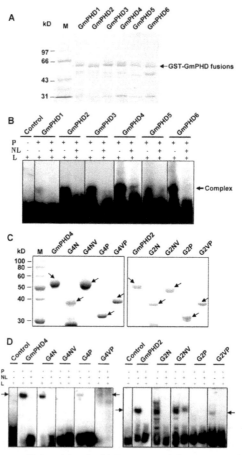

Figure 7. DNA-binding specificity of the GmPHD proteins. (A) Coomassie blue staining of the six GST-GmPHD fusion proteins on SDS/PAGE. Arrow indicates the fusion proteins. Lower bands probably represent the degradation products. (B) Gel shift assay of the six GmPHD proteins. GmPHD proteins (P) were incubated with a radiolabeled probe containing 5 X GTGGAG (L), in the presence (+) or absence (–) of unlabeled probes (NL) in 10-fold excess. Arrow indicates position of the protein/DNA complexes. (C) Coomassie blue staining of various domains of the GmPHD4 and GmPHD2. Arrows indicate the corresponding proteins. (D) Gel shift assay of the GmPHD4, GmPHD2 and their domains. Others are as in (B). Arrows indicate positions of the protein/DNA complexes.

To investigate which domain is responsible for the DNA-binding, the GmPHD4 that showed strong DNA-binding activity was used for the analysis. Different domains of the GmPHD4 were expressed (Figure 7C, left panel) and subjected to DNA-binding assay. Figure 7D (left panel) showed that the N domain had strong DNA-binding activity whereas the NV domain had no binding activity. The PHD domain had weak DNA-binding ability. The VP domain also had slight DNA-binding in addition to the non-specific binding. To further examine if the roles of different domains in DNA binding are also conserved in other GmPHD proteins, the GmPHD2 and its various domains were expressed (Figure 7C, right panel) and compared for DNA-binding ability (Figure 7D, right panel). The N domain of GmPHD2 had strong DNA-binding ability. Presence of the V domain in G2NV did not affect specific DNA-binding but may lead to some non-specific binding. The PHD domain (G2P) showed no DNA-binding while G2VP had weak DNA-binding ability (Figure 7D, right panel). These results indicate that the N domain had the major ability to bind DNA whereas the PHD domain had weak or no DNA-binding ability. The V domain may have substantial influence on the DNA binding ability of both the N and the PHD domains.

Transgenic Plants Overexpressing the GmPHD2 Showed Higher Salt Tolerance

Because the *GmPHD* genes were responsive to multiple stresses, we investigated if the *GmPHDs* are involved in stress responses. The *GmPHD2* was used for further analysis because the encoded protein showed the least homology to the well-studied Alfin1 (Bastola et al., 1998). We generated the transgenic *Arabidopsis* plants overexpressing the *GmPHD2* gene under the control of 35S promoter. Three homozygous lines G2–3, G2–6, and G2–8, with higher *GmPHD2* expression (Figures 8 and 9), were analyzed for their performance under salt stress condition. Figure 8A showed that under normal condition, the germination rate of *GmPHD2*-trangenic seeds was similar to that of the wild type plants. Under NaCl treatment, the germination rate of the transgenic plants was significantly higher than that in the wild type plants. These results indicate that overexpression of *GmPHD2* in *Arabidopsis* enhanced the salt tolerance of the transgenic plants at germination stage.

To evaluate the effects of salt stress on the growth of transgenic plants, 5-day-old seedlings of transgenic and wild type plants were transferred onto the plates containing various concentrations of NaCl. After 2 weeks, we observed severe stressed-phenotype including short roots and compact aerial parts in wild type plants under 150 mM NaCl treatment (Figure 8B). However, the GmPHD2-overexpressing plants had a better growth under the same stress condition. Under normal condition, no significant difference was observed between wild type plants and the transgenic lines (Figure 8B). The salt-stressed plants in Figure 8B were further transferred to soil, and their growth status was compared after 2 weeks. The growth of wild type plants was severely inhibited compared with that of transgenic plants (Figure 8C). The transgenic plants had higher inflorescences and longer roots than those of wild type plants under salt stress condition (Figures 8C, D, E). These results indicate that the GmPHD2 improved the growth of transgenic plants under salt stress.

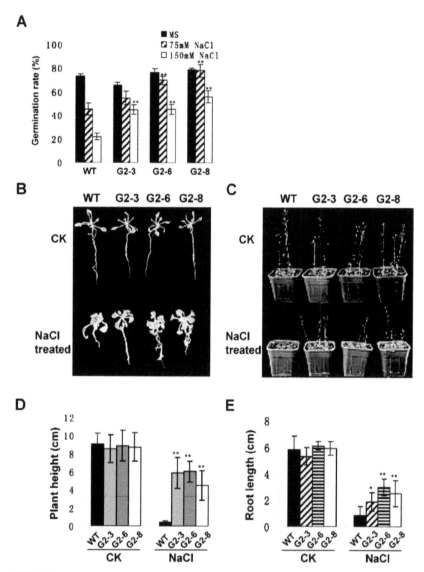

Figure 8. Performance of the GmPHD2-transgenic plants under salt stress. (A) Seed germination under salt stress. The seed germination rate of transgenic lines (G2–3, G2–6, G2–8) was calculated 5 d after sowing. Each data point is the means of three replicates and bars indicate SD. (B) Plant growth in NaCl medium. Five-day-old seedlings were treated on plate without (CK, top) or with 150 mM NaCl (NaCl treated, bottom) for two weeks. (C) Recovery of salt-stressed plants in pots. Seedlings treated with 150 mM NaCl (NaCl treated, bottom) or without NaCl (CK, top) were transferred to pots and grown for two weeks under normal conditions. (D) Comparison of plant height after salt stress treatment. Plant heights in (C) were measured. Values are means±SD (n = 54). (E) Comparison of root length after salt stress treatment. Root length of plants in (C) was measured. Values are means±SD (n = 54). For (A), (D), and (E), "*" and "**" indicate significant difference (P < 0.05 and P < 0.01 respectively) compared to the corresponding WT plants.

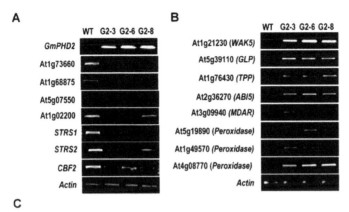

A

WT G2-3 G2-6 G2-8

GmPHD2
At1g73660
At1g68875
At5g07550
At1g02200
STRS1
STRS2
CBF2
Actin

B

WT G2-3 G2-6 G2-8

At1g21230 (WAK5)
At5g39110 (GLP)
At1g76430 (TPP)
At2g36270 (ABI5)
At3g09940 (MDAR)
At5g19890 (Peroxidase)
At1g49570 (Peroxidase)
At4g08770 (Peroxidase)
Actin

C

Locus	Elements (GTGGNG/GNGGTG)
At1g73660	GTGGGG (-2056bp, -)
STRS1	GTGGAG (-2855bp, +; -1690bp, -)
STRS2	GTGGTG (-2623bp, +; -767bp, +)
CBF2	GTGGAG (-2316bp, +), GTGGCG (-1118bp, +)
At1g68875	GTGGTG (-422bp, +; -180bp,+), GTGGAG(-1816bp, -)
At1g02200	GTGGGG (-1406bp,+), GTGGCG (-532bp,+), GTGGTG (-106bp,-), GTGGAG (-154bp,+)
At5g07550	GTGGAG (-1305bp, +), GTGGTG (-1863bp, -; -1554bp,-)
At1g21230	GTGGAG (-1532bp,-), GCGGTG (-1469bp,-;-1284bp,+), GTGGTG (-588bp, +)
At5g39110	GTGGCG (-1566bp, -), GCGGTG (-269bp, +)
At1g76430	GAGGTG (-151bp, +)
At2g36270	GTGGGG (-1013bp, -)
At3g09940	GTGGAG (-78bp, -; -848bp, -), GTGGCG (-812bp, -)
At5g19890	GTGGGG (-452bp, +), GTGGCG (-169bp, -)
At1g49570	GGGGTG (-722bp, -)
At4g08770	GCGGTG (-3122bp, +)

Figure 9. Expression of *GmPHD2*-regulated genes in the transgenic plants. (A) Downregulated gene expression in GmPHD2-transgenic plants (G2–3, G2–6, G2–8) revealed by RT-PCR. Two-week-old seedlings were used for RNA isolation. Actin was amplified as a control. (B) Upregulated gene expression in GmPHD2-transgenic plants. (C) Putative *cis*-DNA elements for GmPHD2 binding in promoter regions of the downregulated and upregulated genes. Numbers indicate the positions upstream the start codon for each gene. "–" indicates that the element was on the antisense strand. "+" indicates that the element was on the sense strand.

GmPHD2-Regulated Genes in Transgenic *Arabidopsis* Plants

Since GmPHD2 has transcriptional suppression activity (Figure 5), it may inhibit gene expressions. Seven stress-responsive genes were examined for their expressions in GmPHD2-transgenic plants. The CBF2/DREB1C is a negative regulator of CBF1/

DREB1B and CBF3/DREB1A expression, and cbf2 mutant showed enhanced tolerance to abiotic stresses (Novillo et al., 2004). The STRS1 and STRS2 encode DEAD-box RNA helicases and mutations in either genes caused increased tolerance to abiotic stresses (Kant et al., 2007). *At1g73660* encodes a putative MAPKKK and negatively regulates salt tolerance in *Arabidopsis* (Gao and Xiang, 2008). These four genes were suppressed in the GmPHD2-transgenic plants (Figure 9A). Three other genes *At1g68875, At5g07550,* and *At1g02200* were also inhibited in the transgenic lines (Figure 9A). *At1g68875* encoded a protein of unknown function; *At5g07550* encoded a glycine-rich protein, and *At1g02200* encoded a putative fatty acid hydrolase with two transmembrane domains.

Eight other genes had higher expression in the transgenic plants in comparison with their expressions in wild type plants (Figure 9B). These genes included *At1g21230* (*WAK5*) encoding a wall-associated protein kinase, *At5g39110* (*GLP*) encoding a germin-like protein, *At1g76430* (*TPP*) encoding a phosphate transporter family protein, At2g36270 (ABI5) encoding an ABA-responsive basic leucine zipper transcription factor, At3g09940 (MDAR) encoding a putative monodehydroascorbate reductase, three peroxidase genes *At5g19890, At1g49570,* and *At4g08770*. The *GmPHD2* gene was also apparently enhanced in the three transgenic lines. These analyses reveal that the *GmPHD2* may improve salt tolerance through affecting stress signal transduction and by scavenging ROS.

Because the GmPHD proteins can bind the GTGGAG element, we then examined if the element or its similarities were present in the promoter region of the regulated genes. Figure 9C showed that in the promoter regions of both the downregulated and upregulated genes, one to four elements were identified. Among the elements from promoter regions of the downregulated genes, the consensus element sequence GTGG(A6/T7/G2/C2)G was found. For the upregulated genes, two consensus element sequence GTGG(A3/T1/G2/C3)G and G(A1/G1/C4)GGTG were identified in their promoter regions (Figure 9C). These elements may be directly or indirectly involved in *GmPHD2*-regulated gene expression. Considering that the *GmPHD2* has transcriptional repression activity, it may bind to the elements and then suppress gene expressions. However, whether the *GmPHD2* can bind to the elements from the downregulated genes needs to be further studied.

Analysis of the Oxidative Stress Tolerance in GmPHD2-Transgenic Plants
Because genes relating to ROS scavenging were identified, we investigated if the transgenic plants overexpressing the *GmPHD2* can tolerate the oxidative stress. Figure 10A showed that the germination rate of wild type plants was dramatically decreased from ~80% to ~23% with the treatments of increasing concentrations of paraquat. However, the germination rates of the transgenic plants were only slightly influenced by the paraquat treatments (Figure 10A). These results indicate that the seed germination process of the GmPHD2-transgenic plants is more tolerant to oxidative stress than that of the wild type plants.

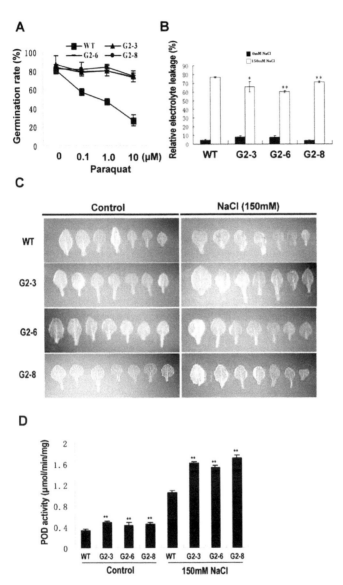

Figure 10. Oxidative stress tolerance of the GmPHD2-transgenic plants. (A) Seed germination under paraquat treatment. The germination rate was calculated 5 days after sowing. Values are means±SD (n = 3, each has 80–100 seeds). (B) Electrolyte leakage in salt-stressed plants. Ten-day-old seedlings were subjected to 150 mM NaCl stress for 5 d on plate. Values are means±SD (n = 4, each has four seedlings). (C) Detection of hydrogen peroxide production in plants. Seedlings in (B) were used to detect the H_2O_2 levels in leaves. H_2O_2 levels were revealed with 3,3'-diaminobenzidine (DAB). Brown color indicates generation of hygrogen peroxide. (D) Peroxidase (POD) activity in salt-stressed plants. Seedlings in (B) were used. Values are means of three replicates and bars indicate SD. For (B) and (D), "*" indicates significant difference (P < 0.05) compared to WT. "**" indicate highly significant difference (P < 0.01) compared to WT.

Salt stress usually caused cell membrane damage and resulted in electrolyte leakage. Both the transgenic plants and wild type plants showed higher relative electrolyte leakage under salt stress in comparison with the untreated plants (Figure 10B). However, the transgenic plants had lower electrolyte leakage than the wild type plants under salt stress. These results indicate that the transgenic plants overexpressing the GmPHD2 are more tolerant to salt stress than the wild type plants.

The GmPHD2 may confer salt tolerance in the transgenic plants through regulation of oxidative stress. We then examined if the hydrogen peroxide level was changed in the plants under salt stress. Figure 10C showed that after salt stress, the three transgenic lines showed no brown color whereas the wild type plants showed brown color, the positive response of DAB staining. These results indicate that the wild type plants have more H_2O_2 than the transgenic plants, suggesting that the transgenic plants are more tolerant to salt stress possibly through inhibition of H_2O_2 accumulation.

Higher expressions of *peroxidase* genes (Figure 9) may result in higher peroxidase (POD) activity. Figure 10D showed that the POD activity in GmPHD2-overexpressing plants were 30% to 45% higher than that in wild type plants under normal growth condition. After salt treatment, the POD activity of the transgenic plants and wild type plants all increased at least three folds, and the increase of POD activity in transgenic plants was higher than that in wild type plants. These results indicate that GmPHD2 enhanced the POD activity in transgenic plants and this may contribute to the salt tolerance by scavenging hydrogen peroxide.

KEYWORDS

- **Alfin1-type PHD finger protein**
- ***Arabidopsis* Protoplast System**
- **Gel Shift Assay**
- **GmPHD proteins**
- **GmPHD-GFP Fusion Proteins**
- **PHD-domain-containing proteins**

AUTHORS' CONTRIBUTIONS

Conceived and designed the experiments: Wei Wei, Jian Huang, Yu-Jun Hao, and Shou-Yi Chen. Performed the experiments: Wei Wei, Jian Huang, Yu-Jun Hao, Hong-Feng Zou, Hui-Wen Wang, and Jing-Yun Zhao. Analyzed the data: Wei Wei, Jian Huang, Yu-Jun Hao, Hong-Feng Zou, Hui-Wen Wang, Xue-Yi Liu, Wan-Ke Zhang1, Biao Ma, Jin-Song Zhang, and Shou-Yi Chen. Contributed reagents/materials/analysis tools: Jing-Yun Zhao, Xue-Yi Liu, Wan-Ke Zhang1, Biao Ma, Jin-Song Zhang, and Shou-Yi Chen. Wrote the paper: Wei Wei, Jian Huang Jin-Song Zhang, and Shou-Yi Chen.

Chapter 7

Cell Wall Biogenesis of *Arabidopsis Thaliana*

Elisabeth Jamet, David Roujol, Helene San Clemente,
Muhammad Irshad, Ludivine Soubigou-Taconnat,
Jean-Pierre Renou, and Rafael Pont-Lezica

INTRODUCTION

Plant growth is a complex process involving cell division and elongation. *Arabidopsis thaliana* hypocotyls undergo a 100-fold length increase mainly by cell elongation. Cell enlargement implicates significant changes in the composition and structure of the cell wall. In order to understand cell wall biogenesis during cell elongation, mRNA profiling was made on half- (active elongation) and fully-grown (after growth arrest) etiolated hypocotyls.

Transcriptomic analysis was focused on two sets of genes. The first set of 856 genes named cell wall genes (*CWGs*) included genes known to be involved in cell wall biogenesis. A significant proportion of them have detectable levels of transcripts (55.5%), suggesting that these processes are important throughout hypocotyl elongation and after growth arrest. Genes encoding proteins involved in substrate generation or in synthesis of polysaccharides, and extracellular proteins were found to have high transcript levels. A second set of 2,927 genes labeled secretory pathway genes (*SPGs*) was studied to search for new genes encoding secreted proteins possibly involved in wall expansion. Based on transcript level, 433 genes were selected. Genes not known to be involved in cell elongation were found to have high levels of transcripts. Encoded proteins were proteases, protease inhibitors, proteins with interacting domains, and proteins involved in lipid metabolism. In addition, 125 of them encoded proteins with yet unknown function. Finally, comparison with results of a cell wall proteomic study on the same material revealed that 48 out of the 137 identified proteins were products of the genes having high or moderate level of transcripts. About 15% of the genes encoding proteins identified by proteomics showed levels of transcripts below background.

Members of known multigenic families involved in cell wall biogenesis, and new genes that might participate in cell elongation were identified. Significant differences were shown in the expression of such genes in half- and fully-grown hypocotyls. No clear correlation was found between the abundance of transcripts (transcriptomic data) and the presence of the proteins (proteomic data) demonstrating (i) the importance of post-transcriptional events for the regulation of genes during cell elongation and (ii) that transcriptomic and proteomic data are complementary.

Plant growth occurs mainly by division and expansion of cells. A meristematic cell might enlarge as much as 5,0000-fold its initial volume. In this process, membrane surface area and amount of cell wall material increase. The primary cell wall plays an essential role since it should allow turgor-driven increase in cell volume by permitting the incorporation of new cell wall material and rearrangement of the existing cell wall. Several plant organs including coleoptiles (Poaceae), internodes (legumes), and hypocotyls (mung bean, sunflower, and *Arabidopsis thaliana*) were used to study cell elongation (Derbyshire et al., 2007a). Environmental signals such as light, temperature, and hormones, regulate hypocotyl growth (Collett et al., 2000; Desnos et al., 1996; Quail et al., 1995; Saibo et al., 2003). *Arabidopsis thaliana* seedlings grown in continuous darkness are a material of choice to analyze the cell elongation process. Indeed, cells of hypocotyls undergo a 100-fold length increase compared to embryo cells (Gendreau et al., 1997). Growth occurs mostly by cell expansion, with little cell division (Gendreau et al., 1997; Raz and Koornneef, 2001; Refrégier et al., 2004; Saibo et al., 2003). Changes in wall thickness during elongation of *A. thaliana* hypocotyls were investigated using cryofield-emission scanning electron microscopy (Derbyshire et al., 2007a). At the germination stage, cell wall thickening occurs and involves high rates of biosynthesis and deposition of cell wall components. During the elongation stage, cell walls undergo remarkable thinning, requiring extensive polymer disassembly and rearrangement.

Many genes are assumed to be involved in cell wall synthesis and rearrangement to support growth of plant cell walls (Cosgrove, 2005). They encode cellulose synthases (CESAs), cellulose synthases-like (CSLs), endo-glucanases, xyloglucan endotransglucosylase/hydrolases (XTHs), and expansins. They belong to multigenic families, but the members of each family involved in elongation of hypocotyl cells were not precisely identified. It is also likely that other genes are important for cell elongation.

In this chapter, the transcriptomes of *A. thaliana* etiolated hypocotyls were compared at two developmental stages, half-grown (yet actively elongating), and fully-grown (after growth arrest). The transcriptome analysis was focused on genes possibly involved in cell wall biogenesis and on genes encoding secreted proteins. Transcript profiling was carried out using CATMA (Complete Arabidopsis Transcriptome MicroArray) (Crowe et al., 2003): (i) to look at the level of transcripts of *CWGs* belonging to families known to be involved in cell wall biogenesis; (ii) to identify genes encoding secreted proteins (*SPGs*) having high or moderate level of transcripts; (iii) to reveal differential gene expression affecting *CWGs* and *SPGs* between half- and fully-grown etiolated hypocotyls; and (iv) to look at the correlation between transcript abundance and protein presence as revealed by a proteomic study performed on the same material (Irshad et al., 2008).

MATERIALS AND METHODS

Plant Material

Arabidopsis thaliana seedlings (ecotype Columbia 0) were grown in continuous dark in Magenta boxes on Murashige and Skoog (Murashige and Skoog, 1962) medium supplemented with 2% sucrose. Etiolated hypocotyls were collected after 5- and 11-days of culture.

Total RNA Extraction

Two RNA extractions from two biological replicates were performed for each sample (5- and 11-day-old hypocotyls). Hypocotyls were cut below the cotyledons and above the crown with sterile scissors. They were ground in liquid nitrogen in a mortar with a pestle. Extraction of total RNAs was performed using the SV Total RNA Isolation kit according to manufacturer's instructions (Promega France, Charbonnières, France). For each RNA extraction, 750 mg of ground hypocotyls were used. Typically, about 110 µg of total RNAs were obtained.

Transcriptome Studies

Microarray analysis was carried out at the Unité de Recherche en Génomique Végétale (Evry, France), using the CATMA array (Crowe et al., 2003; Hilson et al., 2004), containing 24,576 gene-specific tag (GSTs) from *A. thaliana*. RNA samples from the two independent biological replicates were isolated and separately analyzed. For each comparison, one technical replication with fluorochrome reversal was performed for each RNA sample (i.e., four hybridizations in two dye swaps per comparison). The reverse transcription of RNA in the presence of Cy3-dUTP or Cy5-dUTP (Perkin-Elmer-NEN Life Science Products), the hybridization of labeled samples to the slides, and the scanning of the slides were performed as described in Lurin et al. (2004).

Statistical Analysis of Microarray Data

Statistical analysis was based on two dye swaps (i.e., four arrays, each containing 24,576 GSTs and 384 controls) as described in Gagnot et al. (2007). To estimate the transcript level of each gene, a background value was obtained by addition of the average background value to two background standard deviations. The average background value was calculated using a subset of 1,000 non-expressed genes found in the whole CATMA database (http://www.catma.org/database/search.html). The background value was not subtracted from the data presented in this chapter, but was considered for the interpretation of the results. To determine differentially expressed genes, we performed a paired t-test on the log ratios, assuming that the variance of the log ratios was the same for all genes. Spots displaying extreme variance (too small or too large) were excluded. The raw P-values were adjusted by the Bonferroni method, which controls the Family Wise Error Rate. We considered as being differentially expressed the genes with a Bonferroni P-value ≤ 0.05, as described in (Gagnot et al., 2007). We use the Bonferroni method (with a type I error equal to 5%) in order to keep a strong control of the false positives in a multiple-comparison context (Ge et al., 2003).

Data Deposition

Microarray data from this chapter were deposited at Gene Expression Omnibus (http://www.ncbi.nlm.nih.gov/geo/; accession No. GSE14648) and at CATdb (http://urgv. evry.inra.fr/CATdb/; Project RS05-11_Hypocotyls) according to the "Minimum Information About a Microarray Experiment" standards.

Reverse Transcription-Polymerase Chain Reaction (RT-PCR)

The cDNA first strands were obtained from total RNAs using 1 µg total RNAs and SuperScript™ II reverse transcriptase (Invitrogen, Carlsbad, San Diego, CA, USA). As a control, the same amount of pig desmin RNA was added in each sample. Quantitative PCR was performed using a Roche lightcycler system (Roche Diagnostics, Meylan, France) according to manufacturer's recommendations. Using the results from quantitative PCR to determine the number of amplification cycles required to be in a linear range for all genes of interest, semi-quantitative PCR was performed. The amplified fragments were analyzed by electrophoresis in polyacrylamide gels in standardized conditions. In each case, presence of a fragment of the expected size was checked after staining with ethidium bromide.

Bioinformatic Analyses

Sub-cellular localization and length of signal peptides were predicted using PSORT (http://psort.nibb.ac.jp/) and TargetP (http://www.cbs.dtu.dk/services/TargetP/) (Emanuelsson et al., 2000; Nielsen et al., 1997). Prediction of transmembrane domains was done with Aramemnon (http://aramemnon.botanik.uni-koeln.de/) (Schwacke et al., 2003). Molecular masses and pI values were calculated using the aBi program (http://www.up.univ-mrs.fr/~wabim/d_abim/compop.html). Homologies to other proteins were searched for using BLAST programs (http://www.ncbi.nlm.nih.gov/BLAST/) (Altschul et al., 1990). Identification of protein families and functional domains was performed using MyHits (http://myhits.isb-sib.ch/cgibin/motif_scan) and InterProScan (http://www.ebi.ac.uk/InterProScan/) (Quevillon et al., 2005). TargetP, Aramemnon, and InterProScan software were combined to provide the ProtAnnDB friendly user web interface (http://www.polebio.scsv.ups-tlse.fr/ProtAnnDB/) (San Clemente et al., 2009). A MySQL (v4.1) database and PHP5 scripts were used to store, order and extract numeric, and qualitative data.

All the protein families chosen in our *CWG* list were annotated by experts. The GHs and CEs were classified according to the CAZy database (http://www.cazy.org/CAZY/) (Coutinho and Henrissat, 1999) at the Cell Wall Genomics website. (http://cellwall.genomics.purdue.edu/intro/index.html). The GT77 family was annotated according to Egelund et al. (2004). The XTHs and expansions were named according to http://labs.plantbio.cornell.edu/xth/ and http://www.bio.psu.edu/expansins/index.htm respectively. The arabinogalactan proteins (AGPs) and fasciclin-like arabinogalactan proteins (FLAs) were named according to Schultz et al. (2002), Johnson et al. (2003), Van Hengels and Roberts (2003), and Liu and Mehdy (2007). Proteins homologous to COBRA, LRXs, and Hyp/Prorich proteins were annotated according to Roudier et al. (2005), Baumberger et al. (2003), and Fowler et al. (1999) respectively. The lignin toolbox was proposed by Raes et al. (2003). Peroxidases were named as in the PeroxiBase (http://peroxidase.isbsib.ch/index.php) (Bakalovic et al., 2006). Laccases were annotated as in Pourcel et al. (2005) and McCaig et al. (2005). The SKU-like proteins and phytocyanins were described in Jacobs and Roe (2005), and Nersissian and Shipp (2002) respectively. Subtilases are listed at http://csbdb.mpimp-golm.mpg.de/csbdb/dbcawp/psdb.html. Pectin methylesterase inhibitors (PMEIs) were annotated by Dr. J. Pelloux (University of Amiens, France).

DISCUSSION AND RESULTS

Levels of Transcripts of Cell Wall Genes (*CWGs*) During Hypocotyl Elongation

Etiolated hypocotyls were compared at two developmental stages. Five-day-old hypocotyls were approximately half the final size (Figure 1). Growth followed an acropetal gradient. After 5- days, the bottom cells were fully elongated, whereas the top cells were only starting elongation (Refrégier et al., 2004). Eleven-day-old hypocotyls had reached their maximum size (Gendreau et al., 1997). The CATMA was used for mRNA profiling. Since one of the major modifications during cell elongation is the addition and rearrangement of cell wall components, a selection of genes possibly involved in cell wall biogenesis was done. This selection was called *CWGs*. It was mainly based on the knowledge of gene families known to be involved in biogenesis of cell walls, that is synthesis and transport of cell wall components and their assembly or rearrangement in cell walls (see Materials and Methods). Representing 37 gene families, it includes genes encoding proteins involved in substrate generation (nucleotide-sugar inter-conversion pathway, monolignol biosynthesis), polysaccharide synthesis (mainly glycosyl transferases), vesicle trafficking, assembly/disassembly of the wall (glycoside hydrolases (GHs), expansins, carbohydrate esterases (CEs), carbohydrate lyases), structural proteins, oxido-reductases involved in cross-linking of wall components (mainly peroxidases and laccases). Few other gene families encoding cell wall proteins were also included such as AGPs, FLAs, phytocyanins, multicopper oxidases, PMEIs, and subtilases. Only genes annotated by experts were retained (see Materials and Methods).

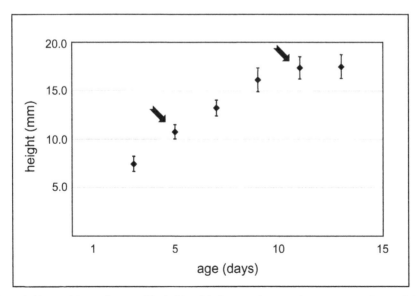

Figure 1. Hypocotyl growth curve. The height of dark-grown hypocotyls was measured every 2 days. Two stages were selected: 5- and 11-days (arrows), corresponding to half- (active elongation) and fully-grown (after growth arrest) etiolated hypocotyls respectively.

Altogether, 1,026 genes were selected among which 856 were analyzed through CATMA, while the remaining genes were not analyzed for technical reasons. Some genes had no *GST* on the microarray, others were not considered because of a poor signal of hybridization to the RNA probe, or of inaccurate duplicates. The level of transcripts was expressed as log2 of the mean signal intensity. Values of log2 below 6.83 were considered as background as defined in Materials and Methods. Three groups of genes were considered after taking into account the dynamic range of CATMA arrays (Allemeersch et al., 2005). Values between 6.83 and 9 corresponded to low level of transcripts (1- to 4-fold the background level), values between 9 and 10 to moderate level (4- to 8-fold the background level) and values higher than 10 to high level (more than 8-fold, and up to 128-fold the background level). Among the 856 genes analyzed, the level of transcripts at one or both stages was below background for 381 genes (44.6%), low for 326 (38.6%), moderate for 62 (7.2%), and high for 82 (9.6%).

Overall, among the analyzed genes, 49.4% of those related to synthesis or transport of cell wall components, and 52.2% of those involved in their modifications in cell walls have detectable level of transcripts. Among the 82 genes with high levels of transcripts at one or both stages, 21 are involved in the synthesis or transport of cell wall components and 35 are involved in modifications of cell wall components. Two separate phases of growth were described in dark-grown *A. thaliana* hypocotyls: an early phase of active synthesis of cell wall polysaccharides up to 3- days after beginning of germination, and a late phase of cell expansion (Refrégier et al., 2004). The former phase results in thicker cell walls which later on become thinner as hypocotyls elongate (Derbyshire et al., 2007a). Our results suggest that both synthesis and rearrangement of cell wall components are required throughout hypocotyl elongation, and even after growth arrest.

Most of the *CWGs* were expected to be transcribed during cell elongation. Genes involved in rearrangement of cell wall components encode GHs such as endoglucanases, XTHs, and beta-galactosidases; CEs such as pectin methylesterases (PMEs); pectin acylesterases; polysaccharide lyases (PLs); expansins of the alpha- or beta-type; and peroxidases. However, 20 genes encoding PMEs and five genes encoding pectin acylesterases have detectable levels of transcripts among which four genes have high levels of transcripts in elongating hypocotyls. This result seems contradictory with previous results showing that a low level of pectin esterification, associated to abundance of PMEs and acylesterases, restricted cell elongation in *A. thaliana* hypocotyls (Derbyshire et al., 2007b). It should be noted that 18 genes encoding PMEIs also have detectable levels of transcripts. The interplay between PMEs and their inhibitors could regulate the activity of PMEs. Fifteen genes encoding proteins possibly involved in oxido-reduction reactions have high levels of transcripts (peroxidases, laccases, phytocyanins, and protein homologous to SKU5). Seven peroxidase genes fall in this category, among which *AT2G37130* (*AtPrx21*) has one of the highest levels of transcripts. The multiple roles of peroxidases during growth and growth arrest were reviewed (Passardi et al., 2004).

Other *CWGs* are also well-represented such as those encoding AGPs, FLAs, and subtilases. Even if AGPs and FLAs were shown to be associated with wood formation in poplar (Lafarguette et al., 2004), their role in cell expansion is not very clear at present.

Likewise, nothing is known about the role of proteases during cell elongation. Finally, the *COBRA* gene (*AT5G60920*) has a high level of transcripts. It has been shown to play an important role in microfibril deposition during rapid elongation and in the orientation of cellulose microfibrils (Roudier et al., 2005).

This work gives clues for understanding the function and possible involvement in multiple processes of members of multigene families either during cell elongation or after its arrest. Indeed, general functions were proposed for most of these gene families, but only scarce information is available for specific members.

Genes Encoding Secreted Proteins with High or Moderate Level of Transcripts in Etiolated Hypocotyls

Most of the gene families described above was already known to be involved in cell wall biogenesis. In order to identify new genes encoding secreted proteins that might be involved in cell expansion, a second selection was carried out. Based on results of proteomic studies, some proteins without a predicted signal peptide were assumed to be secreted. However, the sub-cellular localization of such proteins was never shown in another way (Jamet et al., 2008). For this study, only the genes encoding proteins with a predicted signal peptide were selected. The 2,927 selected genes were ranked by level of transcripts, producing a profile similar to the one obtained with *CWGs* at 5- days; 1,161 genes (39.7%) above background level, 1,295 (44.2%) with a low transcript level, 238 (8.1%) with a moderate level, and 235 (8.0%) with a high level. Same results were obtained at 11- days. From this selection, only genes encoding proteins predicted to be located either outside the cell or in the plasma membrane were retained. Protein families grouped according to their predicted functional domains were already described in cell wall proteomic studies (Boudart et al., 2007; Jamet et al., 2006, 2008): proteins acting on carbohydrates (69 genes); proteases (37 genes); proteins possibly involved in signaling (44 genes); structural proteins (15 genes); proteins possibly involved in oxido-reduction reactions (27 genes); proteins with interacting domains (33 genes); proteins related to lipid metabolism (40 genes); miscellaneous proteins (43 genes); proteins of unknown function (125 genes). Main differences between transcriptomic and proteomic data lie in the genes encoding proteins possibly involved in signaling since they comprise AGPs, FLAs, and plasma membrane proteins that are difficult to isolate, separate, or identify through proteomics (Jamet et al., 2006). In the same way, the group of proteins of unknown function is very important because 48% of them are predicted to have trans-membrane domains. On the contrary, the group of structural proteins is probably under-represented because of the lack of appropriate GSTs for many of them. Indeed, their repetitive amino acid sequences make the design of specific probes difficult. One should note the abundance of proteases that can be assumed to be essential for protein turnover in tissues undergoing rapid elongation followed by elongation arrest within a short time. They may also be involved in signaling (Berger and Altmann, 2000; Tanaka et al., 2001) or in protein maturation (Rautengarten et al., 2008). In addition, there are probably interactions between proteases and protease inhibitors to regulate the proteolytic activities in cell walls. Among the 125 proteins of yet unknown function, 25 have known structural domains. Others share domains with other proteins, such as domains of unknown function

(DUF), or belong to the so-called uncharacterized protein families (UPF). Many are of particular interest, since they are only present in plants.

Among these 433 *SPGs*, only the 69 encoding proteins acting on carbohydrates, and the 12 encoding peroxidases or laccases were shown or assumed to contribute to assembly or rearrangement of cell wall components. It means that this study allowed identifying about 350 genes encoding secreted proteins that are candidates to play roles during growth of *A. thaliana* etiolated hypocotyls. Their functional characterization will be paramount to understand cell wall architecture and assembly during an elongation process.

Are there Variations in the Level of Transcripts Between Half- and Fully-Grown Hypocotyls?

In 5-day-old hypocotyls, apical cells are elongating whereas basal cells are fully elongated. When compared to 11-day-old hypocotyls, where all the cells are fully elongated, the observed differences in levels of transcripts should mainly come from the cells which are at a different developmental stage, namely growing cells. Altogether, 559 genes are differentially expressed between 5- and 11-day-old hypocotyls. Among these genes, 108 encode proteins predicted to be secreted. In addition, 32% of the genes having levels of transcripts modified by a factor two between the two developmental stages encode secreted proteins. The following detailed analysis will be focused on these genes. Sixty-three and 45 genes have a higher level of transcripts in 5- and 11-day-old hypocotyls respectively. The highest difference (8.6-fold increase) was found at 11 days for a gene encoding a glycine-rich protein (GRP, AT2G05440). Conversely, the largest decrease (4.5-fold) was observed at 11 days for a gene encoding a putative Asp protease (AT5G10770). The number of genes of selected families expressed differentially in both samples is represented in Figure 2. For comparison, the number of genes of the same families having high or moderate levels of transcripts is also represented. All the selected gene families are represented by almost the same number of genes at both developmental stages (Figure 2A). However, there are striking differences when the comparison is done with genes showing significant variation in transcript level (Figure 2B).

Several genes encoding GTs, GHs, and PMEs have higher levels of transcripts at 5 days than at 11 days, that is at a time hypocotyls undergo active elongation. Although GTs and GHs are expected to be expressed during elongation, PMEs could play roles during both elongation and growth arrest (Micheli, 2001). The proportion of cells already elongated could also be significant after 5 days of growth. Alternatively, there might be a delay between synthesis of mRNAs, and production of a functional protein. This might be the case for some PMEs that are produced as polyproteins comprising an inhibitor at their N-terminus and an active enzyme at their C-terminus (Micheli, 2001).

Other genes having higher levels of transcripts at 5 days encode peroxidases, proteases, and proteins homologous to GDSL lipases/acylhydrolases. The multiple functions of peroxidases were mentioned above. The role of proteases in cell walls during active elongation has not yet been described. It should be noted that four genes encoding protease inhibitors are up-regulated at 11 days, suggesting complex regulations of proteolytic activities in cell walls after the arrest of hypocotyl elongation.

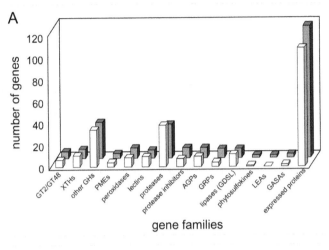

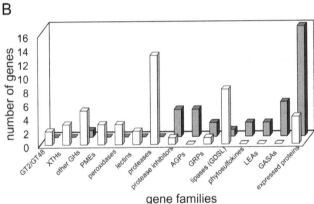

Figure 2. Overview of *SPGs* expressed in 5- and 11-day-old hypocotyls. (A) Genes belonging to several families of *SPGs* with moderate and high level of transcripts are shown: white and grey bars are used for 5- and 11-day-old hypocotyls respectively. (B) The genes of the same families showing significant differences in expression between 5- and 11-day-old hypocotyls are counted: genes with higher level of transcripts at 5 days are represented by white bars; genes with higher level of transcripts at 11 days are represented by grey bars. XTHs stands for xyloglucan endotransglucosylase/hydrolases, GHs for glycoside hydrolases, AGPs for arabinogalactan proteins, LEAs for late embryogenesis abundant, GASAs for gibberellic acid-stimulated *Arabidopsis* proteins, and expressed proteins for proteins with unknown function.

Genes belonging to other gene families have higher levels of transcripts at 11 days. They code for protease inhibitors, AGPs, *GRPs*, late abundant embryogenesis (LEAs) proteins, phytosulfokines (PSKs), and gibberellic acid-stimulated *Arabidopsis* proteins (GASAs). The *AGP* and *FLA* genes are very well-represented in the transcriptome of hypocotyls with 17 *AGP* genes and 13 *FLA* genes having detectable levels of transcripts. The *AGPs* are candidates for cell-to-cell communication (McCabe et al., 1997), and *FLAs* were found to be associated to wood formation in poplar (Lafarguette et al., 2004). Concerning *GRP*s, they were shown to be associated to cell walls of xylem and phloem by tissue printing (Ye et al., 1991). The great increase in the amount of transcripts of *AT2G05440* is consistent with the development of protoxylem elements that contain *GRP*s (Ringli et al., 2001). Nothing is known about the role *LEA*s could play in fully-developed hypocotyls. The PSKs were shown to promote tracheary element differentiation in Zinnia cell suspension cultures, and to play roles during growth (Matsubayashi and Sakagami, 2006). Five *GASA* genes have higher levels of transcripts at 11 days than at 5 days. *Arabidopsis thaliana GASA4* and *GASA5* were shown to play roles in flowering regulation and seed development (Roxrud et al., 2007), as well as in stem growth and flowering time (Zhang et al., 2009). Increase in the level of transcripts of the *GASA* genes in 11-day-old hypocotyls thus points at their role in elongation arrest. Finally, 16 genes encoding secreted proteins of unknown function are transcribed at higher level at 11 days. Exploring the function of these proteins will be one of the major tasks for the future.

These differences in transcript abundance between the two stages of hypocotyl growth should be taken carefully with regard to the possible functions carried out by the proteins, since many other genes from the same families are transcribed in half- and fully-grown hypocotyls, but without significant differences.

Transcriptome Versus Proteome

In order to look for the consistency between levels of mRNAs and presence of the corresponding proteins in cell walls, a proteomic analysis was performed on cell walls (Irshad et al., 2008), and the results were compared to those of this transcriptomic analysis. The cell wall proteomes of 5- and 11-days-old hypocotyls were achieved and a total of 137 proteins predicted to be secreted were identified. When these 137 proteins were compared to the 433 *SPGs* with moderate and high levels of transcripts, only 48 proteins matched (11.8%). Conversely, from the 228 *SPGs* having high levels of transcripts in etiolated hypocotyls, only 28 (12.2%) showed the corresponding proteins (Table 1). It was expected that proteomic profiling identified at least the proteins encoded by the highly-transcribed genes. The great inconsistency between the abundance of mRNAs and the presence of the corresponding proteins was surprising, but several reasons may explain this disparity. It is known that CWP extraction and identification can be challenging (Boudart et al., 2007; Jamet et al., 2006). Many proteins can remain linked to the polysaccharide matrix, such as the structural proteins (Brady et al., 1996; Schnabelrauch et al., 1996), or some peroxidases that might be strongly bound to pectins (Shah et al., 2004). Others are difficult to identify because of their structure, for example highly *O*-glycosylated AGPs, which requires a special deglycosylation step (Schultz et al., 2004). Some proteins contain few linkages sensitive to tryptic digestion, and can escape identification by peptide mass mapping. Finally, low-abundant proteins elude proteomic analyses. For the proteins that were identified without particular problems such as GHs, expansins and proteases, only a few of them correspond to highly-transcribed genes. It indicates that a high level of transcripts is not always correlated with the presence of the protein in sufficient amount to be identified in proteomic approaches.

Table 1. Genes with high levels of transcripts in either 5- or 11-day-old hypocotyls for which the encoded proteins were identified in a proteomic study performed on the same material.

Functional Class	AGI Number	Predicted or known gene function	5-days	11-days	log₂ of ratio 11-days /5-days	p-value
Proteins acting on carbohydrates						
glycoside hydrolase family 16 (xyloglucan endotrans-glycosidases/ hydrolases)	AT2G06850	AtXTH4	13.27	12.62	-0.65	1.96E-01
glycoside hydrolase family 20 (beta-hexosaminidase)	AT3G55260		10.40	9.90	-0.51	1.00
glycoside hydrolase family 31	AT1G68560	AtXYL1	10.97	10.57	-0.41	1.00
carbohydrate esterase family 8 (pectin methylesterase)	AT3G14310	AtPME3	11.66	10.82	-0.84	1.44E-04
alpha-expansin	AT5G02260	AtEXPA9	12.95	12.54	-0.41	1.00
Proteases						
cysteine protease (papainfamily)	AT4G01610		12.70	12.50	-0.20	1.00
aspartic protease (pepsin family)	AT3G54400		10.30	9.85	-0.45	1.00
aspartic protease (pepsin family)	AT5G10770		11.73	9.55	-2.17	0.00E+00
Structural proteins						
proline-rich protein (PRP)	AT1G28290		11.98	11.89	-0.09	1.00
LRR-extensin	AT3G24480	AtLRX4	10.26	10.33	0.08	1.00
Proteins involved in oxidoreduction reactions						
peroxidase	AT1G71695	AtPrx12	10.44	9.48	-0.96	7.29E-07
peroxidase	AT3G21770	AtPrx30	10.39	10.30	-0.08	1.00
early nodulin AtEN20 (protein homologous to blue copper binding Protein)	AT4G12880	plastocyanin	11.56	11.94	0.38	1.00
Proteins with interacting domains						
protein homologous to lectin (curculin-like)	AT1G78850	curculin-like, man nose binding	11.87	11.37	-0.50	1.00
protein homologous to lectin (curculin-like)	AT1G78830	curculin-like, man nose binding	10.44	9.90	-0.54	1.00
protein with leucine-rich-repeat domains (LRRs)	AT3G20820	expressed protein	10.84	10.26	-0.59	1.00
enzyme inhibitor	AT1G73260	inhibitor family 13 (Kunitz-P family)	12.54	13.41	0.86	5.77E-05

In a second step, the level of transcripts of the 137 genes encoding the proteins identified through proteomics was analyzed (Figure 3A). The transcript data of 31 genes was not found in the CATMA experiment since some of them have no GST or were eliminated because of poor signals of hybridization to the RNA probe. The levels of transcripts of the 106 remaining genes were surprising, for 5 days and 11 days, 36.8% and 40.6% respectively had low level of transcript while 17.9% and 19.8% respectively had levels of transcripts below background. However, all the identified proteins are assumed to be the most abundant. This suggests that the transcripts could have short half-lives and/or that the proteins could have a low turnover. This is the case of several genes encoding proteins acting on carbohydrates (*At1g10550, At2g33160, At4g18180, At3g13790,* and *At4g37950*) or oxido-reductases (*At3g49110, At3g50990, At4g25980, At5g64100, At1g30710, At5g44360, At5g44410,* and *At1g01980*). Additional experiments will be necessary to determine the half-lives of the transcripts and of the proteins in etiolated hypocotyls.

The results obtained with the CATMA analysis were confirmed by quantitative RT-PCR analysis. Several genes corresponding to the three cases described were chosen: high or moderate level of transcripts and proteins identified; high level of transcripts and proteins not identified; low or below background level of transcripts and proteins identified. Note that genes having very low levels of transcripts give signals below the sensitivity of the CATMA analysis.

Altogether, these results show that there is not a clear correlation between the presence of CWPs as shown by cell wall proteomic analysis and the amount of transcripts of the corresponding genes. The quality of this correlation may depend on genes and/on environmental conditions. For example, the quantification of soluble proteins of yeast at mid-log phase showed that for a given transcript level, protein levels were found to vary by more than 20-fold, whereas for a given protein level, transcript levels were found to vary 30-fold (Gygi et al., 1999). However, up-regulation of *yeast* genes in response to glucose or nitrogen limitation was found to be controlled at the transcriptional or post-transcriptional level respectively (Kolkman et al., 2006). In *A. thaliana* and rice, changes observed in the soluble proteome in response to bacterial challenge were not strictly correlated to changes in transcript levels (Jones et al., 2004). These results show that quantitative analysis of transcript levels is not sufficient to infer protein levels. Multilevel analysis must take into account the stability of transcripts, their availability for active translation, as well as the stability of proteins, which is certainly essential considering the high number of proteases in cell walls. With regard to transcript stability, data from a recent study aiming at measuring mRNA decay rates in *A. thaliana* cell suspension cultures (Narsai et al., 2007) were used to look for half-lives of gene transcripts identified through proteomics (Figure 3B). It can be seen that more than half of the proteins (64%) identified by cell wall proteomics correspond to genes having transcripts with rather long half-lives (>24 hr). Conversely, no gene corresponding to proteins identified by cell wall proteomics has transcripts with half-lives shorter than 1 hr. This distribution differs from that of transcripts of genes having high or moderate level of transcripts since 48.5% of these genes have half-lives shorter than 6 hr.

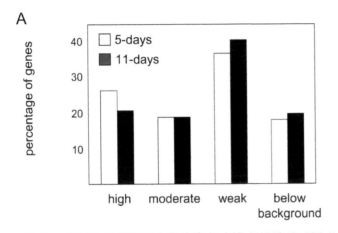

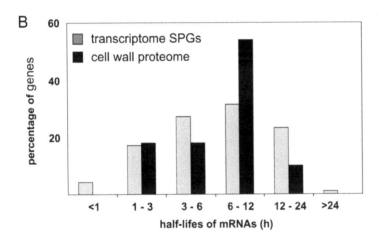

Figure 3. Level of transcripts of genes encoding CWPs identified through proteomics. (A) The levels of transcripts of genes were determined by the CATMA analysis in 5- and 11-day-old hypocotyls (white and dark grey bars respectively). Percentage of genes falling in the following three categories are represented: high transcript level corresponds to log2 values of the mean signal intensity higher than 10, moderate to values between 9 and 10, and weak to values between background (6.83) and 9. (B) Half-lives of mRNAs (in hr) corresponding to *SPGs* having high or moderate levels of transcripts (grey bars) or to proteins identified through cell wall proteomics (black bars). Percentage of genes in each range of half-lives is indicated for each set of genes. Half-lives of mRNAs in cell suspension cultures were from (Narsai et al., 2007).

As a particular case, etiolated hypocotyls of *A. thaliana* at 5 days and 11 days, 28 and 22 genes respectively showed both high level of transcripts, and encoded proteins identified by cell wall proteomics. They might be considered as good markers for cell elongation of dark-grown hypocotyls.

CONCLUSION

The mRNA profiling of the genes potentially involved in cell wall biogenesis (*CWGs*) in etiolated hypocotyls showed that more than half of them present a detectable level of transcripts. All gene families are expressed. The results suggest that both synthesis and rearrangement of wall components are required throughout hypocotyl elongation. The transcriptomic analysis of genes encoding secreted proteins showed that around 350 new genes might be implicated in this process. Understanding the biochemical and biological functions of these genes might reveal new mechanisms of cell wall expansion or of growth arrest, or new functions for the cell wall.

Around 100 genes encoding secreted proteins had significantly different levels of transcripts between growing and fully-elongated hypocotyls. As expected, genes acting on polysaccharides (GTs, GHs) had higher levels of transcripts at 5 days, whereas others encoding PMEs or peroxidases were not supposed to have higher level of transcripts during active elongation. Their function during cell elongation should be re-examined. Several genes encoding proteases also have higher level of transcripts at 5 days and could play roles in protein maturation or turn over. On the contrary, several genes encoding AGPs, protease inhibitors, and proteins homologous to gibberellin regulated proteins had higher levels of transcripts at 11 days. Their functions after the end of the elongation process remain to be found. As expected, some genes encoding GRPs were found to have much higher levels of transcripts in fully-grown hypocotyls at a time lignification is an active process. However, since all these genes belong to multigene families, one cannot rule out the fact that a similar function can be shared by several genes.

Finally, looking into the transcript level of the genes corresponding to the 137 proteins identified by proteomic analysis of the cell walls of half- and fully-grown hypocotyls, 15% were below the CATMA background. On the contrary, only 13% of the genes encoding secreted proteins with high or moderate levels of transcripts corresponded to proteins identified through proteomics. Thus, the comparison between transcript levels and presence of the corresponding proteins suggested that many genes encoding proteins secreted in cell walls are regulated at a post-transcriptional level. In conclusion, transcriptomic and proteomic data appeared to be complementary to describe the regulation of gene activity during the elongation of etiolated hypocotyls.

KEYWORDS

- *Arabidopsis thaliana*
- Cell elongation
- Cell wall biogenesis
- Hypocotyls

AUTHORS' CONTRIBUTIONS

Elisabeth Jamet conceived the study, participated in its design, coordination, analysis of data, and drafted the manuscript. David Roujol and Muhammad Irshad carried out the culture of plants, growth curve determination, RNA extraction and PCR-analysis. Helene San Clemente was involved in bioinformatic analyses. Ludivine Soubigou-Taconnat and Jean-Pierre Renou performed the microarray and statistical analyses of the results. Rafael Pont-Lezica contributed to the analysis of data and to drafting of the manuscript. All authors read and approved the final manuscript.

ACKNOWLEDGMENTS

The authors thank CNRS (Institut des Sciences Biologiques) and Université Paul Sabatier (Toulouse III, France) for financial support. Muhammad Irshad is a fellow of the Higher Education Commission of Pakistan, Islamabad, and of the French government on the behalf of SFERE. They also thank Dr. Hervé Canut (UMR 5546 UPS/CNRS, France), and Dr. Jérôme Pelloux (University of Amiens, France).

Chapter 8

Spindle Assembly Checkpoint Protein Dynamics and Roles in Plant Cell Division

Marie-Cécile Caillaud, Laetitia Paganelli, Philippe Lecomte, Laurent Deslandes, Michaël Quentin, Yann Pecrix, Manuel Le Bris, Nicolas Marfaing, Pierre Abad, and Bruno Favery

---

INTRODUCTION

In eukaryotes, the spindle assembly checkpoint (SAC) ensures that chromosomes undergoing mitosis do not segregate until they are properly attached to the microtubules (MTs) of the spindle.

We investigated the mechanism underlying this surveillance mechanism in plants, by characterizing the orthogolous SAC proteins BUBR1, BUB3, and MAD2 from *Arabidopsis*. We showed that the cell cycle-regulated BUBR1, BUB3.1, and MAD2 proteins interacted physically with each other. Furthermore, BUBR1 and MAD2 interacted specifically at chromocenters. Following SAC activation by global defects in spindle assembly, these three interacting partners localized to unattached kinetochores. In addition, in cases of "wait anaphase", plant SAC proteins were associated with both kinetochores and kinetochore MTs. Unexpectedly, BUB3.1 was also found in the phragmoplast midline during the final step of cell division in plants.

We conclude that plant BUBR1, BUB3.1, and MAD2 proteins may have the SAC protein functions conserved from yeast to humans. The association of BUB3.1 with both unattached kinetochore and phragmoplast suggests that in plant, BUB3.1 may have other roles beyond the SAC itself. Finally, this study of the SAC dynamics pinpoints uncharacterized roles of this surveillance mechanism in plant cell division.

In eukaryotes, the SAC is a sophisticated surveillance mechanism that ensures the fidelity of chromosome segregation during mitosis (Musacchio and Hardwick, 2002; Musacchio and Salmon, 2007). The SAC monitors the interaction between chromosomes and MTs at specialized chromosomal regions, the kinetochores. In response to unattached kinetochores and to kinetochores lacking tension, the SAC is activated and localized to unattached kinetochores. The SAC transmits a "wait anaphase" signal until all chromosomes achieve bipolar attachment. This signal is transmitted through the inhibition of anaphase-promoting complex/cyclosome (APC/C) activity by sequestration of the CDC20 co-factor. The SAC components were first identified through genetic screens in budding yeast and include the mitotic arrest-deficient (MAD) and budding uninhibited by benzymidazol (BUB) proteins (Hoyt et al., 1991; Li and Murray, 1991). In metazoans and yeast, the mitotic checkpoint complex (MCC), which contains the three SAC proteins MAD2, MAD3 (equivalent of BUBR1, for BUB1-related, in higher eukaryotes), and BUB3 together with CDC20, is regarded as the SAC effector

(Fang et al., 1998; Sudakin et al., 2001; Tang et al., 2001). In budding yeast, the SAC is a non-essential device and it only becomes essential in response to "damage" that is perturbations in the kinetochore-MT attachment process (Hoyt et al., 1991; Li and Murray, 1991). On the other hand, in metazoans, the SAC is an essential pathway, the integrity of which is required to prevent chromosome mis-segregation and cell death (Musacchio and Salmon, 2007). In plants, SAC protein homologs have been identified *in silico* (Houben and Schubert, 2003; Lermontova et al., 2008; Menges et al., 2005), but function has been investigated only for MAD2 for which localization to unattached kinetochores has been demonstrated by immunolocalization (Kimbara et al., 2004; Yu et al., 1999).

In this chapter, we investigated how this surveillance mechanism operates in the green kingdom. We demonstrated physical interactions between *A. thaliana* BUBR1, BUB3.1, and MAD2 and their dynamics at unattached kinetochores. In cases of "wait anaphase", plant BUBR1, BUB3.1, and MAD2 proteins were unexpectedly associated with both kinetochores and kinetochore MTs. Our findings suggest that plant BUBR1, BUB3.1, and MAD2 have both the SAC protein functions conserved from yeast to humans and pinpoints uncharacterized roles in plant cell division.

MATERIALS AND METHODS

Sequence Identification and Gene Cloning

Arabidopsis thaliana proteins orthologous to human BUB3, BUBR1, and MAD2 were identified by the OrthoMCL (Li et al., 2003) clustering of six proteomes based on standard parameters. The six proteomes compared were those of *A. thaliana* (TAIR, http://www.arabidopsis.org), *Homo sapiens* (http://www.ncbi.nlm.nih.gov/projects/CCDS/CcdsBrowse.cgi), *Oryza sativa* (http://rice.plantbiology.msu.edu/), *Drosophila melanogaster* (http://flybase.org/), *Caenorhabditis elegans* (http://wormbase.org/) and *Meloidogyne incognita* (http://meloidogyne.toulouse.inra.fr/ (Abad et al., 2008)). Interpro scans (http://www.ebi.ac.uk/interpro) were used to study domain organization. The *A. thaliana* BUBR1, BUB3.1, MAD2, and HTR12/CENP-A coding sequences were amplified by PCR, using specific primers. They were inserted into the pDON207 donor vector and then into the pK7FWG2, or pK7WGF2 for HTR12, plant expression vector and BiFC vectors (pAM-35SS-GWY-YFPc and pAM-35SS-GWY-YFPn), using Gateway Technology (*Invitrogen*).

Promoter Analysis and Histochemical Localization of GUS Activity

For the promoter:GUS fusion, fragments of the 1,365 bp, 1,001 bp, 999 bp, and 1,000 bp immediately upstream from the start codon, for BUB3.1, BUB3.2, MAD2, and BUBR1, respectively, were amplified by PCR, inserted into the pDON207 donor vector and then into the pKGWFS7 plant vector, using Gateway Technology (*Invitrogen*). Wild-type (WS ecotype) *A. thaliana* plants were stably transformed and GUS activity was assayed histochemically, as previously described (Caillaud et al., 2008), on 10 independent transformed plants for each construct. Samples were observed with a Zeiss Axioplan 2 microscope and images analyzed with AxioVision 4.7 (Zeiss).

Yeast Two-Hybrid Split-Ubiquitin Assay

The split-ubiquitin assay was carried out in *S. cerevisiae* strain JD53, as previously described (Deslandes et al., 2003). The *BUBR1*, *BUB3.1*, and *MAD2* coding sequences were inserted into the GW:Cub:URA3 bait vector (pMKZ) and the NuI:GW prey vector, using the Gateway system. Standard procedures were used for yeast growth and transformation. Transformants were selected on 5-fluoroorotic acid (5-FOA) plates containing minimal medium with yeast nitrogen base without amino acids (Difco) and glucose, supplemented with lysine, leucine, uracil (M-HW), and 1 mg/ml 5-FOA.

Nicotiana *benthamiana* Transformation and Cell Cultures

Nicotiana *benthamiana* plants were grown under continuous light for 1 month at 26°C. Infiltration of *Agrobacterium tumefaciens* into tobacco leaves was as described (Voinnet et al., 2003) and plants were analyzed 2 days after infiltration. For tobacco cell culture establishment, *N. benthamiana* leaves were co-cultured 2 days with *A. tumefaciens* in the dark at 26°C, rinsed in a liquid MS medium containing 3% sucrose and 150 mg/l cefotaxime (Sigma). The tissue was blotted dry and placed on regeneration medium (MS medium, 3% sucrose, 1.0 mg/l indole acetic acid, and 0.1 mg/l benzyladenine, Sigma, 0.8% agar), and supplemented with 150 mg/l cefotaxime and 50 mg/l kanamycin. Explants were incubated in a controlled growth chamber at 26°C. All explants were subcultured onto fresh regeneration/selection medium every 10 days. Two explants were used to generate suspension cultures: stably transformed explants were placed on MS medium supplemented with 0.5 mg/l 2,4D (2, 4-dichlorophenoxyacetic acid) and 40 mg/l kanamycin for the induction of callus which was transferred into liquid MS medium supplemented with 1 mg/l 2,4D and 50 mg/l kanamycin. The cultures were incubated at 26°C in the dark with continuous shaking.

Drug Treatments and Microscopy

Optical sections of tobacco leaf epidermal cells or tobacco cell cultures were observed with a ×63 water immersion apochromatic objective (numerical aperture 1.2, Zeiss) fitted to an inverted confocal microscope (Axiovert 200 M, LSM510 META; Zeiss) at 25°C. The GFP and SYTO 82 (Molecular Probes) fluorescence were monitored in Channel mode with a BP 505–530, 488 beam splitters, and LP 530 filters for GFP and 545 nm beam splitters for SYTO 82 (488 nm excitation line). For DAPI staining, cells were first fixed in $1 \times$ PBS + 2% paraformaldehyde in PBS (1 x) supplemented with 0.05% Triton X-100. DNA was stained *in vivo* with the orange fluorescent dye SYTO 82 (2 µM final concentration). Propyzamid (Sigma), Paclitaxel (Sigma), and carbobenzoxyl-leucinyl-leucinyl-leucinal (MG132; kindly provided by M. C. Criqui, IBMP, Strasbourg, France) were used at final concentrations of 50 µM, 50 µM, and 100 µM, respectively. These preparations were stored for no more than 1 month at −20°C. The samples treated with MG132 were collected at different time point to be observed during metaphase arrest by *in vivo* confocal microscopy. For Propyzamid and Paclitaxel treatments, samples were collected 10 min after drug adjunction and used immediately for observation. Digital images were analyzed using LSM Image Browser (Zeiss), imported to Photoshop CS2 (Adobe), and contrast/brightness was uniformly changed. For immunolocalization of β-tubulin, samples were collected 3 hr after MG132 treatment. Cells were first fixed in $1 \times$ PBS + 2% paraformaldehyde supplemented with 0.05% Triton X-100. Immunolabeling was performed according to Ritzenthaler et al. (2003). Cells were incubated overnight with the monoclonal anti-β-tubulin clone TUB 2.1 (Sigma-Aldrich). Two hour incubation at room temperature was performed with Alexa 596 goat antimouse IgG (Molecular Probes, Eugene, OR, USA). The DNA was stained with 1 μg.ml^{-1} 4′,6-diamidino-2-phenylindole (DAPI, Sigma) in PBS 1 x buffer. The GFP and Alexa 596 (Molecular Probes) fluorescences were monitored in Channel mode with a BP 505–530, HFT 488 beam splitters for GFP and LP 530 filters NFT, 545 nm beam splitters for Alexa Red (488 nm excitation line).

DISCUSSION AND RESULTS

As a first attempt to study SAC during the plant cell cycle, candidate *A. thaliana* orthologs of the human essential MCC proteins BUBR1, BUB3, and MAD2 were identified by OrthoMCL (Li et al., 2003) clustering of orthologous proteins from six model eukaryotic species. The six complete proteomes compared included those of plants (*A. thaliana* and *Oryza sativa*), human (*Homo sapiens*), insect (*Drosophila melanogaster*) and nematodes (*Caenorhabditis elegans* and *Meloidogyne incognita*). *Arabidopsis thaliana* BUBR1 (AT2G33560) is a 46 kD protein containing an N-terminal MAD3-BUB1 conserved domain and two KEN boxes conferring substrate recognition by APC/C (Pfleger and Kirschner, 2000). These two KEN boxes are conserved from yeast MAD3 to human BUBR1 and are required for the concerted action of MAD3 and MAD2 in the checkpoint inhibition of CDC20-APC/C (Chan et al., 1999; Hardwick et al., 2000; Sczaniecka et al., 2008). Like the MAD3 proteins of *Saccharomyces cerevisiae* and *Schizosaccharomyces pombe*, *A. thaliana BUBR1* differs from human *BUBR1* by the absence of a C-terminal kinase domain. However, the kinase activity of *BUBR1* has been shown to be dispensable for spindle checkpoint function in *Xenopus larvei* (Chen, 2002). Two *A. thaliana* BUB3 proteins (BUB3.1, AT3G19590; BUB3.2, AT1G49910) were identified. Both are 38 kD proteins containing WD40 repeats, which have been shown to be involved in the association of BUB3 with MAD2, MAD3, and CDC20 in yeast (Fraschini et al., 2001). *Arabidopsis thaliana* BUB3.1 and BUB3.2 are 88% identical. The BUB3.1 is 52% and 22% identical to the human and *S. cerevisiae* BUB3 proteins (Hoyt et al., 1991; Taylor et al., 1998), respectively, over its entire length. *Arabidopsis thaliana* MAD2 (AT3G25980) is a 24 kD protein containing a HORMA domain. It is 44% identical to the human MAD2 protein (Li and Benezra, 1996) and 81% identical to the maize MAD2 protein (Yu et al., 1999), over its entire length.

Arabidopsis BUBR, BUB3.1, and *MAD2* Genes were Expressed in Tissues Enriched in Dividing Cells

We investigated the pattern of expression of the *A. thaliana BUBR1*, *BUB3.1*, *BUB3.2*, and *MAD2* genes during plant development, using *A. thaliana* transgenic lines transformed with the corresponding promoter-GUS reporter gene constructs. Similar patterns of GUS expression were observed for the *BUBR1*, *BUB3.1*, and *MAD2* promoters, both of which directed expression in tissues with a high proportion of dividing cells, early in organ development, in young leaves (Figure 1A), lateral root primordia (Figure 1B), lateral root meristems (Figure 1C), and root meristems (Figure 1D). Individual cells with strong GUS activity were observed in root meristems. In contrast to the cell cycle regulated pattern observed for both *BUBR1*, *BUB3.1* and *MAD2* promoter, no GUS activity in dividing cells was observed for the *BUB3.2* promoter in young leaves (Figure 1A), lateral root primordia (Figure 1B), lateral root meristems (Figure 1C), and root meristems (Figure 1D). These results are consistent with global transcriptome and RT-PCR analysis showing that *BUB3.1*, *BUBR1*, and *MAD2* presented a distinct expression peak at the G2/M boundary in synchronized *A. thaliana* cell cultures that was not observed for BUB3.2 (Lermontova et al., 2008; Menges et al., 2005). Because *BUB3.2* was not a cell cycle regulated gene, we next focused on *BUB3.1* candidate gene.

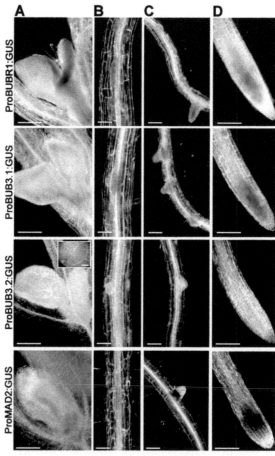

Figure 1. The pattern of expression of *BUBR1*, *BUB3.1*, *BUB3.2*, and *MAD2* during *A. thaliana* development. Promoter:GUS fusions revealed *BUBR1*, *BUB3.1*, and *MAD2* expression in the developing leaves of 7-day-old seedlings (A), in lateral root primordia (B), lateral root meristems (C), and root meristems (D). *BUB3.2* expression was only detected in cotyledons (insert). Bars, 200 µm (A), 50 µm (B), 100 µm (C and D).

BUBR1 and *MAD2* Interact Specifically at Chromocenters

In yeast and humans, *BUBR1*, *BUB3*, and *MAD2* may be found together in large complexes (MCC) (Fraschini et al., 2001; Hardwick et al., 2000; Sudakin et al., 2001). To carry out possible interactions between the cell cycle-regulated *A. thaliana BUBR1*, *BUB3.1*, and *MAD2*, a yeast two-hybrid split-ubiquitin approach was used. It is based on the fusion of the prey and the bait to the N- and C-terminal halves of ubiquitin (Nub and Cub, respectively), which are then able to form a native-like ubiquitin upon interaction (Stagljar et al., 1998). Ubiquitin-specific proteases recognize the reconstituted ubiquitin and cleave off a reporter protein, URA3, linked to the C terminus of Cub and whose degradation results in uracil auxotrophy and 5-FOA resistance. Co-expression of BUBR1:Cub:URA3 with either Nub:BUB3.1 and Nub:MAD2 conferred resistance to 5-FOA, indicating that *BUBR1* interacted with both BUB3.1 and MAD2. The BUB3.1 and MAD2 also interacted (Figure 2A). These interactions were confirmed in a reciprocal bait-prey experiment.

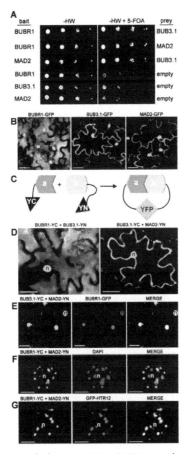

Figure 2. Interactions between *Arabidopsis* BUBR1, BUB3.1, and MAD2 in yeast and in planta. (A) Interactions in the yeast two-hybrid split-ubiquitin system. Dilution series of yeast JD53 cells expressing both bait fusions (BUBR1 or MAD2:Cub:URA3) and prey fusions (Nub:BUB3.1 or MAD2)

Figure 2. *(Caption Continued)*

were grown on yeast medium minus histidine and tryptophan (–HW) but containing 5-FOA, as indicated. Interaction resulted in uracil auxotrophy and 5-FOA resistance. (B) Single-plane images of tobacco epidermal leaf cells infiltrated with *A. tumefaciens* expressing BUBR1:GFP, BUB3.1:GFP, or MAD2:GFP constructs. (C) Principle of *in vivo* bimolecular fluorescence complementation (BiFC). BiFC is based on the fusion of the prey (a) and bait (b) to the N- and C-terminal halves of the yellow fluorescent protein YFP (YN and YC, respectively), forming a functional YFP upon interaction (Hu et al., 2002). (D–G) In *planta* BiFC assay. Single-plane confocal images of epidermal leaf cells infiltrated with *A. tumefaciens* co-expressing (D) BUBR1-YC and BUB3.1-YN or BUB3.1-YC and MAD2-YN, (E) BUB3.1-YC and MAD2-YN (green channel) and BUBR1-GFP (red channel), (F) BUBR1-YC and MAD2-YN, (G) BUBR1-YC and MAD2-YN (green channel) and GFP-HTR12 (red channel) fusion constructs. The merged images show (E) nuclear BUBR1, BUB3.1 and MAD2 colocalization in yellow, (F–G) that BUBR1 and MAD2 interaction colocalized with (F) bright chromocenter spots stained with *DAPI* (blue channel) and with (G) the centromeric marker *GFP-HTR12*. n, nucleus. Bars, 25 μm (B, D, and E), 5 μm (F and G).

To better characterize the physical interactions between BUBR1, BUB3.1, and MAD2, we performed in *planta* localization of these interactions. Following transient expression of the appropriate gene construct in *Nicotiana benthamiana* leaf epidermis, BUBR1 fused to GFP (BUBR1:GFP) was specifically targeted to the nucleus (n = 30; Figure 2B), whereas BUB3.1:GFP and MAD2:GFP were detected in both the nucleus and the cytoplasm (n = 30; Figure 2B). Using bimolecular fluorescence complementation (BiFC; (Hu et al., 2002), we demonstrated a close interaction between BUBR1, BUB3.1, and MAD2. Co-expression of the constructs encoding BUBR1:YC (BUBR1 fused to the C-terminal half of YFP) and BUB3.1:YN (BUB3 fused to the N-terminal half of YFP) resulted in the reconstituted YFP complexes only in the nuclei (n = 20; Figures 2C–D). In addition, BUB3.1 interacted with MAD2 in the nuclei and cytoplasm of epidermal cells (n = 20; Figure 2D). No YFP fluorescence was detected in negative control experiments in which BUBR1:YN, BUB3.1:YN, BUB3.1:YC, or MAD2:YC was produced together with the corresponding vector control (n = 30). Co-expression of the constructs encoding BUB3.1:YC, MAD2:YN, and BUBR1:GFP showed that BUB3.1 and MAD2 interact, and that they colocalize with BUBR1 in the nucleus (n = 20; Figure 2E). Interactions between BUBR1:YC and MAD2:YN were observed exclusively in the nucleus, as bright subnuclear foci (n = 40; Figure 2F). Within the nuclei, fluorescence signals were localized with the core of bright DAPI-stained condensed chromocenters (Figure 2F). Using the centromeric Histone H3 variant from *A. thaliana* GFP:HTR12 (CENH3, AT1G01370) as *in vivo* marker for centromeres (Fang and Spector, 2005; Lermontova et al., 2006; Talbert et al., 2002), we confirmed that BUBR1 and MAD2 interact at interphase centromeres (n = 10; Figure 2G) corresponding to the position on the chromosome at which kinetochore proteins associate.

BUBR1, BUB3.1, and MAD2 Localized to the Kinetochores Following SAC Activation

In metazoan cells, the BUBR1, BUB3, and MAD2 proteins are specifically localized to the kinetochores following the activation of the SAC by global defects in spindle assembly in cells treated with MT poisons (Basu et al., 1998; Chen et al., 1996; Martinez-Exposito et al., 1999; Taylor et al., 1998). The maize and wheat MAD2 proteins are the only plant SAC proteins for which localization to unattached kinetochores has been demonstrated (Kimbara et al., 2004; Yu et al., 1999). By combined direct immunofluorescence of maize MAD2 and CENPC, the identity of the MAD2-positive regions as kinetochores has been demonstrated (Yu et al., 1999).

To gain insight into the spindle checkpoint activation in plant, we profiled the spatial distribution of *A. thaliana* SAC proteins in tobacco cell cultures stably expressing the BUBR1:GFP, BUB3.1:GFP, and MAD2:GFP constructs. At a prometaphase-like stage, following treatment with the microtubule-destabilizing herbicide propyzamid, which prevents the formation of microtubule-kinetochore attachments, the MAD2 fusion protein was found to cluster strongly in bright spots on condensing chromosomes corresponding to unattached kinetochores (n = 20; Figure 3). Similar localization was observed for the BUB3.1 and BUBR1 fusion proteins (n = 20; Figure 3). Thus, the plant BUBR1, BUB3.1, and MAD2 partners identified in this study are all in place at the unattached kinetochores and may therefore fulfill the evolutionarily conserved functions of SAC proteins, delaying anaphase until all the chromosomes are attached to both poles of the spindle.

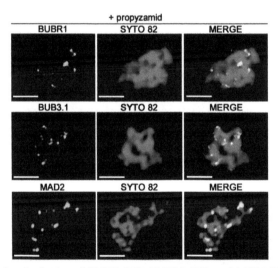

Figure 3. Subcellular distribution of BUBR1:GFP, BUB3.1:GFP, and MAD2:GFP fusion proteins in propyzamid-treated tobacco cells. Single optical section of prometaphase-like arrested cells expressing BUBR1:GFP, BUB3.1:GFP, and MAD2:GFP fusion constructs, 1 hr after propyzamid treatment. In merged images, the yellow color corresponds to BUBR1:GFP, BUB3.1:GFP, or MAD2:GFP (green channel) colocalization with SYTO 82 (red channel). BUBR1:GFP, BUB3.1:GFP, and MAD2:GFP localize *in vivo* to the kinetochores of chromosomes stained with SYTO 82 orange-fluorescent nucleic acid stain. Bars, 5 μm.

We further analyzed plant SAC protein distribution in living cells in cases of delayed anaphase onset. As anaphase initiation requires the ubiquitylation and degradation by the 26S proteasome of key mitotic regulators (Musacchio and Salmon, 2007), such as the separase inhibitor securin and the Cdk1 subunit cyclin B, we studied cells that had been treated with the proteasome inhibitor MG132. The MG132 tripeptide has been shown to be a very efficient proteasome inhibitor in mammalian and plant cell cultures. It preserves metaphase spindles and kinetochore-microtubule (kMT) attachments but inhibits the onset of anaphase (Genschik et al., 1998). As previously reported in plants (Criqui et al., 2000; Genschik et al., 1998), 2 hr after the addition of this molecule to a concentration of 100 µM, tobacco cells arrested in metaphase were found to have highly condensed chromosomes (n = 30; Figure 4A). At this time point, *A. thaliana* BUBR1, BUB3.1, and MAD2 were localized to the sister kinetochores of condensed chromosomes in metaphase arrested cells (n = 20; Figure 4A). In cells in which chromosomes were aligned at the spindle equator, BUBR1, BUB3.1, and MAD2 were found to be present in all the kinetochores (n = 15; Figure 4B). Progressively, much of the BUBR1, BUB3.1, MAD2:GFP-derived fluorescence took on a fibrillar appearance, probably as a result of association with the acentrosomal metaphase spindle apparatus (n = 30; Figure 4B). Three hour after MG132 treatment, the initially diffuse spindle BUBR1, BUB3.1, and MAD2:GFP staining accumulated progressively onto MT-like structures within the spindle (n = 25; Figure 4C). At this time point, bright spots corresponding to kinetochores were also detected for BUBR1, BUB3.1, and MAD2 (n = 25; Figure 4C). To determine if the MT-like SAC fluorescence was in fact MT dependent, we treated BUBR1:GFP cells with the MT-stabilizing agents Paclitaxel. Three hour after MG132 treatment, the adjunction of Paclitaxel dramatically intensified the fibrillar nature of BUBR1:GFP (n > 20). In addition, immunostaining of β-tubulin confirmed that BUBR1 colocalized with spindle MTs when proteolysis is blocked by MG132 (n > 10). Previous reports have provided evidence for the motor-assisted transport of human MAD2 complexes from kinetochores to the spindle poles along MTs (Howell et al., 2000). This mechanism may play an important role in removing checkpoint proteins from the kinetochores and turning off the checkpoint. Based on our observations, plant SAC proteins have an intriguing intracellular distribution, apparently accumulating onto both kinetochores and the spindle MTs in cell arrested in metaphase.

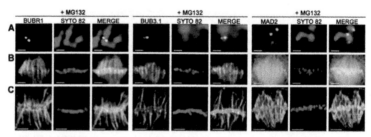

Figure 4. Subcellular localization of BUBR1, BUB3.1, and MAD2 in MG132-treated tobacco cells. Single optical section of cells expressing BUBR1:GFP, BUB3.1:GFP, and MAD2:GFP fusion constructs (green channel) after treatment with 100 µM MG132. Chromosomes in living cells were stained with SYTO 82 (red channel). In merged images, the yellow color corresponds to the colocalization of BUBR1:GFP, BUB3.1:GFP, or MAD2:GFP with SYTO 82. (A) 2 hr after MG132 treatment, BUBR1, BUB3.1, and MAD2 were localized into two bright spots per condensed chromosome, corresponding to kinetochores. (B) When chromosomes were arrested in metaphase, BUBR1, BUB3.1, and MAD2 localized to all the kinetochores of chromosomes arrested in metaphase. A diffuse signal in the metaphase spindle apparatus was also observed for BUBR1, BUB3.1, and MAD2. (C) 3 hr after MG132 treatment, BUBR1, BUB3.1, and MAD2 localized to bright spots corresponding to the kinetochores of chromosomes and staining accumulated onto MT-like structures within the spindle in metaphase arrested cells. Bars, 2 µm (A), 5 µm (B and C).

SAC Inactivation in Normal Cell Division

We then investigated the distribution of *A. thaliana* SAC proteins *in vivo* in normal mitosis conditions, when SAC is inactivated. This was made possible since the expression of the chimeric proteins did not prevent cell cycle progression. In the absence of SAC activation, BUBR1 was found exclusively in nuclei stained with SYTO 82 during interphase (n = 30; Figure 5A). BUB3.1 and MAD2 proteins were localized to the nucleus and gave a weak cytoplasmic signal during interphase (n = 30; Figure 5A). In early prophase, BUBR1, BUB3.1, and MAD2 became localized in the cytoplasm following nuclear envelope breakdown and remained there until the end of metaphase (n = 10; Figure 5A). By telophase, when a new nuclear envelope forms around each set of separated sister chromosomes, *A. thaliana* BUBR1 and MAD2 were again concentrated in the nucleus (n = 15; Figure 5A).

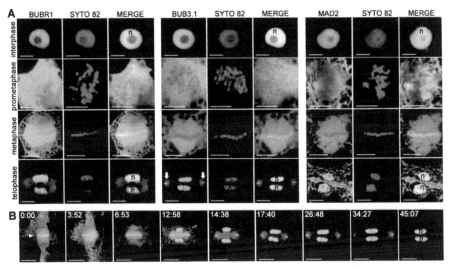

Figure 5. Subcellular localization of BUBR1, BUB3.1, and MAD2 in tobacco cells undergoing normal mitosis. (A) Single optical section of cells expressing BUBR1:GFP, BUB3.1:GFP, and MAD2:GFP fusion constructs (green channel). Chromosomes in living cells were stained with SYTO 82 (red channel). In merged images, the yellow color corresponds to the colocalization of BUBR1:GFP, BUB3.1:GFP, or MAD2:GFP with SYTO 82. By telophase, BUB3.1:GFP was detected in daughter nuclei (n) and in the midline at the cell periphery (arrow), forming a ring around the edge of the newly formed cell plate. (B) Selected frames from a fluorescence time-lapse analysis of the distribution of BUB3.1:GFP during cytokinesis. Single optical section of a cell expressing the BUB3.1:GFP fusion construct (green channel). After chromosome separation, BUB3.1 is localized along the midline of the anaphase spindle (arrowhead). During telophase, BUB3.1 is gradually transferred into the daughter nuclei. During phragmoplast extension from the centre to the periphery of the cell, BUB3.1 localizes with the margin of the expanded phragmoplast. At the end of telophase, BUB3.1 is present at the cell periphery, forming a ring around the edge of the newly formed cell plate. This specific localization at the phragmoplast midline disappeared when the newly formed cell plate completely separated the two daughter cells. At the end of cytokinesis, BUB3.1 was again concentrated in the nucleus. Time is in min:s. Bars, 10 μm.

Overall, our data show that checkpoint proteins are only recruited at kinetochore in case of damage in spindle assembly. During normal mitosis, BUBR1, BUB3.1, and MAD2 staining at the kinetochore was not detected, inconsistent with reports for metazoan cells (Basu et al., 1998; Howell et al., 2004; Musacchio and Salmon, 2007). We propose that organism-specific differences in the behavior of SAC are likely to reflect evolutionary divergence in the mechanics of spindle assembly rather than extensive differences in the pathways of checkpoint signaling. Animal cells undergo an open mitosis in which prometaphase chromosomes are initially free of spindle MTs after nuclear envelope breakdown. High levels of MAD and BUB proteins are present on these unattached kinetochores (Basu et al., 1998; Howell et al., 2004; Musacchio and Salmon, 2007). Plant cells undergo mitosis in which acentrosomal pro-spindle assembly is initiated before nuclear envelope breakdown (Vos et al., 2008). Our data suggest that in plant, kinetochores do not recruit high level of these SAC proteins during normal mitosis which is consistent with the idea that plant chromosomes are continuously linked to MTs.

BUB3.1 Localized to Phragmoplast Midline During Plant Cytokinesis

The BUB3.1 displayed an unexpected distribution during cytokinesis in late anaphase to telophase in plant cells. It first appeared in the centre of the forming cell plate, and was subsequently redistributed to the growing margins of the cell plate as the cell plate grew outwards. Time-lapse analysis showed A. thaliana BUB3.1 was strongly localized to the anaphase spindle midline after chromosome separation in late anaphase (n = 15; Figure 5B). During the final stages of cell division, a cytokinetic apparatus unique to plants, the phragmoplast, was generated. The phragmoplast directs Golgi-derived vesicles to the midline where they fuse to form a cell plate, permitting the separation of the daughter cells. During telophase, a strong BUB3.1:GFP signal was detected in the early phragmoplast midline and in the newly formed daughter nuclei (n = 17; Figure 5B). At the end of telophase, when the phragmoplast was fully expanded, BUB3.1 was observed at the cell periphery, forming a ring around the edge of the newly formed cell plate (n = 20; Figure 5A–B). This signal disappeared when the fully expanded cell plate completely separated the two daughter cells (n = 18; Figure 5B). This BUB3.1 subcellular localization appeared intriguing since metazoan and yeast BUB3 has not been described to be involved in cytokinesis. In animal cells, after sister chromosomes have separated, the remaining non-kinetochore MTs form a structure called the spindle midzone. The spindle midzone was compressed by the ingressing cleavage furrow. This spindle remnant also persists during cytokinesis in plant cells, where it becomes the early phragmoplast. The difference is that, instead of being the focus of constriction, as in animals, the central spindle/early phragmoplast opens out as a ring that directs Golgi-derived vesicles to the midline where they fuse to form a cell plate.

We found that, during outward cytokinesis, BUB3.1 was specifically localized to the phragmoplast midline, at which the cell plate was held by phragmoplast MTs. The BUBR1 and MAD2 proteins did not follow this pattern. Thus, BUB3.1, in addition to its known role in the SAC itself, may have a plant-specific role in late mitosis coordinating phragmoplast expansion. The phragmoplast midline defines the interface between phragmoplast MT plus-ends and the newly formed cell plate. Recent studies indicated that the phragmoplast midline could contain linker molecules that help to stabilize MT plus-ends and connect them to cell plate membranes. This results in optimally organized phragmoplast MTs that deliver the Golgi-derived vesicles to the growing cell plate (Austin et al., 2005).

We hypothesize, that BUB3.1 could be part of a MT plus-end capture complex associated with other "phragmoplast midline proteins" and then may regulate phragmoplast expansion, essential for cytokinesis. An analysis of the cell cycle regulators present in synchronized *A. thaliana* cell cultures showed that BUB3.1 expression was co-regulated with the expression of other cytokinesis-related genes (Menges et al., 2005). In addition, the AURORA-like kinase 1 (Van et al., 2004), the microtubule-associated protein MAP65-3 (Caillaud et al., 2008), the molecular motor kinesin PAKRP2 (Lee et al., 2001) and the CDC27/HOBBIT APC/C subunit (Perez-Perez et al., 2008) have phragmoplast midline distributions similar to that of BUB3.1 during cytokinesis. However, none of these proteins has been reported to be localized to both unattached kinetochores and the phragmoplast midline. The association of BUB3.1 with both these structures suggests that plants may coordinate spindle assembly and cytokinesis through shared machinery. This study provides clues to the possible functional links between the spindle and phragmoplast assembly checkpoints, ensuring failsafe mitosis.

KEYWORDS

- **Budding uninhibited**
- **Mitotic arrest-deficient**
- **Phragmoplast**
- **Spindle assembly checkpoint**

AUTHORS' CONTRIBUTIONS

Conceived and designed the experiments: Marie-Cécile Caillaud, Laetitia Paganelli, Philippe Lecomte, Laurent Deslandes, Pierre Abad, and Bruno Favery. Performed the experiments: Marie-Cécile Caillaud, Laetitia Paganelli, Philippe Lecomte, Michaël Quentin, and Yann Pecrix. Analyzed the data: Marie-Cécile Caillaud, Laetitia Paganelli, Philippe Lecomte, Michaël Quentin, Yann Pecrix, Manuel Le Bris, Pierre Abad, and Bruno Favery. Contributed reagents/materials/analysis tools: Laurent Deslandes and Nicolas Marfaing. Wrote the chapter: Marie-Cécile Caillaud and Bruno Favery.

ACKNOWLEDGMENTS

We thank Marylin Vantard (iRTSV, Grenoble, France), Pascal Genschik, Laurent Pieuchot, Anne Catherine Schmit (IBMP, Strasbourg, France), and Michel Ponchet (IBSV, France) for helpful discussions, Etienne G.J. Danchin for OrthoMCL clustering, Catherine Mura for growing tobacco plants, Mansour Karimi (VIB Ghent, Belgium) for the plant Gateway vectors, Imre E. Somssich (Max-Planck Institut, Köln, Germany) for the split-ubiquitin system, Marie-Claire Criqui (IBMP, France) for MG132 inhibitor and helpful discussions. This chapter is dedicated to the just born Elyas Marfaing.

Chapter 9

Soybeans Iron Deficiency Response

Jamie A. O'Rourke, Rex T. Nelson, David Grant, Jeremy Schmutz, Jane Grimwood, Steven Cannon, Carroll P. Vance, Michelle A. Graham, and Randy C. Shoemaker

INTRODUCTION

Soybeans grown in the upper Midwestern United States often suffer from iron deficiency chlorosis (IDC), which results in yield loss at the end of the season. To better understand the effect of iron availability on soybean yield, we identified genes in two near isogenic lines with changes in expression patterns when plants were grown in iron sufficient and iron deficient conditions.

Transcriptional profiles of soybean (*Glycine max*, L. *Merr*) near isogenic lines Clark (PI548553, iron efficient) and IsoClark (PI547430, iron inefficient) grown under Fe-sufficient and Fe-limited conditions were analyzed and compared using the Affymetrix® GeneChip® Soybean Genome Array. There were 835 candidate genes in the Clark (PI548553) genotype and 200 candidate genes in the IsoClark (PI547430) genotype putatively involved in soybean's iron stress response. Of these candidate genes, 58 genes in the Clark genotype were identified with a genetic location within known iron efficiency Quantitative Trait Locus (QTL) and 21 in the IsoClark genotype. The arrays also identified 170 single feature polymorphisms (SFPs) specific to either Clark or IsoClark. A sliding window analysis of the microarray data and the 7X genome assembly coupled with an iterative model of the data showed the candidate genes are clustered in the genome. An analysis of 5' untranslated regions in the promoter of candidate genes identified 11 conserved motifs in 248 differentially expressed genes, all from the Clark genotype, representing 129 clusters identified earlier, confirming the cluster analysis results.

These analyses have identified the first genes with expression patterns that are affected by iron stress and are located within QTL specific to iron deficiency stress. The genetic location and promoter motif analysis results support the hypothesis that the differentially expressed genes are co-regulated. The combined results of all analyses lead us to postulate iron inefficiency in soybean is a result of a mutation in a transcription factor(s), which controls the expression of genes required in inducing an iron stress response.

Iron is a critical micronutrient for both plant and animal nutrition, serving as a required co-factor for a variety of cellular processes. Iron deficiency anemia is one of the leading human nutritional disorders worldwide, affecting 43% of the population of developing countries (Theil et al., 2004). For most of the world's population, legumes are a major source of dietary iron (Ghandilyan et al., 2006; Theil et al., 2004). Though iron composes 5% of the earth's crust (Mengel et al., 2001) it is largely unavailable to plants, particularly in calcareous soils with a pH greater than 7.5. Calcareous soils are especially prevalent in the upper Midwest of the US (Hansen et al., 2003; Lin et al., 1997) and have been implicated in iron deficiency in soybeans. IDC in soybeans is characterized by interveinal chlorosis of the developing trifoliates (Inskeep and Bloom, 1987) contributing to yield loss directly proportional to the severity of the chlorosis (Inskeep and Bloom, 1987).

Plants have evolved two systems to uptake iron from the soil. These systems are termed strategy I and II (Romheld, 1987; Romheld and Marschner, 1986). Soybeans and other dicots utilize strategy I, in which the rhizosphere is acidified by the release of protons to produce a favorable environment for the release of iron from chelating agents in the soil. A membrane bound reductase reduces iron to the usable Fe^{+2} form. The iron is then transported across the plasma membrane by a specific transporter for distribution and use within the plant. The reduction of the iron from Fe^{3+} to Fe^{2+} has been shown to be the rate-limiting step in IDC (Connolly et al., 2003). Graminaceous monocots utilize strategy II, whereby the roots release chelators called phytosidero-phores to bind Fe^{+3} ions. Once bound, the entire complex is transported into the root where it is uncoupled. The Fe^{+3} ion is reduced to Fe^{+2} and the phytosiderophores are re-released into the soil.

The quantitative nature of IDC makes field studies problematic. Previous studies have identified multiple QTL associated with IDC (Lin et al., 1997, 1998). Many of the same QTL have been identified in both field and greenhouse studies, where plants are grown in a hydroponics system designed specifically to induce IDC (Lin et al., 1998). Growing plants in a controlled greenhouse environment with regulated nutritional availability allows for reproducible induction of iron deficiency stress. In addition, the advent of microarray technology now allows for the identification of individual transcripts whose expression levels are affected by iron availability (O'Rourke et al., 2007a, 2007b). The availability of a whole-genome sequence assembly for the soybean genome has, for the first time, allowed us to genetically position differentially expressed genes induced by iron deficiency.

Genomic studies in many organisms have shown genes in close proximity to one another in the genome are often co-expressed. These co-expressed genes create clusters of expression neighborhoods (Kosak et al., 2007) which are conserved by natural selection (Hurst et al., 2007). A study in *Arabidopsis* showed clusters of up to 20 different genes were coordinately regulated, with a median cluster size of 100 kb (Williams and Bowles, 2004). In rice, approximately 5% of the genome has been associated with co-expressed gene clusters (Ren et al., 2007). Initially co-expressed genes were thought to belong to similar biological pathways (Williams and Bowles, 2004), but further studies have shown co-functionality to be a poor predictor of co-expression (Michalak, 2008). Instead, promoter analysis has found co-regulated genes are often regulated by common transcription factors (Kosak et al., 2007; Michalak, 2008; Oliver et al., 2008). The co-expression of clustered genes may be partially regulated by the interaction of common promoter elements and transcription factors (Oliver et al., 2008). Co-regulated genes often have common transcription factors (Michalak, 2008), so an increase in the number of transcription factor binding sites (TFBSs) in promoter regions would increase the likelihood of the transcription factor binding and aiding in the expression of the gene cluster.

The objectives of our research are to identify a list of candidate genes with a potential involvement in soybean iron deficiency and to associate these genes with the genome sequence to determine any correlation with previously identified QTL. We also wanted to determine whether the changes in candidate gene expression were due to structural or sequence differences in the candidate genes. The results from these analyses confirmed the co-expressed genes were co-localized and possibly coordinately regulated.

MATERIALS AND METHODS

Plant Growth and RNA Extractions

The NILs developed for their characteristic response to limited iron conditions, were developed by the USDA-ARS (USDA, 2004). The iron efficient PI548533 (Oliver et al., 2008) was crossed with iron inefficient T203 (PI54619). Five repeated backcrosses to Clark yielded the iron inefficient line PI547430 (IsoClark). Both the iron efficient Clark and the iron inefficient IsoClark were germinated in sterile vermiculite and transferred to a DTPA buffered nutrient hydroponics system 7 days after planting. Each 10 L hydroponic unit contained 2 mM $MgSO_4$ *$7H_2O$, 3 mM $Mg(NO_3)_2$ *$6H_2O$, 2.5 mM KNO_3, 1 mM $CaCl_2$ *$2H_2O$, 4.0 mM $Ca(NO_3)_2$ *4H2O, 0.020 mM KH_2PO_4, 542.5 μM KOH, 217 μM DTPA, 1.52 μM $MnCl_2$ *$4H_2O$, 4.6 μM ZnSO4 *$7H_2O$, 2 μM $CuSO_4$ *$5H_2O$, 0.20 μM $NaMoO_4$ *$_2H_2O$, 1 μM $CoSO_4$ *$7H_2O$, 1 μM $NiSO_4$ *$6H_2O$, 10 μM H_3BO_3, and 20 mM HCO_3. A pH of 7.8 was maintained by the aeration of a 3% CO2: air mixture. A supplemental nutrient solution containing 16 mM potassium phosphate, 0.287 mM boric acid and 355 mM ammonium nitrate was added daily to maintain proper plant nutrition. Both iron efficient and iron inefficient plants were grown in iron sufficient (100 μM $Fe(NO_3)_3$) and iron limiting (50 μM $Fe(NO_3)_3$) hydroponic conditions. Leaf tissue from the 2nd trifoliate was collected 21 days after planting, or after 14 days in the hydroponics system. Tissue was flash frozen in liquid nitrogen and stored at −80°C until RNA could be extracted. Three independent biological replicates were used as the experimental tissue. The RNA extractions were performed using the Qiagen RNeasy Plant Mini Kit (catalog # 74904). The RNA samples were submitted to the Iowa State University GeneChip® facility to be hybridized and scanned using the Soybean Affymetrix® GeneChip®. Chip data has been uploaded to Gene Expression Omnibus as accession number GSE10730. Hybridization data was visualized using Bioconductor to ensure all hybridizations had normal distributions. The data was then loaded into the Gene Traffic Microarray Analysis program where it was normalized using the invariant set command, using the Clark 100 μM Fe as the control group, and a model based expression index (MBEI) (Li and Wong, 2001) analysis was performed on perfect match probes only. Hybridization statistics were used to determine a two-fold change in expression, consistent across all replicates, reflected a statistically significant (p values and standard errors generated by analysis not shown) difference in expression between genotype and iron concentrations. An analysis of Clark plants grown in iron sufficient and iron deficient conditions showed 835 transcripts differentially expressed at two-fold or greater. IsoClark plants grown in identical conditions showed 200 transcripts that met the criteria for differential expression.

Candidate Gene Annotation

The candidate genes were queried against the SoyBase Affymetrix® GeneChip® Soybean Genome Array Annotation page, publicly available at http://www.soybase.org/AffyChip/. Here, researchers with the USDA-ARS have used BLASTX and TBLASTX (Altschul et al., 1997) to compare the sequences from which all Affymetrix probes were derived to the UniProt database and the *Arabidopsis* genome gene calls (TAIR7, http://www.arabidopsis.org/). The top three UniProt BLAST hits and the *Arabidopsis* best hit GO annotation is reported for each Affymetrix probe set. To assign a putative function and classification to the differentially expressed genes (Table 1) the three UniProt annotations were compared. If all three were identical that annotation was assigned to the gene. If the top three BLAST hits were not in concordance, that sequence was re-examined to determine if one of the annotations was more likely correct than the others. If no annotation could be confidently identified by BLAST analysis with UniProt, the differentially expressed gene was annotated as an unknown. If the gene sequence for the Affymetrix® probe showed no sequence homology to any of the proteins in the UniProt database, the sequence was annotated as No UniProt Hit.

GO Slim Term Analysis

For expressed genes with homology greater than 10E-6 to an *Arabidopsis* gene, custom perl scripts were written to parse and tally each transcript GO slim ID for biological process, molecular function, and cellular process. The same scripts were used to tally GO slim IDs for the entire chip. Differences between the expressed genes and the entire chip were compared using a Fisher exact test (Fisher, 1949). This test was performed to identify the GO slim terms within each of the three GO slim classifications that were over- or under-represented in the lists of differentially expressed genes in relation to their presence on the soybean Affymetrix® chip. A Bonferroni correction (Bonferroni, 1935), using the number of identifiers present on the Affymetrix® chip, was applied to the two-tailed probability value (p-value) of each GO slim identifier. The GO slim identifications with a p-value of less than or equal to 0.05 after the Bonferroni correction were considered statistically over- or under-represented in our list of differentially expressed genes (Tables 2 and 3). This correction is likely to underestimate the number of categories of genes either over-or under-represented on the lists of differentially expressed genes in comparison to their prevalence on the Affymetrix® chip.

Real Time PCR Confirmation

The differential expression observed in the microarray experiment to identify candidate genes was confirmed using semi quantitative Real Time Reverse Transcriptase PCR (sqRT-PCR). Eleven transcripts identified as differentially expressed in the microarray experiment were tested using sqRT-PCR (Table 4). Genes for sqRT-PCR confirmation were chosen based on differential expression levels in the microarray. We tested genes showing both extreme differential expression and those just exceeding the two-fold criteria. Primers were designed from the EST sequence used to construct the Affymetrix probe to produce 250 bp amplicons. The sqRT-PCR was conducted as described by the Stratagene protocol (Catalog #600532) using the Stratagene Brilliant qRT-PCR kit with 25 uL reactions. For each experimental reaction, 200 ng of total RNA was added as initial template along with 125 mM MgCl2 and 100 nM forward and reverse primers. Cycling parameters were as follows: 45 min at 42°C for reverse transcription, 10 min at 95°C to denature reverse transcriptase StrataScript, 40 cycles of 30 sec at 95°C, 1 min at proper annealing temperature, 30 sec at 72°C. All sqRT-PCR reactions were performed in the Stratagene Mx3000P followed by a dissociation curve, taking a fluorescence reading at every degree between 55°C and 95°C to ensure only one PCR product was amplified. As controls, a passive reference dye was added to each reaction to ensure the increase in fluorescence was due to an increase in amplicon and not an artifact of the PCR. Additionally, each sample was run in triplicate and normalized against tubulin amplification to ensure differential expression was not due to differing amounts of initial template RNA added to each sample.

To be considered differentially expressed, samples had to differ in cycle thresholds (Ct) by more than one cycle, which corresponds to the two-fold difference in gene transcripts between the NILs identified by the microarray experiment. The resulting fold change of the sqRT-PCR was calculated from the differences in Ct using the 2ΔCt method (Schmittgen et al., 2000).

SFP Identification and Association with Known IDC QTL on Soybean Genome

The SFPs were identified following the protocol outlined by West et al. (2006). In brief, the microarray data from plants grown under iron sufficient conditions was transformed by robust multichip analysis (RMA) (Irizarry et al., 2003). Custom perl scripts were used to examine each of the 10 individual probes comprising a single perfect match probe. These perl scripts assigned each perfect mach probe set an SFPdev score by subtracting the average hybridization signal from the other 10 probes from the hybridization signal of the probe in question and dividing that by the hybridization signal of the probe being examined ((hyb signal probe 1-(hyb signal probe 1+ hyb signal probe 2 + hyb signal probe 3 + hyb signal probe 4 ... hyb signal probe 10)/10)/ hyb signal probe 1). The SFPdev scores with an absolute value greater than or equal to two on all replicates indicated an SFP.

Statistical Modeling and Cluster Analysis

To determine if gene distribution along the assembled genome could be explained by random chance, a simulation program originally reported by Grant (Grant et al., 2000) was applied to a theoretical genome. A genome of 996,903,313 bp (the combined size of the 7x genome assembly which has been assigned to soybean molecular linkage groups) was partitioned into 1,000,000 bp, 100,000 bp, and 10,000 bp windows resulting in 953 bins, 9,530 bins, and 95,300 bins respectively. The program positioned 760 or 200 genes depending on the genotype being simulated on the genome and determined the number of genes within the window. The simulation was repeated 1,000 times. The mean number of bins with 0–8 genes was calculated for the 1,000 repetitions. A standard deviation for each gene bin size was also calculated. To determine how this compared with our experimental data, the sequences assigned to chromosomes were concatenated together and the sliding window analysis was performed to identify clusters. Chromosomes are designated as shown in http://www.soybase.org website. The difference between the microarray data and the simulated data is calculated in terms of the number of simulated data standard deviations (USDA, 2004). A difference greater than two SD is considered statistically significant. The sign of the difference is indicative of whether there are more or fewer genes than expected.

Motif Identification and Analysis

The consensus sequence used by Affymetrix® to generate the probes on the Soybean GeneChip® identified as differentially expressed between Clark plants grown under iron sufficient and iron deficient conditions were queried against the 7X genome gene calls. The top hit for each differentially gene was used as the gene call for the differentially expressed sequence on the Affymetrix® GeneChip®. The 835 differentially expressed sequences in the Clark genotype correlated with 760 of the predicted genes in the 7X genome release http://www.phytozome.net while the 200 predicted genes from the IsoClark genotype correlated with 200 predicted genes from the 7X genome. Custom perl scripts identified the 500 bases upstream of the start codon for each gene from the 7X genome assembly. The reverse complement of each of the 500 bp promoter regions was also identified. The program MEME (Multiple Em for Motif Elicitation (Bailey et al., 1996)) was run against the 500 base promoter regions of all IDC genes to identify short conserved sequences in the promoter regions of the differentially expressed genes using the-dna-mod anr -evt 1 commands. Identified motifs with E-values<1E-6 were then compared against a modified TRANSFAC database using BLASTN (Altschul et al., 1997) to determine if identified motifs contained any known TFBSs (Table 7).

DISCUSSION

Microarray Analysis

Through a combination of a suite of analyses we have extended the fundamental understanding of the genetics underlying iron uptake and homeostasis in plants, but specifically soybean. Affymetrix gene chip analysis allowed us to identify candidate genes that are induced by iron deficiency in the leaf tissue of two NILs, complementing previous studies done in roots (O'Rourke et al., 2007a, 2007b). The Clark genotype analysis identified 835 differentially expressed genes when grown under iron sufficient and iron insufficient conditions while in IsoClark only 200 were identified. These genes have been aligned with the genomic sequence to determine their location. A sliding window analysis determined the co-expressed genes are clustered in the genome, suggesting co-regulation. The SFP analyses determined the differentially expressed genes are not a result of structural differences in the genes between the two NILs, providing further support that the differentially expressed genes are being co-regulated. Finally, motif analysis identified 11 short conserved motifs in the promoter regions of the candidate genes, which are most likely TFBSs. The cumulative results of all analyses leads us to propose the differential iron response in the NILs is a result of a mutation in the iron inefficient NIL of a transcription factor, or factors, probably encoded within one or more of the previously identified QTL, that prevents the induction of the iron deficiency gene expression responses seen in the iron efficient NIL.

The candidate genes identified with the microarray experiment suggest the Clark genotype is capable of recognizing the iron deficiency and eliciting a change in transcription patterns as a response to the stress. We hypothesize that the iron deficient Clark plant compensates for the lack of iron availability by adjusting its physiological processes to conserve available iron. Alternatively, the IsoClark genotype does not appear to initiate an effective response to the iron deficient conditions. The lack of differentially expressed genes in the IsoClark genotype, when comparing iron sufficient and iron deficient conditions, implies the iron deficient IsoClark plant continues to function as if still in iron sufficient conditions. However, the lack of iron as a cofactor in many of the basic biological processes results in a multitude of biological pathway failures, resulting in chlorotic plants.

In *Arabidopsis*, iron deficiency stress causes an increase in the transcription of electron transport chain components. Specifically, cytochromes are upregulated (Thimm et al., 2001). Our experiment identified 17 genes associated with cytochrome P450 in iron stressed Clark plants. All 17 genes were down-regulated in iron stressed tissue compared to non-stressed tissue, the opposite response as seen in *Arabidopsis* plants (Thimm et al., 2001). Thimm et al. proposed a correlation between iron deficiency stress and in induction of phosphoenolpyruvate carboxylase activity (Thimm et al., 2001). Four genes associated with phosphoenolpyruvate activity were identified as differentially expressed in the Clark genotype by microarray analysis. All four of these genes were down regulated in plants grown under iron stress rather than in iron sufficient conditions. Iron deficiency has also been shown to induce glycolytic activity (Espen et al., 2000). Three enzymes involved in glycolysis; glyceraldehydes 3 phosphate dehydrogenase (G3PD), pyruvate kinase (PK), and fructose 6 phosphate kinase (F6PK) have been shown to be up-regulated in *Arabidopsis* (Thimm et al., 2001) and cucumber (Espen et al., 2000) under iron deficiency stress. Microarray analysis comparing Clark plants grown in iron sufficient and iron stressed conditions only identified a single G3PD and a single F6PK, both of which were down-regulated in the iron stressed tissue compared to iron sufficient tissue. Seven genes associated with PK were identified in our microarray. Again, all seven were down regulated. The down regulation of the three main components of glycolysis suggests soybean, unlike *Arabidopsis*, does not increase non-photosynthetic carbon fixation or phosphoenolpyruvate carboxylase activity under iron stressed conditions. The contrasting results support the hypothesis proposed by Zocchi et al. that soybeans do not follow canonical iron deficiency responses (Zocchi et al., 2007).

Soybean does follow some of the established responses to iron deficient stress conditions. It has been proposed that under iron deficient conditions citrate provides a carbon skeleton for chlorotic leaves to allow for sustained growth and respiration (Abadia et al., 2002). Clark iron deficient stressed plants show a down regulation of citrate lyase (GO: 0008815) in comparison to non-stressed plants. The reduced breakdown of citrate in iron stressed plants lends credence to this hypothesis. Additionally, iron deficient conditions cause decreased activity of lipoxygenases (Thimm et al., 2001). All 13 lipoxygenases identified by microarray analysis in the Clark genotype showed decreased expression in the iron stressed tissue compared to the iron sufficient tissue.

The discrepancies between previously reported literature and the soybean iron deficiency response highlight the complexity of the iron stress response. However, it is important to remember that transcriptional regulation is only one form of regulation. Post-transcriptional modification may be an important component to understanding soybean's iron deficiency response, but that is beyond the scope of this investigation.

GO Slim ID Analysis

The Clark (iron efficient) genotype had an over-representation of genes in GO slim categories specific to iron availability/usage and categories associated with a more general stress response (Table 2). This reinforces the hypothesis that Clark responds specifically to iron but also to a more general stress response. A similar pattern was observed in a cDNA microarray experiment (O'Rourke et al., 2007b). Additionally, the Clark genotype showed a statistically significant number of GO slim IDs that were over-represented related to DNA replication and DNA binding activity. The increased expression levels of genes involved in these processes is probably a result of the DNA repair required to prevent lethal mutations from ROS (Sun et al., 2007), which are more prevalent under conditions of stress (Sun et al., 2007). The DNA binding activity suggests the activity of transcription factors, which lead to dramatic expression changes downstream. However, the down regulation of genes related to translation that is: GO0006412 (translation) and GO: 0006468 (protein amino acid phosphorylation) is indicative that the plant is not synthesizing proteins at a normal rate as it would under optimal growth conditions and is instead reducing the expression of genes involved in cellular processes not imperative to survival.

The IsoClark genotype had many fewer GO categories significantly over represented on our lists of candidate genes (Table 3) in comparison to Clark. Only two of the GO classifications were related to iron GO:0008940 (nitrate reductase activity) and GO:0008382 (iron superoxide dismutase activity). The remaining GO categories show little association to either a general or an iron specific stress response. It appears the IsoClark genotype is unable to recognize or respond to the iron stress. The IsoClark genotype had fewer genes differentially expressed due to iron deficiency and most of the genes that were differentially expressed are not associated with stress related pathways.

Clusters of Co-Expressed Genes

Expression analysis has been used in some model organisms to identify differentially expressed genes that are clustered together within the genome (Hurst et al., 2007; Kosak et al., 2007; Michalak, 2008; Oliver et al., 2008; Ren et al., 2005, 2007; Williams and Bowles, 2004; Zhan et al., 2006). These genomic neighborhoods are thought to be conserved by natural selection (Hurst et al., 2007) but are not entirely explained by co-functionality (Michalak, 2008). The combined use of expression data with known

QTL positions and expression clusters should further narrow the list of candidate genes to identify functionally important differences in the soybean genome affecting iron efficiency.

Co-expressed genes show a non-random distribution throughout the genome (Hurst et al., 2007; Michalak, 2008) where similarly expressed genes are located in clusters. Localized co-expression of genes has been reported in many different species including (but not limited to *Arabidopsis* (Ren et al., 2005; Williams and Bowles, 2004; Zhan et al.,2006), rice (Ren et al., 2007), human (Kosak et al., 2007; Vogel et al., 2005), and yeast (Hurst et al., 2007)). Williams et al. (2004) found genes located nearby in the genome and genes involved in the same pathways are more likely to be co-expressed. The incidence of co-expressed gene clusters has been widely studied (Hurst et al., 2007; Kosak et al., 2007; Michalak, 2008; Vogel et al., 2005; Williams and Bowles, 2004). One proposed explanation is that the co-expressed genes are regulated by a common transcription factor. Grouping these genes creates an increase in the abundance of binding sites specific to that transcription factor (Vogel et al., 2005). A related hypothesis suggests the co-expressed genes are regulated by similar promoter sequences, so a co-expression "neighborhood" would increase the availability of these promoter sequences (Oliver et al., 2008). However, genomic studies have, as of yet, been unable to confirm either of the two hypotheses.

Cluster analysis, as first reported by Grant et al. (2000), was performed to determine if candidate genes identified by the microarray experiment were randomly distributed across the genome. Iterative simulations modeling our data showed our candidate genes were not distributed evenly throughout the genome. Using a sliding window of 1,000,000 bases, we identified more genes in smaller regions of the genome than expected by a random distribution of the differentially expressed with 3–8 candidate genes per 1,000,000 bases (Tables 5 and 6). The same patterning held true when the sliding window was reduced to 100,000 and 100 bases. The statistical significance, from comparing the experimental data to the simulated data, is found in the number of simulated standard deviations (SDs) the experimental data is from the simulated data (Simulation SD column in Tables 5 and 6). When comparing clusters of three or more genes in either Clark or IsoClark, there are only four instances (three and six genes per cluster, Table 7; two and three genes per cluster) where the difference between the experimental data and the simulation study is not statistically different.

In the Clark genotype, with a window of 1,000,000 bases, there were 36 clusters of four genes and 13 clusters of five genes per window identified in the experimental data. There were only 17 clusters of four genes and only four occurrence of five genes clustering together in the simulation study. The difference in SDs is 5.05 and 4.59 respectively, indicating a highly statistically significant difference. The million base window allowed larger gene clusters to be identified in the experimental data (two clusters of seven genes and a single cluster containing eight genes). No clusters of these sizes were identified in the simulation study, further supporting the clustering hypothesis. When the window size is decreased to 100,000 or 100 bases, three genes in a cluster become significantly over represented in the experimental data compared to the simulation study. The microarray experiment identified 22 clusters of three genes per 100,000 bases. No clusters of three or more genes were identified in the simulation study at either window size.

The IsoClark genotype identified fewer candidate genes in the microarray experiment, which reduces the number of gene clusters identified. However, even with a reduced number of candidates, IsoClark still exhibited clustering of co-expressed genes. With a window of 1,000,000 bases, there were eight clusters of three genes identified in the experimental data, but only four clusters are identified in the simulation study. There were seven clusters of four genes per million bases identified in the IsoClark simulation, but none in the simulated data. The retention of clusters, even among so few candidate genes, lends further support that the soybean genome has conserved genomic regions with co-expressed genes.

Individual gene clusters are interesting because so many of them contain multiple copies of similar genes. For example, all six genes in the cluster on chromosome nine encode proline rich proteins while three of the five genes in a cluster on chromosome six encode caffeic acid O-methyltransferase. The co-expression of these genes coupled with their close physical proximity lends further credence to the hypothesis that they are regulated by a common transcription factor.

Single Feature Polymorphisms (SFPs)

Identifying candidate genes for a trait of interest is the most widely used method of analyzing the data provided by microarray experiments. However, mining the hybridization data to identify SFPs provides a high throughput platform for detecting polymorphisms (Kumar et al., 2007). Single Nucleotide Polymorphisms (SNPs) are the most commonly recognized polymorphism, but identification is labor intensive and SNP coverage across the genome is fairly sparse (Borevitz et al., 2007). It has been suggested that there is a greater probability of identifying a causal polymorphism for the trait of study using SFPs than traditional SNPs (Borevitz et al., 2007), perhaps due to better genic coverage.

To date, only three molecular markers (Satt 481, Satt114, and Satt239) segregate with the iron efficiency trait in soybean across multiple populations (Charlson et al., 2005; Wang et al., 2008). The 72 SFPs identified in the Clark genotype and the 98 SFPs identified in the IsoClark genotype relative to Williams 82 in this study have the potential to be developed into molecular markers specific to IDC. Initially, we hypothesized the SFPs would correlate with the differentially expressed genes. However, only one SFP (GmaAffx.41460.1.S1_at) was found in a gene differentially expressed in IsoClark leaf tissue. In *Arabidopsis* this gene is essential for NADH mediated reduction of the plastiquinone pool in respiratory electron transport and is up-regulated under mild heat stress (Wang and Portis, 2007). It is logical that this gene might be differentially expressed in the iron inefficient plant as photosynthesis slows due to a lack of iron serving as electron transporters. The remaining 169 SFPs were not differentially expressed due to iron stress. The majority of the sequences identified containing an SFP have an unknown function and the largest class of annotated SFPs is transcription factors. These SFP polymorphisms may alter transcription and/or translation rates of key genes and proteins, or serving some other regulatory function in soybean iron homeostasis.

Promoter Motifs

Analysis of the 500 bp upstream of the start codon for the predicted genes in soybean http://www.phytozome.net that coordinate with the differentially expressed candidate genes identified 11 conserved motifs (Table 7). These small motifs were notable for both their highly conserved sequences and conserved positions. A comparison of these motif sequences to the TRANSFAC database identified three of the 11 motifs as TF-BSs. It is likely that the remaining eight motifs represent previously uncharacterized TFBSs. One of the three motifs that showed high similarity to a TRANSFAC TFBS was a Myb TFBS. The Myb transcription factors have been implicated in inducing the stress response in plants in response to various abiotic stresses including phosphate stress (Hernandez et al., 2007) and asian soybean rust (Mortel et al., 2007). The identification of the Myb TFBS in the promoter region of candidate genes from the Clark genotype supports the idea that Clark is able to induce both an iron specific stress response and a more generalized stress response.

The identification of a basic helix loop helix (bHLH) TFBS motif in the promoter region of candidate genes from the Clark genotype may be indicative of the iron specific stress response induced in Clark under iron deficient conditions. In *Arabidopsis*, bHLH proteins have been identified as essential components in mediating iron uptake under iron stress conditions. Specifically, *AtbHLH38* and *AtbHLH39* both form heterodimers with AtbHLH29 to regulate iron uptake gene expression under iron deficient conditions in *Arabidopsis* (Youxi et al., 2008). *AtbHLH29* encodes a transcription factor known as FIT (FER like iron deficiency induced transcription factor (Wang et al., 2007)), which dimerizes with either *AtbHLH38* or 39 to induce *FRO2* and *IRT1* gene expression (Youxi et al., 2008). Though the soybean FIT homolog was not identified as differentially expressed due to iron deficiency, it was one of the 780 transcription factors predicted to be encoded within previously identified QTLs. The importance of bHLH transcription factors in regulating iron uptake gene expression makes the identification of a bHLH TFBS in the promoter region of iron deficiency induced genes particularly exciting. This is the first evidence that iron uptake gene expression may be similarly regulated in *Arabidopsis* and soybean.

Iron QTL and the Soybean Genome

The QTL mapping and marker assisted selection have been utilized by plant breeders for decades in the pursuit of crop improvement. This approach has been especially important for quantitative traits such as IDC (Charlson et al., 2005; Diers et al., 1992; Lin et al., 1997; Wang et al., 2008). Only in recent years have scientists been able to utilize microarray technology to examine gene expression on a global scale to identify candidate genes for their trait of interest. The development of the Affymetrix® GeneChip® Soybean genome array (Affymetrix), representing approximately 75% of the predicted genes in soybean (data not shown), means repeatable precision, providing more confidence to the microarray experiments than cDNA arrays.

The availability of the whole-genome soybean sequence has provided the ability to visualize the placement of candidate gene sequences within the genome. This view will allow further insight into soybeans' response to iron deficiency stress. Nineteen QTL regions have been previously identified for IDC, both in field and hydroponic studies (Diers et al., 1992; Lin et al., 1997, 1998). These regions represent approximately 182 cM of genetic information. Our initial hypothesis was that the majority of the genes identified in the microarray experiment would map within known iron QTL regions. However, only 58 of the 835 (7%) candidates in the Clark genotype and 21 of the 200 (10%) in the IsoClark genotype mapped within known QTL regions (Figure 1). Thus, the majority of the candidate genes identified in this study lie outside of regions identified as iron QTL. However, given the evidence of coordinate gene expression, gene clustering and conserved promoter motifs in our data, we have revised our previous hypothesis. We now propose the previously identified QTL regions likely correspond to transcription factors that regulate gene expression during iron stress. While microarray experiments would identify IDC regulated genes whose expression changes in response to a transcription factor, they may not identify the transcription factor itself. In contrast, QTL mapping would identify a mutation in a transcription factor, which is at the top of the signaling pathway. The mutation would affect either the expression of the transcription factor or its ability to bind to target promoters. This hypothesis is supported by our data. The clustering of co-expressed genes suggests they are being coordinately regulated. This is supported by the conserved motifs identified in the promoter regions of candidate genes. Most often, motifs are conserved throughout a previously identified cluster of genes in Clark. It is unlikely these motifs are missing or are altered in the promoter regions of the IsoClark genome. More likely, IsoClark may have a mutation in the transcription factor that controls the expression of these genes. Only by combining QTL analyses, microarray analyses of NILs, and the genome sequence could this conclusion be reached.

RESULTS

Candidate Gene Identification and GO Analysis

The RNA from the second trifoliate of both iron efficient Clark and iron inefficient IsoClark grown under iron limiting conditions (50 μM Fe(NO$_3$)$_3$, iron inefficient plants show severe chlorotic symptoms and iron sufficient conditions (100 μ Fe(NO$_3$)$_3$, no chlorotic symptoms in either genotype) were hybridized to the Affymetrix® GeneChip® Soybean Genome Array. The 835 transcripts were differentially expressed between Clark plants grown under iron sufficient and iron limiting conditions. By comparison 200 transcripts differentially expressed between IsoClark plants grown under the same conditions. Only 18 transcripts were common between the two lists (data not shown). Under iron deficient growth conditions, there were 179 genes differentially expressed between the two NILs. However, an analysis of the data revealed only 21 transcripts met or exceed the two-fold difference required to be considered differentially expressed between Clark and IsoClark genotypes grown under iron sufficient conditions (Table 1). This result confirms the NILs probably differ by only a limited number of genes.

Table 1. Differentially expressed genes between Clark and IsoClark genotypes grown under iron sufficient conditions.

Affymetrix Probe ID	Fold Change	UniProt ID	Plant GOSlim AtHomolog	PlantGOSlim
GmaAffx.93650.1.S1_s_at	-12.383	Q6WE90	No Homolog	
Gma. 12096.1A1_at	-9.9	No UniProt	No Homolog	
Gma.18.1.S1_at	-9.416	Q39819	AT4G10250	response to stress
Gma.17141.1.S1_at	-8.776	Q9ZSA7	AT4G10490	other metabolic
Gma.14554.1.S1_at	-7.959	Q1SJ63		
GmaAffx.62046.1.S1_at	-5.922	Q5CAZ5	AT1G34210	developmental
Gma.2185.3.S1_at	-5.724	Q9FJL3	AT3G25230	response to stress
Gma.10282.1.A1_at	-5.591	Q1T3Y4	AT4G27670	response to stress
GmaAffx.90956.1.S1_s_at	-5.297	No UniProt	AT5G53740	biological process
Gma.11793.1.S1_at	-3.469	No UniProt	No Homolog	
GmaAffx.89665.1.A1_s_at	-3.365	Q9AY32	No Homolog	
Gma. 12660.1.A1_at	-3.173	Q8H2B1	AT1G56300	protein metabolism
Gma.10282.2.S1_at	-2.899	Q1T3Y4	AT4G27670	response to stress
GmaAffx.93424.1.S1_x_at	-2.884	Q9S7H2	AT5G20620	protein metabolism
GmaAffx.56241.2.S1_at	-2.739	Q9SCW4	AT5G62020	transcription
Gma.8636.1.S1_at	-2459	O80982	AT2G26150	transcription
GmaAffx.72322.1.S1_at	-2.32	Q7F1F2	AT5G48570	protein metabolism
Gma.1727.1.S1_at	-2.297	Q1SVQ0	AT2G39730	response to abiotic or biotic stimulus
GmaAffx.5924.1.S1_at	-2.295	Q8L7T2	AT1G52560	response to stress
GmaAffx.74022.1.S1_at	-2.267	Q1RY14	AT4G28480	protein metabolism
GmaAffx.93424.1.S1_s_at	-2.261	Q9S7H2	AT5G20620	protein metabolism

The GO slim categories that were over-represented in our lists of differentially expressed genes were identified for both the Clark and IsoClark comparisons. Transcripts with GO slim classifications that are over-represented on our list of differentially expressed genes should be representative of the processes and pathways being affected in both the iron efficient and iron inefficient plants. The Clark genotype had 488 out of 835 unique transcripts with GO slim IDs. Of the corresponding GO slim IDs, 24 were over-represented in our list of differentially expressed genes (Table 2) in comparison with the entire chip. The over-represented GO slim categories could be further divided into 14 biological process IDs, nine molecular function IDs, and one cellular component processes (Table 2). Of the 200 differentially expressed genes in the IsoClark genotype, 49 had corresponding *Arabidopsis* GO slim IDs. Of these, 21 genes had GO annotations that were over-represented. These GO categories fell into two molecular function categories and three biological process categories (Table 3).

Table 2. GO slim terms over represented in candidate genes from comparison between clark plants grown in iron sufficient and iron deficient hydroponics solutions.

GO Slim ID	GO Term Description	Number of Genes with GO ID	Bonferroni Corrected P-Value
Go:0000004 BP	Unknown Function	128	0
GO:0006270 BP	DNA Replication Initiation	10	0
GO: 0009611 BP	Response to Wounding	36	0
GO:0009695 BP	Jasmonic Acid Biosynthesis	24	0
GO:0006826 BP	Iron Ion Transport	9	1.2E-07
GO:0006879 BP	Iron Ion Homeostasis	10	3.6E-07
GO:0010039 BP	Response to Iron Ion	9	5.28E-06
GO:0009617 BP	Response to Bacterium	8	0.000018
GO:0006275 BP	Regulation of DNA Replication	4	0.00303
GO:0006972 BP	Hyperosmotic Response	4	0.00303
GO:0030397 BP	Membrane Disassembly	8	0.004381
GO:0008299 BP	Isoprenoid Biosynthesis	10	0.006706
GO:0009408 BP	Response to Heat	20	0.014334
GO:0019373 BP	Epoxygenase P450	5	0.045066
GO:0008199 MF	Ferric Iron Binding	9	0
GO:0008094 MF	DNA-dependent ATPase Activity	10	1.29E-06
GO:0016165 MF	Lipoxygenase Activity	12	0.000128
GO:0047763 MF	Cafeate O-Methyltransferase	8	0.001014
GO:0030337 MF	DNA Polymerase Processivity Factor	4	0.001426
GO:0005544 MF	Calcium Dependent Lipid Binding	8	0.001965
GO:0009978 MF	Allene Oxide Synthase	5	0.017949
GO:0046423 MF	Allene Oxide Cyclase	4	0.020157
GO:0008815 MF	Citrate (Pro-3S) Lyase	5	0.029946
GO:0009346 CC	Citrate Lyase	5	0.002807

Table 3. GO slim terms over or under represented in candidate genes from comparison between IsoClark plants grown in iron sufficient and iron deficient hydroponics solutions.

GO Slim	GO Term Description	Number of Genes with GO ID	Bonferroni Corrected P Value
GO:0000004 BP	Unknown Function	49	0
GO:0006809 BP	Nitric Oxide Biosynthesis	4	0.0006102
GO:0010025 BP	Wax Biosynthesis	5	0.00736524
GO:0019953 BP	Sexual Reproduction	5	0.01223184
GO:0008940 MF	Nitrate Reductase	4	0.00006192
GO:0008382 MF	Iron Superoxide Dismutase	3	0.01245882

Examining the GO terms associated with the candidate genes provides further insight into the disparity of the number of differentially expressed genes between genotypes. The IsoClark (inefficient) genotype does not appear to induce the full complement of genes induced in Clark in response to the iron deprivation stress. The most prevalent GO term in all three classifications for both genotypes was "unknown function" (Tables 2 and 3). However, the Clark (efficient) genotype also had a high proportion of GO terms (and thus, transcripts) specifically related to iron availability and usage, (ferric iron binding (GO:0008199), iron ion transport (GO:0006826), and iron ion homeostasis (GO:0006879)) that were over-represented on our lists of candidate genes responding to iron stress. In addition, Clark genes encoded a number of GO terms not specifically related to iron, but which are associated with a more general stress response (GO:0009611 response to wounding, GO:00099 jasmonic acid biosynthesis, and GO:0009408—response to heat).

Real Time PCR Confirmation

The differential expression observed through sqRT-PCR analysis mirrored, in direction, the expression differences observed in the microarray study (Table 4). The difference in expression levels seen in between the sqRT-PCR and the microarray experiment is most likely due to cross hybridization. Multiple members of the same gene family may hybridize to the same spot on the microarray, while the sqRT-PCR experiment is designed to amplify only single members of the gene family. The sqRT-PCR experiments confirmed the iron deficient plants had lower expression levels of the transcripts than the iron sufficient plants, replicating the results seen in the microarray data.

Table 4. Semi quantitative real time PCR results.

Affymetrix Probe Set	Annotation	Conditions of Differential Expression	Differential Expression in Microarray	Diff Express in sqRT-PCR
Gma.13296.3.S1_at	Lipid Transfer	CSSvCSD	190.34	2.62
Gma.17724.3.S1_at	GDSL motif	CSSvCSD	6.26	8.46
Gma.17825.1.A1_at	GDSL motif	CSSvCSD	4.69	10.26
GmaAffx.89896.1.S1_at	Heat Shock Protein	CSSvCSD	23.45	2.04
GmaAffx.93268.1.S1_at	Heat Shock Protein	CSSvCSD	19.91	1.36
GmaAffx.51733.1.A1_at	Ribonuclease T2	CSSvCSD	-13.35	-2.98
GmaAffx.88242.1.S1_at	SnRNP protein	CSSvCSD	3.58	11.18
Gma.16500.1.S1_at	Lipoxygenase	CSSvCSD	11.36	16.9
Gma.3705.1.S1_at	Nitrate Reductase	CSDvlCSD	-3.53	-1.101
Gma.9609.1.S1_at	Reductase	CSDvlCSD	2.26	2.23
GmaAffx.36066.1.S1_at	Replication Factor	CSDvlCSD	2.63	3.58

Positioning Candidate Genes on the 7X Genome Assembly

Sequencing of the soybean genome by the Department of Energy, Joint Genome Institute currently has produced 7X sequence coverage of the genome http://www.phytozome.net website, which has been assembled by USDA-ARS researchers into 20 draft pseudo chromosomes based on marker homology, allowing us to place our candidate genes on specific chromosomes (Figure 1).

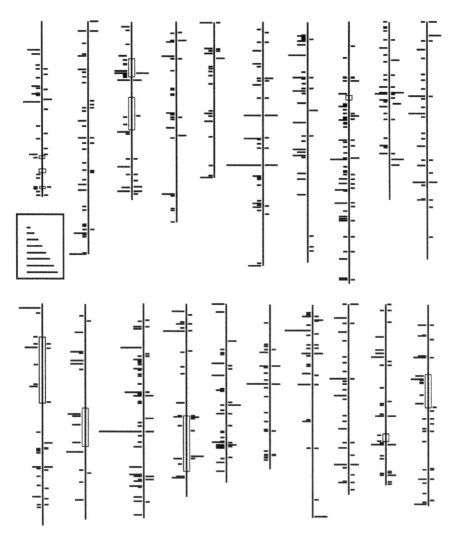

Figure 1. Differentially expressed genes on soybean chromosomes. Genes identified as differentially expressed in the microarray experiment have been aligned on the 7X build of the soybean genome, assembled into chromosomes. Each horizontal line represents one gene; longer lines represent multiple genes. Lines to the left of the chromosome are genes in Clark; lines to the right of the chromosome are genes in IsoClark. Open boxes on the chromosomes represent previously identified iron QTL regions. Clusters of differentially expressed genes are apparent throughout the genome.

The sequences of transcripts identified as differentially expressed by microarray analysis (see above) were obtained from the Affymetrix® website http://www.affyme-trix.com. These sequences were then queried against the 7X soybean genome using BLASTN (Altschul et al., 1997) and an e-value cutoff of 10E-50 to ensure a high sequence similarity between the aligned sequences. The same parameters were used to compare the sequences of SFPs to the 7X genomic sequence assembly. Markers used in previous iron QTL studies were also identified on the pseudo chromosomes to delineate known iron QTL regions (Figure 1). The iron efficiency QTL were scaled to the 7X build and used to determine if any of the candidate genes from the microarray experiment were encoded within the iron QTL regions. Fifty-eight genes in the Clark genotype and 21 genes in the IsoClark genotype were located within previously identified iron QTL (Figure 1).

Sequences of the delineated iron QTL regions were analyzed using FGENESH http://www.softberry.com website using *Arabidopsis* as the training model to identify gene structure predictions. The identified gene sequences were mined from the genome sequence and compared to known transcription factors in *Arabidopsis* http://datf.cbi.pku.edu.cn/download.php (Guo et al., 2008) rice, soybean, barley, and medicago http://planttfdb.cbi.pku.edu.cn/index.php (Guo et al., 2008). This comparison identified 780 predicted genes within the previously identified QTL regions that show high sequence similarity (10E-50) to known transcription factors. One of these, within a QTL region on chromosome 12, showed a 100% identity to the *Arabidopsis FIT* gene (*AT2G28160*).

Cluster Analysis

The gene distribution simulation randomly placed genes across the assembled genome. A second analysis assumed 36% of the genome was heterochromatic as proposed by Singh and Hymowitz (Singh and Hymowitz, 1988) and reflected in the distribution of predicted gene locations http://www.phytozome.net. This analysis further constrained the algorithm, reducing the probability of candidate genes exhibiting clustering within the genome. If the candidate genes were randomly located throughout the genome, we would expect the experimental results to closely mirror the simulated data study. However our results, for both clustering analyses, strongly indicate that the differentially expressed genes exhibit clustering of two or more genes within 1,000,000 bp, 100,000 bp, and 10,000 bp in the genome (Tables 5 and 6).

Table 5. Clusters of candidate genes on 7X build of soybean genome from Clark genotype assuming 36% of the genome is heterochromatic.

	1,000,000 base bins			
# of Genes	Stimulation Mean	Stimulation SD	Experimental Data	Sds from Sim Mean
0	523	7.94	569	5.79
1	221	11.92	188	-2.77
2	137	9.34	99	-4.07
3	56	6.37	44	-1.88
4	17	3.76	36	5.05
5	4	1.96	13	4.59
6	1	0.90	2	1.11
7	0	0.41	2	4.88
8	0	0.14	1	7.14

Table 5. *(Continued)*

	100,000 base bins			
# of Genes	Stimulation Mean	Stimulation SD	Experimental Data	Sds from Sim Mean
0	8885	6.47	8926	6.34
1	671	12.50	486	-14.8
2	42	6.05	98	9.26
3	2	1.29	22	15.5
4	0	0.22	2	9.09
5	0	0.03	1	33.33
	100 base bins			
# of Genes	Stimulation Mean	Stimulation SD	Experimental Data	Sds from Sim Mean
0	95243	2.16	94678	-2.62
1	751	4.31	593	936.66
2	5	2.16	71	30.55
3	0	0.10	7	70
4	0		0	
5	0		1	

Table 6. Clusters of candidate genes on 7X build of soybean genome from IsoClark genotype assuming 36% of the genome is heterochromatic.

	1,000,000 base bins			
# of Genes	Stimulation Mean	Stimulation SD	Experimental Data	Sds from Sim Mean
0	757	4.98	783	5.22
1	165	8.81	118	-5.33
2	33	4.40	38	1.14
3	4	2.02	8	1.98
4	0	0.66	7	10.60
	100,000 base bins			
# of Genes	Stimulation Mean	Stimulation SD	Experimental Data	Sds from Sim Mean
0	9359	2.13	9325	-15.96
1	236	4.21	178	-13.78
2	5	2.07	28	11.11
3	0	0.25	4	16
	100 base bins			
# of Genes	Stimulation Mean	Stimulation SD	Experimental Data	Sds from Sim Mean
0	95754	0.72	95128	-869.44
1	245	1.44	199	-31.94
2	1	0.72	22	29.17
3	0	0.03	1	33.33

The candidate genes do not show a high concordance with known iron QTL regions, but do serve to identify additional genomic regions of IDC transcriptional importance (Figure 1). The largest cluster contained eight candidate genes located within 1 MB on chromosome 6 (Figure 1). There was also a cluster of seven genes on chromosome 2 (Figure 1). Chromosome 2 contained six clusters of four or more genes within 1,000,000 bases, as did chromosome 13 (Figure 1). None of these clusters were located within known iron QTL. However, another cluster of seven genes falls on chromosome 12 (Figure 1), which has three known iron QTL, together spanning 24 cM of the linkage group. Chromosome 7 contained the most gene clusters, eight separate clusters, each with four candidate genes (Figure 1). Again, chromosome 7 has not been previously shown to contain regions of genetic importance to soybean IDC. A number of the gene clusters contain multiple copies of genes encoding proteins with similar functions. A cluster of six candidate genes on chromosome 9 all encode proline rich proteins while three of the genes in a cluster of five on chromosome 6 encode caffeic acid O-methyltransferases. The close physical proximity of co-expressed genes of the same function provides further support for the clustering of co-expressed and co-functional gene clusters.

The identification of these gene clusters on chromosomes not previously known to be involved in soybean IDC opens new regions of genetic interest to investigate in future studies. The majority of the candidate genes identified were not within the iron QTL, nor were the largest clusters of differentially expressed genes. Genes affecting chlorosis and yield loss may not be confined to the previously identified QTL regions. However, the iron QTL must contain sequence of importance to IDC. It is likely a transcription factor(s) controlling the expression of genes required to induce an iron stress response is encoded within the QTL regions. We have identified 780 predicted transcription factors, including the soybean homolog to the *Arabidopsis* FIT gene, within the previously identified QTL. This, or another of these transcription factors may be responsible for inducing a cascade of gene expression changes due to limited iron conditions. The transcription factor/factors may also affect the expression of the canonical iron genes, such as IRT and FRO, none of which are encoded within the previously identified QTL, nor identified as differentially expressed in our microarray experiment.

Motif Analysis

Previous research has demonstrated that genes clustered in close proximity in a genome may be coordinately regulated. To determine if clusters of IDC genes were coordinately regulated, we examined 500 bases from the 5' untranslated regions (putative promoters) of all differentially expressed IDC genes and used these as input into the MEME software program. As an internal control, sequences were not analyzed as members of clusters; rather, all sequences were analyzed as a single large group. If IDC genes were coordinately regulated, MEME could also be used to independently identify potential gene clusters. In total, the putative promoters of 835 iron deficiency induced genes from the Clark genotypic comparison and 200 genes from the IsoClark comparison were analyzed using MEME. There were no motifs found using MEME in the IsoClark (inefficient) promoter regions. All motifs identified by MEME were found in the promoter regions of genes differentially expressed due to iron deficiency in the Clark (efficient) genotype. Twenty-one motifs with E-values more significant than $10E^{-6}$ were identified by MEME analysis. Following visual inspection, this number was reduced to 11. Motifs were eliminated if they contained repetitive sequence or had lower significance E-values. The 11 motifs were identified in 248 IDC genes, representing 129 of the clusters of two or more genes as mentioned above. One mechanism by which genes can be coordinately regulated is through the action of transcription factors that bind to the promoter to regulate gene expression. Therefore, the 11 motifs identified above were compared to known TFBSs in the TRANSFAC (Matys et al., 2003) database (Table 7). Three motifs showed significant sequence similarity (99% identity) to known TFBSs (Table 7). These 3 sites bind a helix-loop-helix transcription factor (bHLH), an elongation factor (EF2), and a Myb transcription factor. These binding sites were identified in the promoter regions of 42, 40, and 28 iron responsive genes respectively. Both helix loop helix and Myb transcription factors are known to be involved in regulating the iron stress response and general stress responses in other plant species.

Table 7. Motifs in promoter regions of differentially expressed genes identified by MEME and their similarity to transcription factor binding sites (TFBSs) in the TRANSFAC Database.

Identified Motif Sequence	# Seqs with Motif	Motif E-Value	TRANS FAC Hit ID	TRANS AC Binding Site Sequence	TRANS FAC Annotation
CATCCAACGGC	29	1.2E-1	M00227	TCCAACGGC	Myb
CCCGCCACGCGCCAC	48	5.1E-26	M00187	GCCACGTGCC	Helix Loop Helix
TGGCGGGA	50	5.8E-13	M00024	TGGCGGA	Elongation Factor 2
CCAAACCC	50	2.7E-5	No Hit		
CCACCACCACC	48	3.8E-16	No Hit		
ACACAACACAC	45	2.2E-10	No Hit		
AAAATAAAAATAAAA	9	2.27E-7	No Hit		
AATAAAAAAATAAAA	8	1.51E-7	No Hit		
AGCTAGCTAGC	6	1.47E-7	No Hit		
AGCGAGCGAGC	4	6.23E-8	No Hit		
AGCAAGCTAGC	3	2.47E-7	No Hit		

SFP Analysis

The SFP analysis identified 170 SFPs, 72 SFPs were unique to the Clark genotype, and 98 unique to the IsoClark genotype. A number of the Affymetrix® sequences found to contain an SFP perfectly matched more than one genomic location, giving a potential of 208 predicted genes with an SFP. Only one of the 170 SFPs identified in this study (GmaAffx.41460.1.S1_at) was encoded within a gene identified as differentially expressed between IsoClark plants grown under iron sufficient and iron deficient conditions. The remaining SFPs were not in differentially expressed genes in either Clark or IsoClark genotypes. This suggests most of the SFPs are encoded in regulatory elements, which would not necessarily be differentially expressed. GO slim ID analysis, as previously described, was performed with the gene sequences containing SFPs. Of the genes containing SFPs, 20% had an unknown biological process annotation. The most prevalent group with known annotations was related to transcriptional regulation. Genes involved with electron transport, ATP binding, ligases, and transferases were also identified as over-represented by their GO IDs (data not shown).

CONCLUSION

The use of near isogenic soybean lines, microarray analysis, SFP identification, and the sequence of the soybean genome has allowed us to identify individual genes lying within known iron efficiency QTL whose expression levels are affected by iron availability. We have also identified 11 conserved motifs in the promoter sequence of genes differentially expressed due to iron deficiency stress. The 58 differentially expressed genes identified in Clark and 21 in IsoClark, located within known QTL regions, are the first genes identified by microarray analysis within QTL regions specific to iron deficiency stress. The conserved motifs throughout the promoter regions of the differentially expressed genes in the Clark genotype provide compelling evidence that the differential iron response is likely due to the differential expression or binding of a transcription factor. Co-expressed genes clustered either by physical proximity (Tables 5 and 6) or through shared promoter motifs (Table 7) provide new regions of genetic interest in the study of IDC in soybean. Additionally, both types of clustering suggest the control of soybeans' iron deficiency response is regulated by the differential expression of a transcription factor or a mutation within the transcription factor, which affects its ability to bind to target promoter regions. This implies the eight transcription factors differentially expressed in Clark under iron deficiency stress which are located within known iron QTL regions are likely candidate genes for the QTL. An analysis of the 780 transcription factors predicted within the IDC QTL regions, specifically the FIT homolog, additional 52 bHLH transcription factors, and the other 50 genes in Clark and 21 genes in IsoClark that map within the QTL regions may further elucidate the response induced in soybean due to iron deficiency stress. Additionally, the conserved motifs identified by MEME in the promoter regions of iron deficiency induced genes can be used to mine the soybean genome for additional genes potentially affected by IDC, but which are not represented on the soybean Affymetrix® GeneChip®.

KEYWORDS

- **Basic helix loop helix**
- **Iron deficiency chlorosis**
- **Quantitative Trait Locus**
- **Robust multichip analysis**
- **Single feature polymorphisms**
- **Single Nucleotide Polymorphisms**

AUTHORS' CONTRIBUTIONS

Jamie A O'Rourke carried out the sample collection and preparation, RT PCR validation, data analysis, and drafted the manuscript. Rex T. Nelson wrote custom Perl scripts and performed various bioinformatic analyses. Steven Cannon, Jeremy Schmutz, and Jane Grimwood performed the genome sequence alignments and assembly. David Grant performed the genome cluster analysis. Carroll P Vance provided comments, suggestions, and revisions to the manuscript. Michelle A Graham performed various bioinformatic analyses, did the SFP identification, GO category analysis, wrote custom Perl scripts for various analyses, and provided editorial assistance. Randy C. Shoemaker conceived the study, coordinated the design of the project, and drafted the manuscript. This research was funded in part by the North Central Soybean Research Program.

Chapter 10

Global Characterization of *Artemisia annua* Glandular Trichome Transcriptome

Wei Wang, Yejun Wang, Qing Zhang, Yan Qi, and Dianjing Guo

INTRODUCTION

Glandular trichomes (GTs) produce a wide variety of commercially important secondary metabolites in many plant species. The most prominent anti-malarial drug artemisinin, a sesquiterpene lactone, is produced in GTs of *Artemisia annua*. However, only limited genomic information is currently available in this non-model plant species.

We present a global characterization of *A. annua* GT transcriptome using 454 pyrosequencing. Sequencing runs using two normalized cDNA collections from GTs yielded 406,044 expressed sequence tags (ESTs) (average length = 210 nucleotides), which assembled into 42,678 contigs and 147,699 singletons. Performing a second sequencing run only increased the number of genes identified by ~30%, indicating that massively parallel pyrosequencing provides deep coverage of the *A. annua* trichome transcriptome. By BLAST search against the NCBI non-redundant protein database, putative functions were assigned to over 28,573 unigenes, including previously undescribed enzymes likely involved in sesquiterpene biosynthesis. Comparison with ESTs derived from trichome collections of other plant species revealed expressed genes in common functional categories across different plant species. The RT-PCR analysis confirmed the expression of selected unigenes and novel transcripts in *A. annua* GTs.

The presence of contigs corresponding to enzymes for terpenoids and flavonoids biosynthesis suggests important metabolic activity in *A. annua* GTs. Our comprehensive survey of genes expressed in GT will facilitate new gene discovery and shed light on the regulatory mechanism of artemisinin metabolism and trichome function in *A. annua*.

Secreting GTs are major site for biosynthesis and accumulation of a wide range of plant natural products. These plant natural products often function to protect the plants against insect predation (Ranger and Hower, 2001; Wagner et al., 2004), and contribute to the flavor and aroma of plants. Many of the natural products also have pharmacological effects, such as the analgesic drug morphine, the anticancer compound taxol, and the anti-malarial drug artemisinin. Artemisinin, a sesquiterpene lactone, is currently recognized as one of the most prominent anti-malarial treatment (Duke and Paul, 1993). A complete understanding of the artemisinin biosynthetic pathway and its regulatory mechanism holds the key to efficient metabolic engineering for increased artemisinin yield. In the past decades, research efforts have been dedicated to identification of enzymes and intermediate compounds leading to artemisinin production. Many genes encoding enzymes participate in the pathway have been cloned and functionally characterized (Bertea et al., 2005; Bouwmeester et al., 1999; Chang et al., 2000; Mercke et al., 2000; Ro et al., 2006; Teoh et al., 2006; Zhang et al., 2008). However, little is known about the regulatory aspects of sesquiterpene metabolism. This is partly due to the fact that *A. annua* is a non-model plant with limited genomic information available, and sequencing of limited number of randomly selected cDNA clones often have insufficient coverage of less abundant transcripts, including important regulatory transcription factors (TFs). In addition, genes uniquely or preferentially expressed in trichomes may be under-represented in non-tissue-targeted EST sequencing projects. A comprehensive survey of genes expressed in GT will facilitate new gene discovery and contribute significantly to elucidating the terpenoid pathway regulation and trichome function in *A. annua*.

Whole genome or transcriptome sequencing enables functional genomic studies based on global gene expression. The newly developed high throughput pyrosequencing technology allows rapid production of sequence data with dramatically reduced time, labor, and cost (Huse et al., 2007; Margulies et al., 2005; Moore et al., 2006; Weber et al., 2007; Wicker et al., 2006). So far, most applications of pyrosequencing have involved analysis of genomic DNA (Poinar et al., 2006). Published reports on 454 pyrosequencing of transcriptomes have been mostly restricted to model species with genomic or comprehensive Sanger EST data available (Bainbridge et al., 2006; Cheung et al., 2006; Emrich et al., 2007; Weber et al., 2007). Previous studies (Cheung et al., 2006; Weber et al., 2007) using genome or Sanger EST sequences for mapping and annotation of 454 ESTs were not able to accomplish *de novo* assembly of their 454 ESTs. We here present the global transcriptome characterization of *A. annua* GT, the so called biofactory for the production of artemisinin and other plant secondary metabolites. We assigned putative function to 28,573 unigenes, including previously undescribed enzymes likely involved in sesquiterpene biosynthesis. We verified the expression of 32 selected unigenes and novel transcripts in GTs using semi-quantitative RT-PCR. These 454 ESTs were linked to metabolic process specific in GTs and form the basis for further investigation.

MATERIALS AND METHODS

Plant Materials

Artemisia annua seeds were purchased from Youyang, Sichuang province of China. Seeds were sown into commercial potting mixture for germination. The germinated plantlets were grown under natural light conditions in the greenhouse located at The Chinese University of Hong Kong. Flower buds were collected for trichome isolation before flowering.

Isolation of Glandular Trichomes

Trichome cells were gently abraded from the surface of flower buds using glass beads and a commercial cell disrupter (BioSpec Products). The isolated secretary cells were separated from other cells and tissue fragments in the mixture by sequentially passing through a 40 µm and a 30 µm nylon sieves. Glandular cells were finally collected in 30 µm meshes with minimum contamination of non-glandular trichomes (NGTs) (Figure 1).

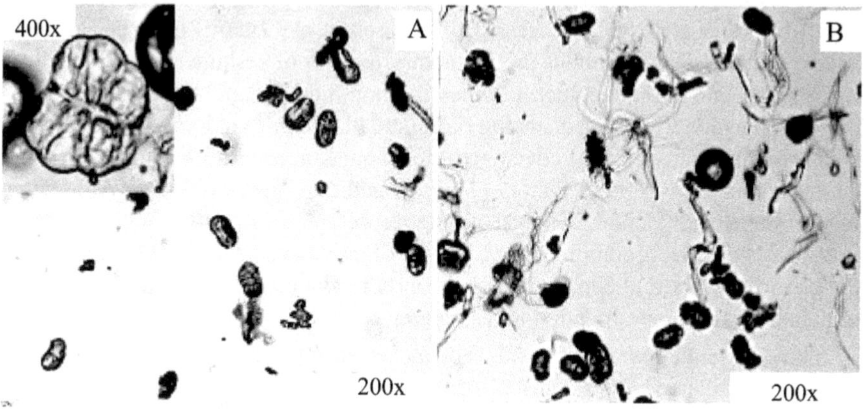

Figure 1. Isolation of glandular trichomes. (A) Extracted glandular trichomes (B) Crude extracts containing both glandular and non-glandular trichomes.

RNA Extraction, cDNA Synthesis, and Normalization

Total RNA was extracted from GTs isolated from 30 g flower buds following the standard protocol of RNeasy Plant Mini Kit (Qiagen). The cDNA was synthesized using the BD SMARTM PCR cDNA Synthesis Kit (Clontech). First-strand cDNA synthesis was performed with oligo(dT) primer as described in the provided protocol using 500 ng total RNA. Double-strand cDNA was prepared from 2 µl of the first-strand reaction by PCR with provided primers in a 100 µL reaction. cDNA was purified using Qiagen QIAquick PCR purification spin columns. Normalization was performed using TRIM-MER cDNA normalization kit (EVR_GEN) to decrease the prevalence of abundant transcripts before sequencing. Approximately 1 µg of normalized double-stranded cDNA was used for 454 pyrosequencing.

454 Pyrosequencing, Data Pre-Process, and Assembly

Approximately 1 µg of the adaptor-ligated cDNA population was sheared by nebulization and DNA sequencing was performed following protocols for the Genome Sequencer GS FLX System (Roche Diagnostic). Reads generated by the FLX sequencer were trimmed of low quality, low complexity (poly(A)), and adaptor sequences using the SeqClean software http://compbio.dfci.harvard.edu/tgi/. The cleaned sequences were subject to CAP3 program (Huang and Madan, 1999) for clustering and assembly using default parameters.

Gene Annotation Using GO Terms

After assembly, the resulting contigs and singlets were aligned with NCBI non-redundant protein database using BLAST2go software with a cut-off e-value of 1e-10. The GI accessions of best hits were retrieved, and the Gene Ontology (GO) accessions were mapped to GO terms according to molecular function, biological process, and cellular component ontologies http://www.geneontology.org/.

Semi-Quantitative RT-PCR Analysis

To verify the presence of pyrosequencing ESTs in GTs, we totally selected 35 unigenes and novel transcripts for RT-PCR analysis. Total RNA were extracted from GTs, non-glandular hairy trichomes, leaves, and hairy roots respectively. The first-strand cDNA was synthesized from 10 μL (about 1 μg) total RNA using SuperScript™ II Reverse Transcriptase (Invitrogen) with Oligo(dT)12–18 Primer. PCR was performed using 0.5 to 2 μL of the cDNA in a total of 50 μL reaction volume. The PCR conditions were 2 min at 95°C, 30 sec at 95°C, 30 sec at 4756°C, 1 min at 72°C for 30 cycles, followed by 5 min at 72°C. These conditions were chosen because none of the samples analyzed reached a plateau at the end of the amplification (i.e., they were at the exponential phase of the amplification). Actin was used as a loading control, and loading was estimated by staining the gel with ethidium bromide. Expression analysis of each gene was confirmed in at least two independent RT-reactions using forward and reverse primers.

DISCUSSION

As the sole plant source for artemisinin production, the *A. annua* has been studied extensively for the past decades. Like most other non-model plant species, it has lacked genetic and genomic resources necessary for mechanistic study. Although a precise estimate of transcriptome coverage is unattainable without full genomic sequence, we appear to have recovered a significant portion of the *A. annua* GT transcriptome. Novel transcripts detected highlights the hypothesis-expanding aspects of 454 deep pyrosequencing approach, which potentially facilitate the understanding of GT metabolic function. The assembled sequence data also provided a rich source of information for further investigation.

Two consecutive pyrosequencing runs identified a large number of genes expressed in GTs. In data analysis, approximately 85% of the pyrosequencing assemblies did not align to any ESTs available in GenBank. This high proportion could reflect the specialized cell type that was sampled or perhaps the greater complexity of the *A. annua* genome. Because our priority goal in this study is gene discovery, we therefore chose normalized cDNA population to reduce oversampling of abundant transcripts and to maximize coverage of less abundant transcripts present in the sample. The average contig length was fairly short (~334 bp), and only 62% of the sequence reads assembled into contigs, leaving 147,699 singletons.

Genes involved in plant secondary metabolism have frequently been identified by EST approach (Bao et al., 2002). The lower cost and greater sequence coverage offered by pyrosequencing makes it possible to identify more candidate genes involved in plant natural product biosynthetic pathways, especially those with low abundance and often missed by conventional EST projects. For non-model species with little or no genomic data available, such as *A. annua*, pyrosequencing offers rapid characterization of a large portion of the transcriptome and therefore provides a comprehensive tool for gene discovery. However, one limitation of pyrosequencing is that one must rely on RACE PCR in order to obtain full-length sequence data for a given gene of interest.

Comparison between our GT 454 ESTs with conventional ESTs generated from trichomes of other plant species revealed likely common function in NGTs and GTs. In addition, some unigenes corresponding to enzymes in sesquiterpene biosynthesis were found to be highly expressed in both trichome types in our RT-PCR analysis. Although it has been suggested that GTs are the site for synthesis and accumulation of plant secondary metabolites, it will be interesting to further investigate the different functional roles of NGTs in artemisinin biosynthesis.

RESULTS

Sequencing and Assembly of 454 Pyrosequencing ESTs

Totally 406,044 ESTs (minimal size >50 bp) averaging 210 bp were generated from two consecutive pyrosequencing runs. Cleaning (removal of primer, polyA tail, etc.) of the raw sequences resulted in a total of 386,881 high quality reads with an average length of 205 nucleotides totalling 85 Mb. After clustering and assembly using TGICL CAP3 clustering tools (Huang and Madan, 1999; Pertea et al., 2003), these reads were assembled into 42,678 contigs and 147,699 singletons. The average length for contigs and singletons are 334 bp and 191 bp respectively. The contigs and singletons are collectively referred to as unigenes. The length distribution of unigenes and their component reads are summarized in Tables 1 and 2.

Table 1. Length distribution of assembled contigs and singletons.

Nucleotides Length (bp)	Contigs	Singletons
50-99	276	22,730
100-199	2,534	41,936
200-299	19,220	82,169
300-399	11,568	863
400-499	4,940	1
500-599	1,991	0
600-699	980	0
700-799	529	0
800-899	296	0
900-999	142	0
1,000-1,499	173	0
1,500-1,999	22	0
> 2,000	7	0
Total	42,678	147,699
Maximum length	2,366 bp	411 bp
Average length	334 bp	191 bp

Table 2. Summary of component reads per assembly.

Number of Reads	Number of Contigs
2 to 10	39,112
11 to 20	2,142
21–30	585
31–40	289
41–50	164
51–100	250
101–150	63
151–200	27
> 200	46

Pyrosequencing Provides Deep Coverage of the *A. Annua* Trichome Transcriptome

The contigs were searched against the NCBI non-redundant (NR) protein database using the BLASTX algorithm. Among the 190,377 contigs and singletons, 29,577 (15.5%) had at least one significant alignment to existing gene model in BLASTX searches (E-value cutoff, e-^{10}). A majority (84.5%) of the pyrosequencing assemblies did not match any known sequences in the existing database and thus likely represent novel (E-value cutoff, pts sion of 17 transcripts identified in this study). Performing a second sequencing run increased the number of genes identified by approximately 30% (Table 3), suggesting that two pyrosequencing runs detect a substantial fraction of genes expressed in GTs and provide deep coverage of the *A. annua* trichome transcriptome.

Table 3. Summary of BLAST hits from two pyrosequencing runs.

Pyrosequencing Run	NCBI Database Unique Hits
1st run	266,976
2nd run	289,467
Total	357,843

Characterization and GO Annotation of Novel Transcripts

The 357,843 sequences that had matches with protein sequences in the NCBI protein database http://www.ncbi.nlm.nih.gov/sites/entrez?db=protein could be condensed into 29,577 clusters based on their top protein hits. Each 454 contig was assigned a putative gene description and a GO classification based on the "best hit" BLASTX search (bitscore > 45, e-value < 110), using the "inferred from sequence similarity" (ISS) level of evidence (Ashburner et al., 2000). The unigenes were classified into three major functional categories: biological process, molecular function, and cellular component, according to the standard GO terms (GO; http://www.geneontology.org). The assigned functionality of genes covers a broad range of GO categories. The top 20 most highly represented GO categories are illustrated in Figure 2. Under the category of biological process, transport, transcriptional control, and metabolic process were among the most highly represented categories, indicating the important metabolic activities in *A. annua* GTs. Other categories include photosynthesis, secondary metabolism (lignin, flavonoid, and isoprenoid biosynthesis process) and primary metabolism (fatty acid, glycolysis, carbohydrate process, etc.).

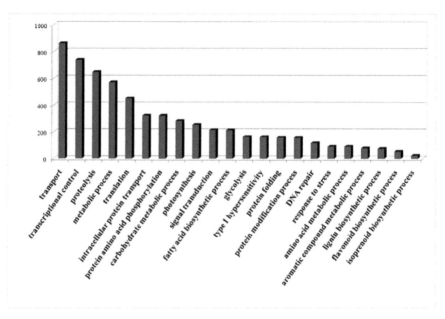

Figure 2. Top-ranked GO categories (molecular function) of assembled pyrosequencing ESTs.

Comparison of 454 Sequence Contigs to Trichome ESTs from Other Plant Species

TrichOME http://trichome.noble.org/trichomedb/ website is a publicly available database of genes and metabolites expressed in plant trichomes. It currently contains 37,017 conventional ESTs derived from eight plant species, including *Medicago sativa, Humulus lupulus, Mentha × piperita, Nicotiana benthamiana, Ocimum basilicum, Solanum habrochaites, Solanum lycopersicum,* and *Solanum pennellii*. A BLASTX search against TrichomeDB showed that only 17,372 (9%) of our 454 contigs had best blast hits (e-value < 1e-10) to 8,095 EST clusters with unique descriptions. Thus 454 sequencing has revealed many transcripts not previously detected in *A. annua*. ESTs homologous to photosynthesis-related proteins (chlorophyll a/b binding protein, ribulose bisphosphate carboxylase small subunit) are among the top 10 most highly expressed transcripts.

The top ranked common molecular function of ESTs identified from all nine plant species are listed in Table 4. Regulation of metabolic process, metabolic process, oxidation reduction, and transport categories has the highest number of contigs. Trichomes are known to be active in photosynthesis, as well as for their roles in storage and secretion of toxic compounds for example, heavy metals (Choi et al., 2001; Kupper et al., 2000), which requires the function of transporters. In our assembled pyrosequencing EST collections, we identified a large number of contigs homologous to ABC transporter, which is one of the most important families of membrane transport proteins that may play critical roles in the transmembrane transport of secondary metabolites in plants. The large amount of transporters can be linked to the secretion and transport function of GTs.

Table 4. Shared common GO terms (biological process) in all trichome EST databases.

GO ID	GO Term	No. of Unigenes
GO:0006464	Positive regulation of protein metabolic process	147
GO:0006730	Metabolic process	36
GO:0008152	Positive regulation of metabolic process	28
GO:0055114	Oxidation reduction	21
GO:0006006	Glucose metabolic process	21
GO:0006334	Nucleosome assembly	16
GO:0006412	Positive rgulation of biosynthetic process	14
GO:0006096	Positive regulation of glycolysis	5
GO:0006810	Transport	5
GO:0006869	Positive regulation of lipid transport	2

Representation of Genes-Related to Secondary Metabolism

Numerous sesquiterpene and monoterpene compounds have been identified in *A. annua* leaves, stems (Bouwmeester et al., 1999; Ma et al., 2007, 2008) and isolated GTs (Bertea et al., 2006). The genes corresponding to enzymes involved in the biosynthesis of major sesquiterpenes have been cloned and characterized (Bertea et al., 2006; Bouwmeester et al., 1999; Cai et al., 2002; Hua and Matsuda, 1999; Jia et al., 1999; Ma et al., 2007, 2008; Matsushita et al., 1996; Mercke et al., 1999; Picaud et al., 2005). To investigate the trichome function in secondary metabolism, the annotated unigenes were searched for enzymes participate in terpenoids biosynthesis. Unigenes corresponding to all the known enzymes in the terpenoids 2C-methyl-D-erythritol 4-phosphate (MEP) and mevalonate (MVA) pathway were identified. In higher plants, terpenoids precursor isopentenyl diphosphate (IPP) can be produced from both MVA and MEP routes, which is then converted to its isomer DMAPP (Figure 3) (Lichtenthaler et al., 1997). The cytosolic MVA terpenoids pathway, which starts from acetyl-CoA and proceeds through the intermediate MVA, provides the precursors for sterols and ubiquinone (Disch et al., 1998). The plastidial MEP pathway, which involves a condensation of pyruvate and glyceraldehyde-3-phosphate, is used for the synthesis of isoprene, carotenoids, abscisic acid, and the side chains of chlorophylls and plastoquinone (Arigoni et al., 1997; Hirai et al., 2000; Milborrow and Lee, 1998; Schwender et al., 1997). Although the subcellular compartmentation allows both pathways to operate independently, there is ample evidence that cross-talk exist between these two pathways (Laule et al., 2003; Piel et al., 1998).

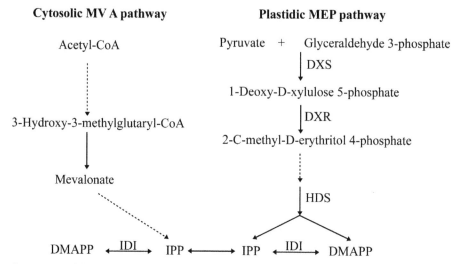

Figure 3. Simplified graphical representation of terpenoid biosynthetic pathway in *A. annua*. DXR: deoxy-D-xylulose-5-phosphate synthase; DXP: 1-deoxy-Dxylulose-5-phosphate reductoisomerase; HDS: 1-hydroxy-2-methyl-2-(E)-butenyl 4-diphosphate synthase; IDI: isopentenyl diphosphate/dimethylallyl diphosphate isomerase. DMAPP: Dimethylallyl Diphosphate. IPP: isopentenyl diphosphate.

Unigenes encoding the MEP and MVA pathway enzymes and all the sesquiterpene artemisinin pathway enzymes were present in our pyrosequencing collection. It is noteworthy that although the sequences were derived from normalized cDNA collections, unigenes corresponding to MEP pathway enzymes were two-fold more abundant as compared with MVA pathway transcripts. This likely suggests that the MEP pathway may serve as a major route for DMAPP/IPP production in the *A. annua* trichomes. The MEP pathway has previously been shown to provide precursors for both mono- and sesqui-terpene biosynthesis in snapdragon flowers (Dudareva et al., 2005). In a recent report on hops, the ESTs encoding MEP pathway enzymes are also found more abundant than those of MVA pathways (Wang et al., 2008).

Except for those well characterized terpenoid pathway genes, other unigenes annotated as sesquiterpene synthase and monoterpene synthase were identified. Three unigenes (*Contig02039*, *Contig16267*, *Contig14765*) annotated as sesquiterpine synthases were selected for RACE PCR to retrieve the full length cDNAs. Sequence analysis indicated that the conserved sesquiterpene synthase functional domain exists in all three genes (Figure 4). Further functional characterization of these enzymes will be reported elsewhere.

Furthermore, large amount of unigenes annotated as phenylpropanoids and flavanoids pathway enzymes were present in the assemebled pyrosequencing EST collection, indicating the metabolic function of GTs in *A. annua* secondary metabolism.

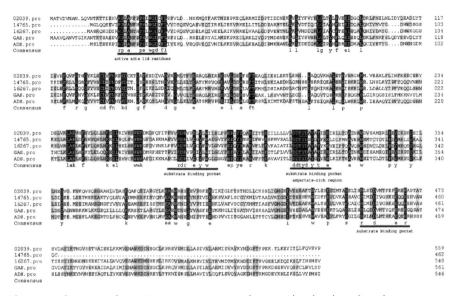

Figure 4. Alignment of putative sesquiterpene synthases with other homologs from *A. annua* (accession no. DQ447636 and AY006482). Identical amino acids are highlighted. The functional motifs are underlined.

RT-PCR Validation

A set of 17 contigs were selected for semi-quantitative RT-PCR analysis to confirm their expression (Figure 5A). The selected contigs encode enzymes involved in artemisinin biosynthesis, and putativetranscription factors. The PCR experiments were conducted on four pools of cDNAs derived from (1) GTs, (2) NGTs (3) leaves, and (4) roots. The results demonstrate that all of the novel transcripts detected among the 454-ESTs are indeed expressed in GTs, including those with low expression levels. This suggests that deep pyrosequencing is effective in revealing the expression of many rare transcripts, for example TFs. Most of the tested contigs were also expressed in leaf and NGT cDNA pools, except for one contig40477, which was only expressed in GTs and roots. Interestingly, three contigs likely encode enzymes needed in sesquiterpene biosynthesis were also strongly expressed in non-glandular type of trichomes. This raises the question as to whether GT is the sole site for the biosynthesis of artemisinin and other sesquiterpenes in *A. annua*.

The RT-PCR was also used to confirm the expression of novel transcripts and singletons. A set of 18 novel transcripts and singletons was randomly selected to test if they are indeed expressed in GT (Figure 5B). Of the 20 primer pairs, 13 produced RT-PCR products that were of the correct size and whose sequence matched the sequences from which the primers were designed. Based on these results, we conclude that many of the novel transcripts and singletons detected among the 454-ESTs are not due to the sequencing artifacts. This result provides further evidence for the value of tissue specific 454 sequencing for gene discovery.

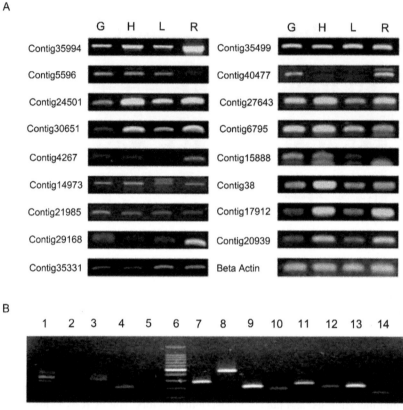

Figure 5. Semi-quantitative RT-PCR analysis of selected unigenes and novel transcripts. (A). Expression of selected contigs in different tissue types. G: Glandular trichome, N: Non-glandular tirhcome, L: Leaf, R: Root. bHLH family proteins: Contig35994 and Contig5596; WD family proteins: Contig24501 and Contig30651; Myb family proteins: Contig14973, Contig21985, Contig29168, Contig35331, Contig35499, and Contig40477; Terpene synthases: Contig27643; amorpha-4,11-diene synthase (*Artemisia annua*) ABM88787: Contig 6795; sesquiterpene cyclase (*Artemisia annua*) AAG24640: Contig15888 (3R)- linalool synthase (*Artemisia annua*) AAF13356; WRKY Proteins: Contig38, Contig17912, Contig20939. (B) Expression of novel transcripts and singletons in GT. Lane 1-5: singletons S122859, S078690, S091943, S154166, and S174533; Lane 6: DNA marker; Lane 7-14: novel transcripts C3719, C13021, C15708, C1441, C20920, C29103, C445, and C14916.

CONCLUSION

In conclusion, we describe the global analysis of GT in *A. annua* using massively parallel pyrosequencing. Mining the pyrosequencing ESTs resulted in the identification of many contigs likely involved in terpenoid biosynthesis and trichome function. Functional characterizations of selected genes are being carried out. These pyrosequencing data form the basis for further characterization of the molecular mechanism of GT function in *A. annua*. The results also highlight the value of using tissue-specific high throughput pyrosequencing technology for gene discovery in non-model plants. Access to all EST contigs obtained in this study is facilitated through a file available in the supplemental data.

KEYWORDS

- *Artemisia annua*
- Glandular trichomes
- Terpenoids biosynthesis
- Unigenes

AUTHORS' CONTRIBUTIONS

Wei Wang carried out the trichome isolation, RT-PCR, and participated in the sequence analysis and drafted the manuscript. Yejun Wang and Qing Zhang carried out EST assembly, data annotation, and bioinformatics analysis. Yan Qi participated in the trichome isolation and sequence analysis. Dianjing Guo conceived of the study and participated in its design and coordination. All authors read and approved the final manuscript.

ACKNOWLEDGMENTS

We thank Mr. Patrick Lau from the core facility in the Faculty of Science at CUHK for performing the 454 pyrosequencing. The work was fully supported by a grant from the Research Grants Council of the Hong Kong Special Administrative Region, China (Project no. CUHK 4603/06M).

Chapter 11

Tissue-Specific Expression of the Adjacent Genes in *Arabidopsis thaliana*

Hernán G. Bondino and Estela M. Valle

INTRODUCTION

Transcription initiation by RNA polymerase II is unidirectional from most genes. In plants, divergent genes, defined as non-overlapping genes organized head-to-head, are highly represented in the *Arabidopsis* genome. Nevertheless, there is scarce evidence on functional analyses of these intergenic regions. The At5g06290 and At5g06280 loci are head-to-head oriented and encode a chloroplast-located 2-Cys peroxiredoxin B (*2CPB*) and a protein of unknown function (*PUF*), respectively. The 2-Cys peroxiredoxins are proteins involved in redox processes, they are part of the plant antioxidant defense and also act as chaperons. In this study, the transcriptional activity of a small intergenic region (351 bp) shared by *At5g06290* and *At5g06280* in *Arabidopsis thaliana* was characterized.

Activity of the intergenic region in both orientations was analyzed by driving the β-glucuronidase (*GUS*) reporter gene during the development and growth of *Arabidopsis* plants under physiological and stressful conditions. Results have shown that this region drives expression either of *2CPB* or *PUF* in photosynthetic or vascular tissues, respectively. The *GUS* expression driven by the promoter in *2CPB* orientation was enhanced by heat stress. On the other hand, the promoter in both orientations has shown similar down-regulation of *GUS* expression under low temperatures and other stress conditions such as mannitol, oxidative stress, or fungal elicitor.

The results from this study account for the first evidence of an intergenic region that, in opposite orientation, directs *GUS* expression in different spatially-localized *Arabidopsis* tissues in a mutually exclusive manner. Additionally, this is the first demonstration of a small intergenic region that drives expression of a gene whose product is involved in the chloroplast antioxidant defense such as *2CPB*. Furthermore, these results contribute to show that *2CPB* is related to the heat stress defensive system in leaves and roots of *Arabidopsis thaliana*.

A promoter region of an eukaryotic protein-encoding gene usually consists of a core promoter region of around 50 bp nucleotides adjacent to the transcription initiation site, and multiple distal DNA regulatory elements to control transcription efficiency. There are several key genetic elements within a core promoter: the TATA box, an initiator element, the downstream promoter element usually found in TATA-less promoters, and the TFIIB-recognition element (Novina and Roy, 1996; Smale and Kadonaga, 2003). The TATA boxes are usually located about 25–30 bp upstream of the transcription start site (TSS), while the less conserved initiator elements span the TSS. These sequences contribute to an accurate transcription initiation and to the TATA-containing promoters strength. In *Arabidopsis* core promoters, the TATA box is located between 50 and 20 relative to the TSS and, instead of the initiator element around the TSS, the YR rule (Y:C or T; R:A or G) applies to most of them. Another element is the pyrimidine patch (Y Patch), although its role is still unknown. These three elements are orientation-sensitive (Yamamoto et al., 2007). Other promoter elements found in *Arabidopsis* and rice are regulatory element groups (REGs), which appear upstream of the TATA box (20 to 400), and exist in an orientation-insensitive manner (Yamamoto et al., 2007).

Transcription initiation by RNA polymerase II is unidirectional from most genes. However, several reports indicate that divergent transcription is likely a common feature for active promoters (Beck and Warren, 1988; Seila et al., 2009).

Divergent genes, defined as non-overlapping genes organized head-to-head in opposite orientation, represent a 36.5% of the total gene pairs when separated by less than 1 kb in the *Arabidopsis* genome (Krom and Ramakrishna, 2008). Nevertheless, there is scarce evidence on functional analyses of the intergenic regions between those gene pairs. Previous findings of head-to-head oriented genes sharing an intergenic region with putative bidirectional promoters were reported in *Brassica napus* (Keddie et al., 1994), *Capsicum annuum* (Shin et al., 2003), and by computational analysis in rice, *Arabidopsis*, and black cottonwood (Dhadi et al., 2009). Large-scale studies of expression data in *Arabidopsis* revealed that neighboring genes in the genome are co-expressed (Williams and Bowles, 20040, and that the lengths of the intergenic sequences have opposite effects on the ability of a gene to be epigenetically regulated for differential expression (Colinas et al., 2008). Two recent papers have shown activity of larger intergenic regions in rice (1.8 kbp) and *Arabidopsis* (2.1 kbp), functioning as bidirectional promoters of chymotrypsin protease inhibitor (Singh et al., 2009) and chlorophyll a/b-binding protein (Mitra et al., 2009) genes, respectively. These systems were assessed in a heterologous background using onion epidermal cells (Singh et al., 2009), and also in stable transgenic plants, the latter intended to be used for genetic engineering-based crop improvement (Mitra et al., 2009).

All divergent gene pairs are potential sources of bidirectional promoters. To define the function of the corresponding intergenic regions and their transcriptional regulation is of great interest for plant molecular biologists.

In this study, a divergent promoter of a protein-encoding gene pair (*At5g06290* and *At5g06280*) with an intergenic region of 351 bp was analyzed. The At5g06290 and At5g06280 loci encode a *2CPB*, which are a chloroplast-located protein (Dietz et al., 2002), and a *PUF*, respectively http://www.arabidopsis.org. The 2-Cys peroxiredoxins are proteins involved in redox processes, and their functions are related to the antioxidant defense of the plant (Baier and Dietz, 1999), photosynthesis, abiotic stress response, and possibly chloroplast-to-cytosol signalling (Dietz, 2007). In yeast, peroxiredoxins could act as molecular chaperons, increasing resistance to heat stress (Jang et al., 2004). The expression pattern of the *At5g06290* and *At5g06280* was tested by fusing the intergenic region in opposite orientation to *GUS* reporter gene during the development and growth of *Arabidopsis* plants as well as during stress situations.

MATERIALS AND METHODS

Plant Material and Growth Conditions

Arabidopsis thaliana ecotype Columbia (Col-7) was synchronously germinated at 4°C for 48 hr and grown in soil-vermiculite mixture (2:1 v/v) in growth chambers at 2022°C, under long day conditions (16 hr light/8 hr darkness). The light intensity was set at 130 µmol m^{-2} s^{-1}.

When assaying stress treatments, *Arabidopsis* plants grown photoautotrophically on agar medium containing 0.5 X Murashige and Skoog (MS) salts (Sigma-Aldrich).

Stress Treatments

Arabidopsis plants were cultivated on agar supplemented with the stress agent: osmotic stress (100 mM mannitol), salt stress (50 mM NaCl), oxidative stress (0.1 µM methyl viologen), or fungal elicitor (1.3 mg/ml autoclaved cellulase, Onozuka R-10, Yakult Honsha, Tokio, Japan). For cold (4°C) and high (37°C) temperature stresses, the plants were grown for 10 days on MS agar without supplements under control conditions and then the temperature treatment was applied for 2 days. For higher light intensity (800 µmol m^{-2} s^{-1}), the plants were grown for 10 days and the treatment was applied for 6 hr.

DNA Constructs

The intergenic region with the 5' UTR regions of the genes *At5g06280* and *At5g06290* was isolated by PCR from an *A. thaliana* DNA CTAB preparation (Stewart and Via, 1993) using the primers 5'-CGCGGATCCAGTCTTTCTTCTTCTTTTTTTTTG-3' and 5'-CGCGGATCCTGACTCTGTTCTCTCTCTCTATC-3'. The PCR product was subcloned into pGEM-T Easy Vector (Promega, Madison, USA). The DNA sequencing was used to confirm that no spurious mutations were introduced during amplification. The fragment was excised with BamHI, and the 530 bp fragments were cloned into the BamHI site of pBI101.1 to create the plasmids pBI280 and pBI290. The orientation of the fragment was analyzed by PCR with primers that hybridize in the pBI101.1 plasmid (5'-ACAGTTTTCGCGATCCAGAC-3' and 5'-TTATGCTTCCG-GCTCGTATG-3') and the primers previously described. *Escherichia coli* strain DH5α was used for plasmid construction. *Agrobacterium tumefaciens* strain GV3101 pMP90 was transformed with plasmids by electroporation, and *Arabidopsis* (Col-7) plants were transformed by floral dip infiltration (Clough and Bent, 1998) with the plasmids pBI101.1, pBI280, or pBI290.

Histochemical Localization of *GUS* Activity

The *GUS* activity was localized by staining the tissues with 0.5 mg of 5-bromo-4-chloro-3-indolyl-b-D-glucuronic acid (X-Gluc; Gold Biotechnology, St Louis, MO, USA) per ml in X-Gluc buffer containing 50 mM sodium phosphate (pH 7.2), 10 mM EDTA, 0.33 mg/ml potassium ferricyanide and 0.001% Tween 20. The tissues were vacuum-infiltrated for three rounds of 1 min each, and staining reactions proceeded overnight at 37°C. Chlorophyll was removed by soaking in ethanol. The photographs were taken with a binocular microscope Leika MZ16F.

Analysis of *GUS* Activity

Quantitative analysis of *GUS* activity was performed on whole aerial part using the *GUS* activity assay (Weigel and Glazebrook, 2002), the experiment was made twice, each treatment had three biological replicates and each replicate was a pool of 10 *Arabidopsis* plants, except the high light treatment which had four biological replicates.

Production of 4-methylumbelliferone (MU) was measured using a DTX 880 Multimode Detector (Beckman Coulter, Fullerton, CA). Protein concentrations of the samples were determined using Bradford reagent (Bradford, 1976) and BSA as a standard. The amount of MU was determined from a standard curve, and *GUS* activity was expressed as nmol MU/min/mg protein. The empty vector transformed plants shown a basal activity of 0.22 ± 0.08 nmoles MU/min/mg protein.

Immunoblot Analysis

To measure the protein levels of *2CPB*, 100 mg of tissue were ground to a fine powder in liquid N2 and then homogenized with 0.2 ml of buffer (25 mM Hepes (pH 7.5), 0.6 M mannitol, 0.462 mg/ml dithiothreitol, 2 mM EDTA, 0.175 mg/ml phenylmethylsulphonyl fluoride and 1% (w/v) polyvinylpolypyrrolidone). The homogenates were centrifuged at 15,000 g for 20 min, and the supernatant protein concentration was determined utilizing BSA as a standard protein as described by (Bradford, 1976). The supernatant was mixed with sample buffer 10× (250 mM Tris-HCl (pH 6.8), 10% SDS, 0.5% bromophenol blue, and 20% glycerol), boiled for 5 min, and separated in a 12% SDS-PAGE as described earlier (Laemmli, 1970). The gels were stained with Coomassie Brilliant Blue R-250. For immunoblotting, the proteins were transferred to nitrocellulose membranes using a Mini Trans-Blot cell (Bio-Rad, CA, USA) at 100 mA for 100 min. The membranes were treated with polyclonal antibody raised against rapeseed 2-Cys peroxiredoxin (Caporaletti et al., 2007). Signals on the membranes were visualized with alkaline phosphatase-conjugated goat anti-rabbit IgG (SIGMA, St Louis, MO, USA).

The signal intensities were quantified from the immunoblot using the Gel-Pro Analyzer software (Media Cybernetics Inc, Silver Spring, MD) and normalized to the intensities observed in control conditions. A representative example from three independent experiments is shown.

Promoter Sequence Analysis

The promoter sequence was analyzed using publicly available databases, PlantCARE http://bioinformatics.psb.ugent.be/webtools/plantcare/html/(Rombauts et al., 2003) and PLACE http://www.dna.affrc.go.jp/PLACE/signalscan.html (Higo et al., 1999), which are databases of plant *cis*-acting regulatory elements; AthaMap http://www.athamap.de/index.php (Steffens et al., 2004), which provides a genome-wide map of potential transcription factor binding sites in *Arabidopsis thaliana*; and Plant Promoter Database (ppdb) http://www.ppdb.gene.nagoya-u.ac.jp (Yamamoto and Obokata, 2008), which is based on species-specific sets of promoter elements, rather than on general motifs for multiple species.

Arabidopsis Promoters Length Analysis

Annotation data for the *Arabidopsis thaliana* genes was downloaded from The *Arabidopsis* Information Resource (TAIR) FTP server ftp://ftp.arabidopsis.org/Maps/seqviewer_data/sv_gene.data. The analysis was performed on 27,141 genes after filtering out pseudogenes and transposon-related genes ftp://ftp.arabidopsis.org/Maps/gbrowse_data/TAIR8/TAIR8_GFF3_genes_transposons.gff from 31,762 annotated genes. Start and stop positions of the transcription units along with information on the strand that encodes an mRNA were extracted. Microsoft Office Excel was used to calculate the distances between the 3' ends of the nearest neighbor genes and the distances between 5' ends of the neighbor genes. The overlapping genes were analyzed only in the graph corresponding to the 3' ends of the nearest neighbor genes and the resulting distances among them were less than zero (shown in Figure 6C, inset).

DISCUSSION

With the availability of complete genome sequences for a number of organisms, functionality of intergenic regions has attracted more attention. Computational analysis has shown that divergent gene pairs with intergenic regions less than 1 kb are quite abundant in the sequenced eukaryotic genomes of both plants and animals (Krom and Ramakrishna, 2008; Trinklein et al., 2004). The interest in studying intergenic region functionality is increasing not only to better understand divergent transcription, but also to use them as a new toolkit to manipulate genomes (Venter, 2007). In plants, particularly, very few reports about this matter are available. An example of such investigations in plants in which data from computational assistance and bidirectionalization were integrated to construct a synthetic transcriptional unit for high-level reporter-gene expression in response to specific elicitors was reported, thus yielding exciting results (Chaturvedi et al., 2006). In this study, it has been found that the region shared by two divergent genes in the chromosome 5 of *Arabidopsis thaliana* (*At5g06280* and *At5g06290*) functions as a promoter in both orientations. In addition, this study was able to demonstrate that tissue and developmental expression patterns differed between *PUF* and *2CPB*. Head-to-head genes from other organisms such as human, mouse, and rat genomes statistically tend to perform similar functions, and gene pairs associated with the significant co-functions seem to have stronger expression correlations (Mittler et al., 2004). In this case, the gene products of *At5g06280* and *At5g06290* are both presumably located in the chloroplasts, although it is unknown if their functions are related. Thus, it is known that *2CPB* is located in the chloroplasts and prevents oxidative damage of chloroplast proteins (Baier and Dietz, 1999). The transcript increase of *2CPB* was correlated with chlorophyll distribution and also accumulated in plants with decreased catalase activity and upon heat stress (Mittler et al., 2004). Downregulation of *2CPB* was observed upon pathogen infection, ozone, and cold (Dietz et al., 2006; Goulas et al., 2006). Instead, the role of *PUF* remains unknown until today, and presumably it would be a chloroplast-located protein as predicted by ChloroP analysis (Emanuelsson et al., 1999).

When searching for *At5g06280* and *At5g06290* potential orthologues, it has been found that this head-to-head gene organization was not conserved among other genomes (data not shown); pointing out that most probably their gene products are not functionally related. In humans, analysis of genome-wide expression data demonstrated that a minority of bidirectional gene pairs are expressed through a mutually exclusive mechanism (Trinklein et al., 2004). In this study, the tissue-specific expression of both genes directed by the divergent promoter has shown unidirectional activity for *PUF* in petiole and vascular bundles and unidirectional activity in the opposite direction in different tissues for *2CPB*. The higher expression of *2CPB* in the leaf mesophyll, but not in vascular bundles, is coincident with its function in the redox processes of chloroplasts (Dietz et al., 2006). Taken together, these results suggest that the directionality of the promoter activity may be regulated to some degree in a tissue-specific manner. In fact, a *cis*-motif associated to vascular bundle expression (AACA) (Scarpella et al., 2005) was found several times in the *PUF* direction of transcription.

Furthermore, it has been demonstrated that the divergent promoter shared by *PUF* and *2CPB* responded to temperature stress. In relation to this, the higher *2CPB* levels in the leaf and root caused by heat treatment of *Arabidopsis* seedlings would indicate a role of this protein in temperature stress. In yeast, peroxiredoxins could alternatively function as peroxidases and molecular chaperons, increasing resistance to heat stress (Jang et al., 2004). It is well known that exposure of plants to high temperature leads to the production of *Hsps*. The yeast heat shock factor 1 binding sequence nTTCn (or nGAAn) (Yamamoto et al., 2005) was found highly represented in the intergenic region of this study. Therefore, it is tempting to speculate that high temperature could stimulate *2CPB* similarly to *Hsp* genes. Remarkably, the *PUF* expression was repressed similarly to *2CPB* by several stress conditions.

In silico analysis of this promoter using ppdb revealed that it is a TATA-less promoter in both orientations. In plant genomes putative bidirectional promoters have TATA boxes underrepresented (Dhadi et al., 2009). A recent study (Walther et al., 2007) suggested that TATA box-containing genes have longer intergenic upstream regions and increased variation across species because their upstream regulatory potential is greater and, therefore, more amenable to change and modulation. The TATA box appears to be responsible for promoter unidirectionality in most cases, whereas having no TATA boxes appears to be a novel mechanism of regulation by bidirectional promoters compared to unidirectional promoters. This analysis also revealed that in a short region of this promoter (28 bp) (Figure 5B), four different *cis*-elements are overlapped. They are: one heat shock element (CCAAT box), a Y Patch found in the majority of *Arabidopsis* promoters but with unknown function (Yamamoto and Obokata, 2008), and three binding sites of homeodomains-leucine zipper transcription factors, some of them being able to bind in both directions (Higo et al., 1999; Steffens et al., 2004). These *cis*-elements would be leading the transcription of *2CPB*, specially ATHB1, which is involved in differentiation of the palisade mesophyll cells, and ATHB5, which in turn is involved in the control of leaf morphology development (Steffens et al., 2004). Upstream of this region there are three AACA elements in the +/ 25 bp region of *PUF* TSS (Figure 5A). This is a negative regulatory element in vascular promoters, which represses activity in other cell types (Scarpella et al., 2005) suggesting that, in the intergenic region under analysis, this *cis*-element would be preventing *PUF* transcription in mesophilic cells. The expression of *PUF* in vascular bundle of midribs could be activated by ATHB2, which has a homeodomain too, and by the Y Patch that is located in the 28 bp region above mentioned. The *2CPB* and *PUF* putative promoter regions mentioned have an element of response to heat near them, which could explain the heat stress experiments. It was not possible to find any abiotic stress element overrepresented in the 530 bp region analyzed, suggesting that the expression pattern observed in Figure 4 could be the result of the complex interaction of the transcription factors that bind the 28 bp region. Overall, results obtained from this study indicate that the multiple stress responsiveness of the intergenic region would reside within the 351 bp.

When length is considered, the short promoter shared by *2CPB* and *PUF* belongs to a minority group of putative bidirectional promoters present in the *Arabidopsis* genomes. In fact, *Arabidopsis* genome has a bimodal distribution of distances between the 5' ends of genes on opposite strands, peaking the smaller group of gene pairs at 323 bp. This is the first intergenic region functionally studied of this small group of *Arabidopsis* promoters. Plants are sessile organisms and, during their growth, they occasionally are affected by adverse environmental conditions; therefore, they may rely more strongly on elaborate transcriptional response programs to survive. Then, it is highly possible that other intergenic regions of similar lengths and regulatory features could be found in plants.

RESULTS

Functional Analysis of the Intergenic Region Between *At5g06280* and *At5g06290* in *Arabidopsis* Plants During Their Development and Growth

To test functionality of the intergenic region shared by the divergent genes *At5g06280* and *At5g06290* during *Arabidopsis* life cycle, the DNA fragment was fused to *GUS* in both orientations (*Prom280:GUS* and *Prom290:GUS*, respectively). Accordingly, we cloned a 530 bp DNA fragment (the 351 bp intergenic region and the 5' untranslated regions) upstream of *GUS* in the binary vector pBI101.1. The constructs were introduced into wild-type *Arabidopsis* plants by floral dip, multiple transgenic plants were obtained, and more than three independent lines were examined for each construct throughout development. The *GUS* staining was performed in *Arabidopsis* plants during life cycle (Figure 1, stages 1.0–6.9 according to (Boyes et al., 2001)). Interestingly, *Prom280:GUS* plants have shown staining almost exclusively in the petiole and vascular bundle of midrib in all the leaves (Figures 1C, 1E, 1G, and 1I), sepals (Figure 1K), but not in the cotyledons (Figures 1A), while *Prom290:GUS* plants have shown staining mainly in the leaf mesophyll (Figure 1B, 1D, 1F, 1H, and 1J), sepals (Figure 1L), and siliques (Figure 1M and 1N). It is worth noticing that stronger *GUS* staining was observed for *Prom290:GUS* plants (it was visualized even after three hr of staining) in comparison with *Prom280:GUS* plants at all growth stages (data not shown). Results indicate that the intergenic region between *At5g06290* and *At5g06280* directs *GUS* expression in a spatially exclusive manner depending on the promoter orientation during *Arabidopsis* development and growth (Figure 1).

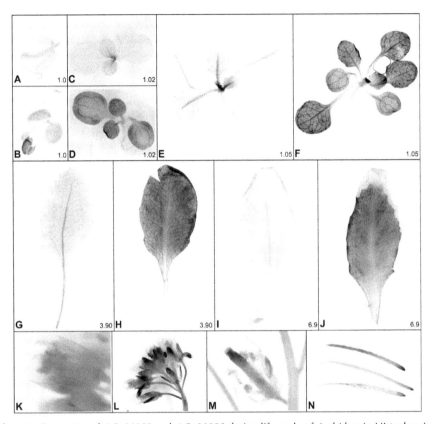

Figure 1. Expression of *At5g06280* and *At5g06290* during life cycle of *Arabidopsis*. Histochemical detection of *GUS* in *Arabidopsis* plants from *Prom280:GUS* and *Prom290:GUS* lines of different ages. The *Arabidopsis* growing stages (according to (Boyes et al., 2001)) are indicated at the right bottom corner of the pictures. The *GUS* activity are seen in *Prom280:GUS* line (C, E, G, I) in the petiole and vascular bundle of midrib in all the leaves, but not in the cotyledons (A). The *Prom290:GUS* line (B, D, F, H, J) has evidenced staining in mesophillic tissue of the leaves at all stages. (K) Open flower of *Prom280:GUS* line showing staining of the vascular tissues of the sepals. (L) Open flower of *Prom290:GUS* line with stained sepals. (M) Senescent flower of *Prom290:GUS* line. (N) Siliques of *Prom290:GUS* line showing the stained style and stigmatic tissue. Siliques of *Prom280:GUS* line were not stained at all (data not shown).

As *2CPB* is a chloroplastic protein (Dietz et al., 2002), we analyzed the putative intracellular location of *PUF* using ChloroP 1.1 Server (Emanuelsson et al., 1999) and the deduced amino acid sequence of *At5g06280*. The prediction results have shown that *PUF* (156 residues) is likely to be a plastidic protein, because it has an amino-terminal extension indicative of chloroplast transit peptide (score 0.506). For comparison, *2CPB* score was 0.598 using this web tool.

Response of *Prom280:GUS* and *Prom290:GUS* Plants to Various Stresses

Different stress conditions lead to the production of reactive oxygen species (ROS) as a consequence of membrane and protein damage (Apel and Hirt, 2004). The expression of 2-Cys peroxiredoxins are reported to be redox regulated (Horling et al., 2001). Therefore, it was decided to test the response of *Prom280:GUS* and *Prom290:GUS* plants to various environmental stresses. Firstly, the effect of temperature treatment in 10-day old *Arabidopsis* seedlings was analyzed. Plants of both transgenic lines were incubated for 48 hr at 37°C or 4°C and, after the treatment, they were submitted to *GUS* staining procedure. Figure 2 shows that leaves from both plant lines were stained stronger under heat stress (Figures 2C and 2D), maintaining the same tissue specificity to the control condition (Figures 2A and 2B). In addition, the root tips were stained in the case of *Prom290:GUS* plants (Figure 2J). In both plant lines the *GUS* staining pattern was conserved under cold stress (Figures 2E and 2F), although the expression levels were weaker than control conditions as revealed by quantification of the *GUS* staining intensity (Figures 2G and 2H). Furthermore, no expression was detected in the plants carrying the vector without the intergenic region (empty vector) (Figure 2I). Further analysis of *PUF* and *2CPB* expression using the response viewer of Genevestigator software http://www.genevestigator.ethz.ch/ (Zimmermann et al., 2005) is presented in Figure 3. Under several cold treatments, the aerial part of *Arabidopsis* plants have evidenced decreased expression of *PUF* and *2CPB*, while under heat conditions, the plants have evidenced enhanced expression of both genes (Figure 3). Similar responses were observed in the expression of *GUS* in *Prom280:GUS* and *Prom290:GUS* plants submitted to temperature stress (Figure 2). Additionally, in roots of *Prom290:GUS* plants, the expression of *2CPB* is markedly increased by heat stress (Figure 2J), which is consistent with data obtained from roots under the same stress treatment (Figure 3).

To confirm the effect of heat treatment on the induction of *2CPB*, 10-day old wild-type *Arabidopsis* plants were submitted for 2 days at 37°C, and the total protein of leaves and roots were extracted and analyzed by SDS-PAGE and immunoblotting. The total protein pattern has shown slight differences between control and treated plants in the leaf or root tissues, especially in higher molecular masses larger than 66 kDa. Immunoblot analysis of these tissues has shown induction of *2CPB* in both leaves and roots after heat treatment. These data indicate that heat treatment was able to increase not only *2CPB* protein level in root and leaf of wild-type plants, but also *GUS* activity in the same tissues as observed in *Prom290:GUS* plants (Figures 2D and 2J).

Other sources of ROS are biotic and abiotic stresses. The effect of different stress conditions on the expression levels of *Prom280:GUS* and *Prom290:GUS* plants were evaluated, and the results are presented in Figure 4. The *GUS* expression was similarly reduced in both *Arabidopsis* lines under oxidative stress caused by methyl viologen (MV), a redox cycling compound, fungal elicitor, and mannitol. Additionally, *GUS* expression was down-regulated by high light and NaCl in *Prom280:GUS* lines, while in *Prom290:GUS* lines were unaffected. It is worth mentioning that the expression of *GUS* in *Prom280:GUS* lines was 10 times lower than in *Prom290:GUS* plants when calculated per mg of protein of the aerial parts of the plant. This could be due to a dilution effect of *GUS* activity specifically located in vascular bundles of the leaf in *Prom280:GUS* lines, in comparison with the whole leaf expression pattern of *GUS* in *Prom290:GUS* plants.

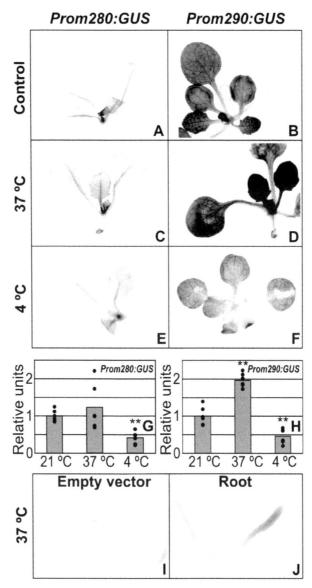

Figure 2. Expression of *GUS* in *Prom280:GUS* and *Prom290:GUS* plants in response to temperature treatments. Ten-day-old plants were grown on MS agar plates photoautotrophically at 21°C, and incubated for 48 hr at 37°C and 4°C. The *GUS* staining of *Prom280:GUS* and *Prom290:GUS* plants before (A and B) and after transferring the plants to higher (37°C, C and D) or lower (4°C, E and F) temperature conditions. The *GUS* activity was quantified in whole aerial part using 4-methylumbelliferone (MU) as substrate, and results are reported in a relative scale (G and H). Control of *Prom280:GUS* line was 6.52 ± 1.04 nmoles MU/min/mg protein, and control of *Prom290:GUS* line was 76.56 ± 18.84 nmoles MU/min/mg protein. Each point represents a single replicate. Asterisks (**) indicate significant differences between treatments and controls according to Student's t-Test at P < 0.005. Plants transformed with pBI101.1 (I). Roots of *Prom290:GUS* line after heat treatment (37°C) (J).

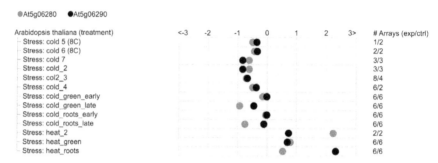

Figure 3. Expression levels of the genes after heat or cold stress as shown by Genevestigator. Response viewer of Genevestigator software shows that *At5g06280* and *At5g06290* genes decrease their expression levels in all cold stress experiments and increase their levels with heat stress treatments.

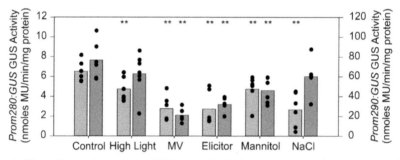

Figure 4. Effect of several stresses on *GUS* expression in *Prom280:GUS* and *Prom290:GUS* lines. Ten-day-old *Arabidopsis* seedlings were grown on MS agar at 21°C and incubated for 6 hr to high light (800 μmol m⁻² s⁻¹). For other stress conditions, *Arabidopsis* plants were cultivated on MS agar supplemented with 50 mM NaCl, 0.1 μM MV, 100 mM mannitol or 1.3 mg/ml elicitors. The *GUS* activity was quantified in whole aerial part using MU as substrate. Each point on the bar represents a single replicate (blue bars for *Prom280:GUS* line and red bars for *Prom290:GUS* line). Asterisks (**) indicate significant differences between treatments and controls according to Student's t-Test at $P < 0.05$. Note the scale difference between *Prom280:GUS* and *Prom290:GUS* lines.

These results suggest that *PUF* and *2CPB* are stress-responsive genes, although they are not always affected in the same way by the same stress conditions.

In Search of *cis*-Elements in the Promoter of *PUF* and *2CPB*

In silico analysis of the divergent promoter was performed looking for *cis*-elements using the ppdb (Yamamoto and Obokata, 2008), PlantCARE (Rombauts et al., 2003), PLACE (Higo et al., 1999), and Athamap (Steffens et al., 2004) web tools. Analysis revealed no TATA box available. The elements distribution in the 530 bp region is shown in Figure 5. We identified binding sites for four homeodomain-leucine zipper transcription factors: ATHB1, which was reported to be involved in differentiation of the palisade mesophyll cells and leaf development (Aoyama et al., 1995; Henriksson et al., 2005); ATHB2, which is responsive to far-red light (Carabelli et al., 1996); ATHB5, which is a transcription factor involved in the regulation of light-dependent developmental phenomena (Henriksson et al., 2005); and transcription factors similar to ZmHox2a, which have the homeodomains ZmHOX2a(1) and ZmHOX2a(2) (Kirch et al., 1998). Furthermore, a Y Patch near *PUF* TSS, and seven REGs near *2CPB* TSS were identified; however, their functions are still unknown. An AACA element, which was described as a negative regulatory element in vascular promoters that re- presses activity in other cell types (Scarpella et al., 2005), were identified in seven positions. Lastly, a CCAAT box, present in the promoter of heat shock protein (*Hsp*) genes (Rieping and Schoffl, 1992), was found four times, and the nCTTn element present in the promoters of several *Hsp* genes (Yamamoto et al., 2005) was found 23 times. This analysis displayed no other overrepresented *cis*-element in the promoter region under study.

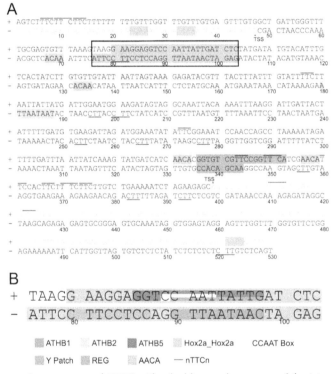

Figure 5. Intergenic sequence and 5'UTRs. The double strand sequence of the intergenic and the 5'UTR regions of *2CPB* and *PUF* are shown in A. The *cis*-elements found in the analyzed region are indicated at the bottom of the figure. More details of the 28 bp region (enclosed) are shown in B. The plus (+) strand is upstream of *At5g06290* and the minus strand is upstream of *At5g06280*. The 5'UTRs are shown in light green. Asterisks indicate the TSS. No TATA boxes have been found. The ATHB1 is the binding site of the transcription factor ATHB1, which is involved in differentiation of the palisade mesophyll cells and leaf development. The ATHB2 is the binding site of the transcription factor ATHB2, which is an element of response to far-red light. The ATHB5 is the binding site of the transcription factor ATHB5, which is involved in the regulation of light-dependent developmental phenomena. The Hox2a_Hox2a is the binding site of proteins with the homeodomains ZmHOX2a(1) and ZmHOX2a(2). The CCAAT box is found in the promoter of *Hsp* genes. The Y Patch is a direction-sensitive plant core promoter element that appears around TSS. The REG is a direction-insensitive element that is preferentially found around 20 to400 bp relative to TSS. The AACA is a negative regulatory element in vascular promoters that repress activity in other cell types. The yeast heat shock factor 1 binding sequence nTTCn is underlined in the minus strand and overlined in the plus strand.

Distribution of Distances between Genes and their Nearest Neighbors in *Arabidopsis* Genome

To further characterize this 351 bp promoter on genome-wide scale, the distribution of intergenic regions of similar lengths into the *Arabidopsis* genome was studied. For that purpose, the distribution of distances between *Arabidopsis* genes and their nearest neighbors in the same and opposite strands were explored. The distances between the TSS of the nearest gene neighbors for each of the 27,141 genes predicted (see Materials and Methods) after filtering out genes annotated as pseudogenes and transposons were calculated. The distribution of distances between 5' ends of genes on opposite strands is bimodal, which could be deconvoluted in two peaks centred at 323 bp (around 140 gene pairs between 300 and 350 bp length) and 2.5 kbp (Figure 6A). This type of distribution was not present in all around 14,000 genes with the nearest neighbors on the same strand (Figure 6B), or when the distances were calculated between the 3' ends of the genes on opposite strands (Figure 6C). Noticeably, only 4.3% of the gene pairs with 5' ends on the same strand are closer than 1,000 bp (Figure 6B), while 75% of the gene pairs with 3' ends on opposite strands are closer than 1,000 bp, with 1,234 of them having overlapping regions (Figure 6C, inset). We designated the region between the two non-overlapping 5' ends of genes located on opposite strands as a putative bidirectional promoter. This analysis shows that out of 6,438 divergent gene pairs (Figure 6A), 2,469 are putative bidirectional promoters of less than 1,000 bp in the *Arabidopsis* genome. Most of the head-to-head oriented genes (98%) have predictably shown non-overlapping bidirectional promoters, and only 874 (13.8%) gene pairs are less than 323 bp in length.

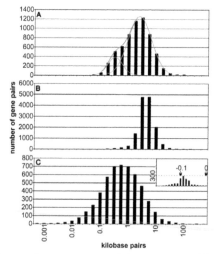

Figure 6. Distribution of distances between genes and their nearest neighbors in *Arabidopsis*. (A) The distribution of distances between 5' ends of genes on opposite strands is deconvoluted in two peaks showing a peak centered at 323 bp and another at 2,518 bp, the 38.4% of the genes pairs are closer than 1,000 bp. (B) Distances between 5' ends of genes on the same strand showing that 4.3% of gene pairs are closer than 1,000 bp. (C) Analysis of the distribution of distances between 3' ends of genes on opposite strands showing that 75% of the gene pairs are closer than 1,000 bp. Inset: indicate the distribution of the overlapping genes.

CONCLUSION

In this report, it has been shown that a 351 bp intergenic region between head-to-head oriented *At5g06290* and *At5g06280* directs genes expression in different *Arabidopsis* tissues in a mutually exclusive manner. Gene products of these loci are a chloroplast-located *2CPB* involved in the antioxidant defense, and a *PUF*. This is the first report of an intergenic region that drives expression of a gene involved in the chloroplast antioxidant defense. These results also show that *2CPB* is induced by heat stress in the leaves and roots, suggesting a function for this protein in the heat stress defensive system of *Arabidopsis thaliana*.

KEYWORDS

- *Arabidopsis thaliana*
- 2-Cys peroxiredoxin B
- Glucuronidase
- Protein of unknown function
- RNA polymerase

AUTHORS' CONTRIBUTIONS

Hernán G. Bondino has made substantial contributions to the conception and design of experiments, acquisition of data, analysis and interpretation of data, as well as drafting the manuscript; Estela M Valle has been involved in the design of the experiments, analysis, and interpretation of data, as well as drafting the manuscript. Both authors have approved the final manuscript.

ACKNOWLEDGMENTS

We thank Dr. Ricardo Wolosiuk, Instituto Leloir, Argentina for the generous gift of 2-CPB antisera.

The work described in this chapter was performed with the financial support of the Agencia Nacional de Promoción Científica y Tecnológica (ANPCyT) and the Consejo Nacional de Investigaciones Científicas y Técnicas (CONICET) from Argentina.

Chapter 12

Plant Physiological Response to Elevated Night Temperature

Abdul R. Mohammed and Lee Tarpley

INTRODUCTION

Global climate warming can affect functioning of crops and plants in the natural environment. In order to study the effects of global warming, a method for applying a controlled heating treatment to plant canopies in the open field or in the greenhouse is needed that can accept either square wave application of elevated temperature or a complex prescribed diurnal or seasonal temperature regime. The current options are limited in their accuracy, precision, reliability, mobility, or cost and scalability.

The described system uses overhead infrared heaters that are relatively inexpensive and are accurate and precise in rapidly controlling the temperature. Remote computer-based data acquisition and control (DAC) via the internet provides the ability to use complex temperature regimes and real-time monitoring. Due to its easy mobility, the heating system can randomly be allotted in the open field or in the greenhouse within the experimental setup. The apparatus has been successfully applied to study the response of rice to high night temperatures. Air temperatures were maintained within the set points ±0.5°C. The incorporation of the combination of air-situated thermocouples, autotuned proportional integrative derivative temperature controllers and phase angled fired silicon controlled rectifier (SCR) power controllers provides very fast proportional heating action (i.e., 9 ms time base), which avoids prolonged or intense heating of the plant material.

The described infrared heating system meets the utilitarian requirements of a heating system for plant physiology studies in that the elevated temperature can be accurately, precisely, and reliably controlled with minimal perturbation of other environmental factors.

Global climate warming can affect functioning of crops and plants in the natural environment. Increased night temperatures have been implicated in decreased crop yields throughout the world and are predicted to warm more than the day time temperatures in the future (Houghton et al., 2001). The effects of high night temperatures are diverse, including, for example, increased coincidence of intervals of unusually high night temperature with sensitive reproductive stages eventually resulting in poor seed set and a decline in vegetative reserves due to increased respiration and alteration in phenology, or, in the case of natural populations, altered quantity, and seasonal distribution of reproductive units.

Precise and accurate control of the temperatures and immediately surrounding small populations of plants is a primary purpose of an apparatus for control of high night temperature, but reliability is also needed to avoid short-term deviations of the tissue temperatures beyond physiologically normal ranges. Short-term deviations of tissue temperature have often been shown to affect plant function, often with effects carried beyond the period of exposure. For example, sublethal heat shock, with short-term tissue temperature increases in the range observed in otherwise well-controlled infrared (IR) heating studies, can induce the synthesis of heat shock proteins, and other physiological changes that are important for thermotolerance (Sun et al., 2002). Another plant-physiological feature that is easy to inadvertently alter during a plant population warming study is vapor pressure deficit (VPD), which can lead to decreased leaf water potential for plants under some growing conditions. Decreased leaf water potential can trigger alterations to plants similar to those observed in sublethal heat shock, for example, the synthesis of heat shock proteins, and other physiological changes that are important for abiotic stress tolerance (Sun et al., 2002). One means to alter the VPD is to alter the absolute humidity (Kimball, 2005). Global climactic change models predict that absolute humidity will change with global warming (Houghton et al., 2001) indicating that the ability to maintain absolute humidities while altering temperature would be an additional prerequisite for a heating system designed for study of various plant physiological responses to elevated temperature.

Plant physiological experimentation employs both square-wave manipulation of environmental variables as well as ambient +/ some proportion or degree of the quantity of an environmental factor, for example, average seasonal temperature $+x°C$ (e.g., (Reddy et al., 2001)) depending on the study objectives, thus the inclusion of computer-based DAC via the Internet is highly desirable to facilitate the study of plant physiological responses to various aspects of high night temperature. The responses by plants and plant populations are multiple, so demand the ability to clearly separate these responses via well-replicated studies, often of fairly subtle temperature changes, thus cost and scalability warrant consideration. The apparatus should avoid unintended effects on the local environment.

Current apparatuses used to study the effects of high night temperatures are limited in ability to carefully control the elevated temperature, conduct replicated study of populations of plants, or minimize perturbation of other environmental factors. Greenhouses, growth chambers, phytotrons, open-top chambers (OTC), and naturally-lit plant growth chambers (known as Soil-Plant-Atmosphere-Research (SPAR) units) are usually used in controlled environmental studies. Greenhouses generally have higher humidity, lower wind speed, and lower light intensity compared to outside. Moreover, greenhouse coverings typically transmit two-thirds to three-fourths of the available sunlight (Allen et al., 1992). In artificially lit growth chambers, the temperature is well controlled, however plants are subjected to an artificial light environment. The phytotron has similar light conditions as that of artificially lit growth chambers and also has smaller rooting volumes, which might restrict the partitioning of carbohydrates to roots (Thomas and Strain, 1991). The OTC requires a high flow rate of air in and out of the OTC to control the temperature and the humidity (Allen et al., 1992). However, many studies have reported higher daytime and night temperatures in OTCs compared to neighboring unenclosed areas (Adros et al., 1989; Fangmeier et al., 1986). The SPAR units are one of the best in controlling the environmental factors (Reddy et al.,

2001), however the cost and lack of mobility of the units makes them site-specific. In most of the above-mentioned facilities, the climatic conditions are unrealistic and fail to couple changes in light, temperature, and other factors resulting in poor simulation of natural environmental conditions (Tingey et al., 1996). In contrast, the use of an IR heating system can be employed in ways that do not alter other natural environmental conditions such as light intensity, humidity, and wind speed, while precisely controlling temperature.

The use of IR heating for study of plant—and ecosystem response to global warming has been increasing during the last 10–15 years. For example, Harte and Shaw (1995) and Harte et al. (1995) have conducted a long-term study of the effect of added heat to plots located in a montane community. The IR radiation warms the vegetation similar to that of normal solar heating and is energetically efficient because it heats the vegetation directly without having to overcome a boundary layer resistance if the air were to be heated (Kimball, 2005). An improvement in IR heater control was made by Nijs (Nijs et al., 1996), who varied the heat output in order to maintain a constant 2.5°C difference in canopy temperature compared to the control plots. Free Air Temperature Increase (FATI), as coined by Nijs (Nijs et al., 1996), is based on modulated IR radiation and increases temperature in a controlled fashion, without enclosing the plants. More recent reports on the use of IR heating systems for ecosystem warming include Luo et al. (2001), Shaw et al. (2002), Wan et al. (2002), Noormets et al. (2004), and Kimball et al. (2008). All the above-mentioned studies have primarily used IR heating to study the effect of warming at the plant population level, mostly with the intent to estimate possible ecosystem effects of global warming. In contrast, we sought to develop an IR-based system allowing the study of plant physiological responses to high temperature. The purposes of this chapter is to explicitly describe the controlling capabilities of the presented IR heating system, provide results indicating that the described apparatus meets the criteria indicated above for use in study of plant physiological response, and to provide additional example results to further characterize the system and illustrate successful application of the apparatus in study of plant physiological response to high night temperature.

DISCUSSION AND RESULTS

Trade names and company names are included for the benefit of the reader and do not imply any endorsement or preferential treatment by the authors or Texas AgriLife Research.

Infrared (IR) Heaters

The IR heaters, purchased from Omega (RAD 3113 BV/208, OMEGA Engineering, Inc. Stamford, Connecticut, USA) are housed in a rigid aluminum housing which is 77.8 cm in length and 9.4 cm in width. The aluminum housing is equipped with interlocking connectors, mounting clamp, conduit connector, polished aluminum reflector, and single radiant (RAD) elements. The single RAD element is a rod-shaped heating element (1 cm diameter and 57.8 cm long) mounted at the focal point of the polished aluminum reflector. The working voltage of the heating element is 120 volts and has a power of 1100 watts. The Incoloy (an iron-nickel alloy) sheath is 9.5 mm diameter.

The operating wavelengths of the IR heaters are >1,200 nm, and the IR heater output is negligible <1,200 nm (Omega, 2008). Hence, there is no significant emission of photo-morphogenic wavelengths. Stranded, insulated, nickel-plated copper wire is used for connecting the heaters to the power controllers. For protection of the IR heaters and personnel, a grill (GR-3, OMEGA Engineering, Inc.) is provided for each IR heater. A detailed description of IR heaters is provided in *The Electric Heaters Handbook* (Omega, 2008). Previous studies have described ways to weatherproof similar IR heaters (Kimball et al., 2008).

Power Semi-Conductor Controllers

A SCR power controller is an output device used for fast heat switching, to control variable resistance heaters and to switch higher amperage electric heaters. In the setup, the IR heaters are controlled using power semi-conductor controllers (SCR71P-208-030-S60, OMEGA Engineering, Inc.) to enable proportioning heating action instead of on/off action. These power controllers are single-phase models, phase-angle firing, 208 volt, 30 amp with a 60-sec soft-start option. These power controllers use "phase-angled fired" proportional control, which eliminates thermal shock, and extends the working life of the heating elements. The "phase-angled fired" proportional SCR control also provides a smooth, rapid (in milliseconds), and controlled heating process. This is potentially advantageous for vegetation warming studies because it provides a nearly continuous adjustment of the heater output thus minimizing the risk of prolonged or intense heating of the plant material with unexpected deviations away from the set point. The power controller receives a 4–20 mA process output signal from a temperature controller. This signal is processed by the electronics in the power controller to switch the heaters at sub-cycle intervals, resulting in a smooth radiation output. A detailed description of power semi-conductor controllers is provided in *The Electric Heaters Handbook* (Omega, 2008).

i-Series Temperature Controllers

The basic function of a temperature controller is to compare the actual temperature with the set point temperature and produce an output which will maintain that set point temperature. The i-Series temperature controllers (CNi16D53-C24, OMEGA Engineering, Inc.) used in the present setup include digital panel meters and single loop autotuned proportional integral derivative (PID) control mode controllers that are simple to configure and use, while providing tremendous options including direct connectivity to an Ethernet network with the ability to serve web pages over a local area network (LAN) or the Internet. The i-Series temperature controllers are Deutsche Industrial Norm (DIN) compatible for easy incorporation into industrial mounting and control systems. The instrument utilizes chip on board (COB) and surface mount technology (SMT) assembly techniques and automation and provides the ability to program and set the temperatures and alarms. The presented system uses 1/16 DIN Omega-series controllers with dual display, analog output, 0–10 Vdc or 0–20 mA, at 500 ohm max, and relay. The embedded Internet and serial communication interface allows remote DAC and flexibility in programming via OLE for Process Control (OPC) software on remote PCs. The i-Series temperature controller has both RS-232 and RS-485 serial communication interface, which allows multiple temperature controllers to be connected through a single industrial server. A detailed description and applications of i-Series temperature controllers are provided in *The Electric Heaters Handbook* (Omega, 2008).

i-Server

The temperature controllers are connected to an i-Server (EIS-2-RJ, OMEGA Engineering, Inc.) using a RJ45 serial port. The RS-485 interface standard used for connecting the temperature controllers to the i-Server provides distances up to 1,200 m, data rates up to 10 Mbps, up to 32 line drivers on the line, and up to 32 line receivers on the same line (Park et al., 2003). The i-Server takes a dynamically assigned the Internet Protocol (IP) address from a Dynamic Host Configuration Protocol (DHCP) server on the network. This DHCP client capability is a valuable and unique feature of the i-Server that makes it extremely easy and simple to use on almost any Ethernet network. The i-Server connects to an Ethernet network with a standard RJ45 connector. In addition, the i-Server can be used to create a virtual tunnel on an Ethernet/Internet network simulating a local point-to-point serial connection between a serial device and a PC. The serial devices will function over the Ethernet network or the Internet as if they were connected directly to a PC. The COM port on the i-Server simulates a local COM port on the PC. The i-Series temperature controllers and i-Servers connected to an Ethernet network or the Internet makes it possible to monitor and control a process from any remote place. A detailed description of the i-Server and its applications are provided in *The Electric Heaters Handbook* (Omega, 2008).

OLE for Process Control (OPC) Server Software

The OPC servers are hardware drivers that are written to a common standard, OPC. The OPC compliant programs (OPC Clients) are available for Distributed Control System (DCS), Supervisory Control and Data Acquisition (SCADA), and Human Machine Interface (HMI). Previously each software or application developer was required to write a custom interface to exchange data with hardware field devices. The OPC eliminates this requirement by defining a common, high performance interface. The OPC specification is a non-proprietary technical specification that defines a set of standard interfaces based upon Microsoft's OLE/COM technology. A complete description and specifications of OPC servers and OPC clients are available at the OPC Website (OPC Foundation).

Thermocouples

The temperature input can be provided through thermocouple, resistance temperature detector (RTD), or process voltage/current. Thermocouples were used in the presented system to provide flexibility in the system, that is use of IR (non-contact) thermocouples, rapid response air temperature thermocouples, or hypodermic needle-type (internal temperature of desired plant part) thermocouples as desired. In the presented system, the temperature controllers receive input from the rapid response air thermocouples (GTMQSS-040E-12, OMEGA Engineering, Inc.), which were attached to the temperature controllers by thermocouple wire (304-T-MO-032, OMEGA Engineering,

Inc.). The thermocouples are low noise thermocouple probes with type "T" grounded-junction probe with a Teflon-insulated extension wire and subminiature male connector termination. The thermocouple wire is also a type "T" wire, MgO insulation, 0.076 cm cable and the sheath material of the wire is 304 stainless steel (Omegaclad thermocouple wire, *The Electric Heaters Handbook*, Omega, 2008).

Setup and Working of IR Heating System

The thermocouple is attached to the i-Series temperature controller by thermocouple wire, which is a type "T" grounded-junction probe with a Teflon-insulated extension. The i-Series temperature controller also communicates with the power controller and i-Server. The i-Series temperature controller communicates with the power controller by electrical wire connections and with the i-Server through an RS-485 interface via a RJ45 serial port. The power controllers are connected to the IR heaters by stranded, insulated, nickel-plated copper wire. The i-Server communicates with the Ethernet/Internet via a RJ45 serial port. The OLE software is installed on a PC connected via the Ethernet/Internet. The temperature can be set at predetermined set points using i-Series temperature controllers, which can be accessed from a remote distance through a PC via the Internet and i-Server. Sophisticated temperature regimes can be applied through use of the OLE software, as can data acquisition. The model of the setup and the actual setup are shown in Figure 1.

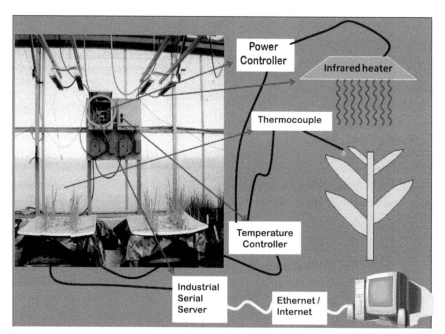

Figure 1. Picture showing cartoon and actual setup of IR heating system. Air temperatures can be set at predetermined set points using i-Series temperature controllers. When the temperature is below the set point as determined by the readings from the thermocouples, a signal from an i-Series temperature controller is sent to a power controller, which in turn controls the heater output to maintain the temperature very near the set point.

Plant Culture and Temperature Treatments

Three experiments were conducted in the greenhouse at the Texas AgriLife Research and Extension Center at Beaumont, Texas, USA. "Cocodrie," a common U.S. rice (*Oryza sativa* L.) cultivar of tropical japonica background, was used in all the experiments. The average ambient night temperature during the reproductive period of the rice growing season at the location varied between 26 and 28°C (Mohammed and Tarpley, 2009). Hence, the ambient and elevated night temperatures were set at 27 and 32°C (ambient plus 5°C), respectively. This is a large temperature difference relative to most vegetation warming studies and is also a square-wave treatment requiring the maintenance of constant temperature over long periods of time, thus challenged the heating system's ability for accurate, precise, reliable heating without causing plant physiological artifacts.

Plants were grown in 3-L pots that were placed in a square box (0.84 m²), 10 pots per box. The boxes were lined with black plastic (thickness = 0.15 mm; FILM-GARD, Minneapolis, Minnesota, USA) that served as a water reservoir. Pots were filled with a clay soil (fine montmorillonite and thermic Entic Pelludert) (Chen et al., 1989). At 20 days after emergence (DAE), the boxes were filled with water to approximately 3 cm above the top of the soil in each pot. A reflective foam cover (Cellofoam Sheathing/ Underlayment, Cellofoam. North America Inc., Conyers, Georgia, USA) was placed over the water surface to prevent direct IR heating of water. A three-way split application of nitrogen was used as described by (Mohammed et al., 2007). Nitrogen was applied in the form of urea and ammonum sulfate, and phosphorus in the form of P_2O_5. At planting, urea-N was applied at the rate of 113.5 kg ha⁻¹ along with 45.4 kg ha⁻¹ phosphorus (P_2O_5). The second and third nitrogen fertilizations (both 79.5 kg ha⁻¹ nitrogen in the form of ammonium sulfate) were applied 20 DAE and at the panicle-differentiation stage.

Plants were subjected to elevated night temperature through the use of the nearly continuously controlled IR heaters, which were positioned 1.0 m above the topmost part of the plants. This involved controlled heating of small unenclosed areas of the greenhouse. Air temperatures were controlled at predetermined set points (27°C and 32°C). The night temperatures were imposed from 2000 hr until 0600 hr starting from 20 DAE until harvest. There were three experiments presented in the present study. The assignment of heat treatments to greenhouse location was random within each experiment. The greenhouse was maintained at 27°C night time temperature and within this, plants of the HNT treatment were subjected to elevated night time temperature through the use of nearly continuously controlled (sub-second response) IR heaters. In each experiment, there were four sets (replications) of IR heaters, two IR heaters in each set. The night temperature and humidity were independently monitored using standalone sensor/loggers (HOBOs, Onset Computer Corporation, Bourne, Massachusetts, USA) in both the ambient and the HNT regimes. In each temperature regime, 1 m below the IR heaters, there were four HOBOs, one HOBO per replication. In addition, under the HNT regime, there were two additional HOBOs per replication placed at 0.75 m and 1.25 m below the IR heaters to measure the temperature at different levels below the heaters. The HOBOs were set to record temperature and humidity at 15-min interval. Hence, the value for a temperature or humidity for a night (2000–0600 hr) is an average of 40 data points. In experiment-I, plants grown under two sets of IR heaters (randomly selected) were exposed to a wind velocity of 2.2 m s^{-1} using industrial fans with speed controls (Super Fan, Mobile Air Circulator, Air Vent Inc., Dallas, Texas, USA). The wind speed was measured using a wind speed meter (ADC™•WIND™, The Brunton Company, Riverton, Wyoming, USA). In addition to this main study (consisting of three experiments), a preliminary study (one experiment) was done wherein the HNT was maintained at 32°C using greenhouse heaters. In the preliminary study, temperature and humidity were recorded using HOBOs. The data for humidity under IR versus greenhouse heating was analyzed using a paired t-test.

Performance of Infrared Heating System

There was no difference between the experiments (locations within the greenhouse) for ambient as well as elevated night temperatures measured using the HOBOs.

The IR heaters provided accurate night (2000–0600 hr) temperatures during the cropping season (emergence to harvest). The average night temperatures were 27.3 and 31.8°C for the ambient (27°C) and ambient + 5°C (32°C) temperature treatments, respectively (Figure 2A), indicating the ability of the described apparatus to maintain a large temperature differential for an unenclosed space. For most of the time of heat exposure (82%), night temperature was held within 0.5°C for 32°C treatment (precision) and the minimum and maximum recorded temperatures at any point during the 32°C treatment were 30.0 and 32.9°C (reliability) (Figure 2B). Similar results of accurate control of temperatures as imposed by the usage of IR heaters are reported in previous studies (Kimball et al., 2008; Nijs et al., 1996). However, a previous study reported short episodes of tissue temperature increases up to 14 °C above set point (Kimball et al., 2008). We have not observed any "thermal shock" type rise in tissue temperature using the optimized conditions described here. The stability of the tissue temperature in the present study can be attributed to the combination of two factors: (1) the smooth, very rapid modulation of heater output provided through the combined use of the phase-angled fired SCR power controllers with the autotuned PID temperature controllers; which prevented thermal shock not only of the heating elements, but also of the target vegetation; and (2) the use of the fast response, low noise air-sensing thermocouples, which were subject to very little temperature buffering of the target and system heating response. A previous study had reported very little change in the surrounding air temperature, although canopy temperature differences were achieved (Nijs et al., 1997). A difference between the above-mentioned study and other reported IR vegetation warming studies compared to the present study is our use of air temperature, instead of canopy temperature for controlling system response.

The IR heating system had more precision, accuracy, and reliability in maintaining set point temperature compared to greenhouse heating (Figures 3A, B, C). The paired t-test results indicated no differences between absolute humidity under IR and greenhouse heating (Figure 3A, B). Moreover, the humidity during the cropping season was 14.3 and 14.4 gm m^{-3} under the high night and ambient night temperature treatments, respectively, suggesting that the VPD was not altered to any unnatural extent through a change in absolute humidity during the IR heating. The ability to maintain the same absolute humidity in the presented study will provide flexibility in studying plant physiological response to elevated temperature in various ways, and was possibly due to the heating of unenclosed areas with light, but nearly constant, wind providing some mixing of the air.

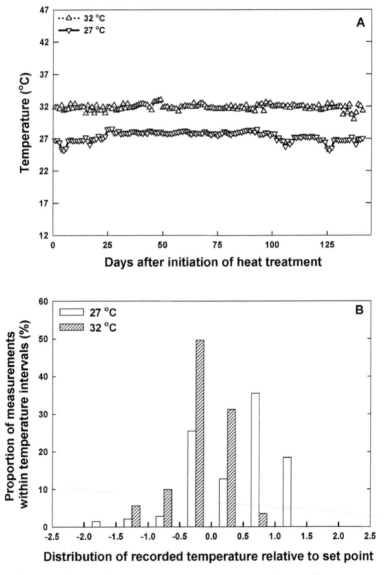

Figure 2. Ability of IR heating system to maintain temperatures at set points. The temperatures were monitored using standalone sensor/loggers (HOBOs). The ambient night temperature was set at 27°C and high night temperature at 32°C (A). Distribution of recorded temperature to the set point for ANT and HNT treatments (B).

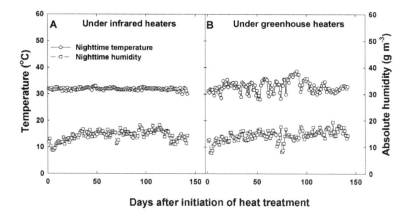

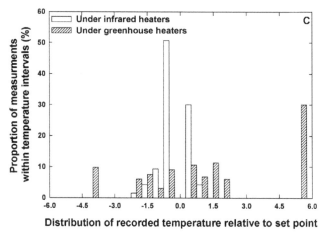

Figure 3. Temperature and humidity under IR heating system and greenhouse conditions. The night temperatures and humidities were monitored using standalone sensor/loggers (HOBOs) under IR heating system (A) and greenhouse conditions (B). The night temperature was set at 32°C. Distribution of recorded temperature to the set point for IR and greenhouse heating systems (C).

The accuracy of the IR heater in maintaining the set point temperature greatly decreased with a wind velocity of 2.2 m s⁻¹ (Figure 4A). At 2.2 m s⁻¹, the IR heaters were off by 3°C. Similar results of decreased IR heater thermal radiation efficiency with increase in wind speed have been reported in previous studies (Kimball, 2005; Kimball et al., 2008). However, the decrease in efficiency with wind speed can be adequately estimated (Kimball, 2005). In the presented setup, the IR heaters were able to maintain the set point temperatures when mounted 1 m above the canopy (Figure 4B), however, further increasing the mounting distance above the canopy also decreased the ability of the IR heaters to maintain the set point temperatures. Similar results of decrease in the ability to maintain the set point temperatures with increase in mounting distance were reported by Kimball (2005).

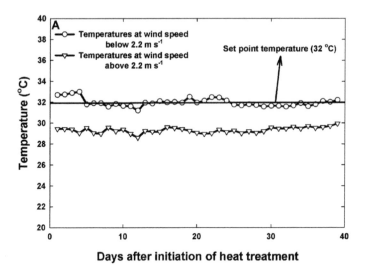

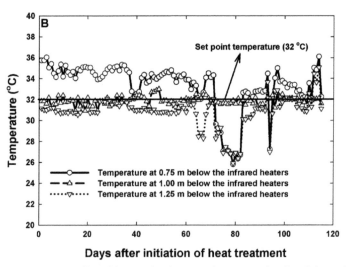

Figure 4. Temperatures as affected by wind velocity and mounting height of the IR heaters. To determine the ability of the IR heating system to control the temperatures, the IR heating lamps were exposed to wind velocities above and below 2.2 m s⁻¹ (A) and mounted at different heights (B).

CONCLUSION

The described IR heating system with the phase-angled-fired SCR power controller, autotuned PID temperature controllers, and fast response, low noise, air temperature thermocouples meets the utilitarian requirements of a heating system for plant physiology studies in that the elevated temperature can be accurately, precisely, and reliably controlled, and can be scaled in replicated study of populations of plants with minimal perturbation of other environmental factors. Changes to the physiology that can alter plant tolerance to abiotic stresses, such as "thermal shock" events or unusual alteration to the VPD due to change in the canopy to air temperature difference or change in the absolute humidity, are avoided. The combination of the lack of effect on other environmental factors and lack of unintended effects on the plant physiology indicate that the presented apparatus is specifically suitable for study of plant physiological response to high night temperature. The described IR heating system was able to maintain constant set point temperature, provided the heaters were not too high above the vegetation. Furthermore, wind speeds of or above 2.2 m s^{-1} decreased the efficiency of this IR heating system. This IR heating system can be used in conductance of studies evaluating plant physiological response to high nighttime temperature.

KEYWORDS

- **Aluminum housing**
- **Global warming**
- **Infrared heating system**
- **Thermocouple**

AUTHORS' CONTRIBUTIONS

Abdul R. Mohammed and Lee Tarpley conceived the project, designed experiments, and prepared the manuscript. Abdul R Mohammed conducted the experiments and developed modifications to the instrumentation. Lee Tarpley designed the instrumentation and acquired funding. Abdul R Mohammed and Lee Tarpley read and approved the final manuscript.

ACKNOWLEDGMENTS

The authors thank the Texas Rice Belt Warehouse for providing a graduate fellowship to Abdul R Mohammed during his studies for the Ph.D. degree, as well as the Texas Rice Research Foundation for partial financial support during the term of this project. We would also like to thank Mr. Robert Freeman for his help with electrical setup and power connections of the IR heating system.

COMPETING INTERESTS

The authors declare that they have no competing interests.

Chapter 13

Auxin Binding Protein 1 for Root Growth

Alexandre Tromas, Nils Braun, Philippe Muller, Tatyana Khodus, Ivan A. Paponov, Klaus Palme, Karin Ljung, Ji-Young Lee, Philip Benfey, James A. H. Murray, Ben Scheres, and Catherine Perrot-Rechenmann

INTRODUCTION

In plants, the phytohormone auxin is a crucial regulator sustaining growth and development. At the cellular level, auxin is interpreted differentially in a tissue- and dose-dependent manner. Mechanisms of auxin signaling are partially unknown and the contribution of the Auxin Binding Protein 1 (ABP1) as an auxin receptor is still a matter of debate.

Here we took advantage of the present knowledge of the root biological system to demonstrate that ABP1 is required for auxin response. The use of conditional ABP1 defective plants reveals that the protein is essential for maintenance of the root meristem and acts at least on the D-type cyclin/retinoblastoma pathway to control entry into the cell cycle. The ABP1 affects Plethora gradients and confers auxin sensitivity to root cells thus defining the competence of the cells to be maintained within the meristem or to elongate. The ABP1 is also implicated in the regulation of gene expression in response to auxin.

Our data support that ABP1 is a key regulator for root growth and is required for auxin-mediated responses. Differential effects of ABP1 on various auxin responses support a model in which ABP1 is the major regulator for auxin action on the cell cycle and regulates auxin-mediated gene expression and cell elongation in addition to the already well known transport inhibitor response 1 (TIR1)-mediated ubiquitination pathway.

The plant hormone auxin plays crucial roles in plant development. While one F-box protein mediated signal transduction route has been discovered, mechanisms of auxin signaling are still partially unknown. Effects of differential accumulation of auxin have been closely analyzed in *Arabidopsis* roots, where auxin mediates stem cell specification, maintenance of the root meristem, patterning and growth. At the cellular level, auxin is interpreted differentially in a tissue- and dose-dependent manner. Auxin concentrations that promote cell expansion in shoot tissues inhibit cell elongation and promote cell division in roots suggesting that in addition to the importance of auxin distribution and local auxin concentration, differences of cell responsiveness also play critical roles. In the presence of auxin, Aux/IAA transcriptional repressor proteins are recruited by the F-box protein TIR1 within the SCF[TIR1] complex, polyubiquitinylated, and degraded via the 26S proteasome (Badescu and Napier, 2006; Mockaitis and Estelle, 2008). The TIR1 binds auxin and acts as an auxin receptor mediating rapid

Aux/IAA protein degradation and subsequent auxin response factor (ARF)-dependent activation of transcription (Dharmasiri et al., 2005; Kepinski and Leyser, 2005; Tan et al., 2007). Auxin responses, however, involve another putative auxin receptor, the auxin binding protein1 (ABP1) (Napier et al., 2002). This protein was isolated based on its capacity to bind auxin and is involved in a set of early auxin responses such as rapid activation of ion fluxes at the plasma membrane (Napier et al., 2002). Previous efforts to characterize ABP1's role during plant development have been hampered by the embryo-lethality of the null abp1 mutant in *Arabidopsis* (Chen et al., 2001). Developmental map of gene expression in *Arabidopsis* revealed that ABP1 (*At4g02980*) exhibit a fairly constant expression in almost all tissues throughout vegetative plant development suggesting that its role is not restricted to embryo development (Schmid et al., 2005; Winter et al., 2007). Using conditional ABP1 *Arabidopsis* lines, we recently showed that ABP1 is required for post-embryonic shoot development acting on various cellular responses in a context-dependent manner (Braun et al., 2008). It remains, however, unknown whether auxin is required for ABP1-driven downstream responses and what is ABP1's role in plant root growth.

Primary root growth is sustained by cell division within the root meristem, which ensures the continuous production of new cells that elongate and differentiate. Accumulating evidence indicates that auxin controls cell identity, cell division, and cell expansion in a dose-dependent manner (Grieneisen et al., 2007; Leyser, 2005; Tanaka et al., 2006; Vanneste and Friml, 2009). The primary root exhibits a longitudinal gradient of cell differentiation overlapping an instructive gradient of auxin (Grieneisen et al., 2007). Thus, we used the *Arabidopsis* root as a model to dissect the role of ABP1 in the auxin mediated control of growth.

MATERIALS AND METHODS

Plant Lines and Growth Conditions

The Col-0 ecotype of *Arabiodopsis thaliana* was used for construction of all transgenic plants. Markers, mutant, overexpressor and RNAi lines were introduced by crosses in ABP1 conditional lines (Braun et al., 2008) and double homozygotes selected from F3 progeny were used for all observations. Controls were either wild-type Col0 or plants expressing GUS under the ethanol inducible system (AlcA:Gus) to guarantee that the observed feature is not related to the ethanol system. Lines used were: CYCB1::DboxCYCB1;1:GUS (Colon-Carmona et al., 1999), DR5::GUS (Ulmasov et al., 1997), promoter::GFP lines A8, Q12, S04, S17, S32 (Lee et al., 2006), SHR::GFP (Helariutta et al., 2000), SCR::GFP (Sabatini et al., 1999), promoter::CFP lines PLT1, PLT2 (Aida et al., 2004), *shr-2* (Fukaki et al., 1998), 35S::PLT2-GR (Galinha et al., 2007), RCH1::RBR RNAi ("rRBr") (Wildwater et al., 2005), and 35S::CYCD3;1 which is in Landsberg ecotype (Dewitte et al., 2003). Seeds were germinated under sterile conditions on plates containing 1/2 Murashige and Skoog (MS) basal salt mixture, buffered at pH 5.7 with 2.5 mM 2-(N-morpholino) ethanesulfonic acid, and containing 0.9% vitro agar (Kalys, St Ismier, France). For selection of transgenic plants or segregating plants issuing from crosses with marker lines either kanamycin or hygromycin was used. In all cases, plates were incubated in a vertical position at 22°C under constant light at 99 µmol. m^{-2} x s^{-1} intensity.

Ethanol Induction and Treatments

The ethanol induction was performed as described (Braun et al., 2008). Inductions were performed either immediately after stratification of the seeds or after 1–4 days of culture for shorter exposure to ethanol vapors as indicated.

For chemical treatments, seedlings were either grown on half MS and transferred on media containing 1–100 nM NAA for the indicated time or grown on 10 µM NPA. The root auxin sensitivity assay was performed by growing seedlings on plates containing appropriate concentrations of indole 3 acetic acid (IAA). The root length was measured after 4 days of growth. Elongation is expressed relative to the mean root elongation of the same genotype on medium without auxin with standard deviation calculated on at least 40 samples each.

Root Imagery

Whole mount microscopic analysis of roots were performed on fresh material stained with 5 µM FM4–64 for 10 min and roots were observed using an inverted confocal microscope TCS SP2 (Leica microsystems, Heidelberg, Germany). Roots of rRBr and rRBr, SS12K plants were fixed on acetic acid and methanol. Periodic acid-Schiff's reagent was used to reveal starch granules and tissues were stained with propidium iodide before observation. The pictures were shaped and assembled using Photoshop (Adobe) without treatment. Quantitative measurements were realized with ImageJ.

The β-glucuronisade (GUS) assays were performed as described (Malamy and Benfey, 1997), samples were stained overnight at 37°C. The GUS-stained seedlings were observed without clearing with a Multizoom AZ100 microscope (Nikon Corporation Instruments Company, Japan).

Immunodetection

Immunolocalization was performed on 4-day-old seedlings. Immunolocalization in roots was performed as described (Ueda et al., 2004). Labeling was performed with rabbit anti-PIN1, guinea pig anti-PIN2, and rabbit anti-PIN4 antibodies at 1:500, 1:400, and 1:500 dilutions, respectively. Alexa Fluor 488-conjugated goat anti-rabbit and Alexa Fluor 555-anti-guinea pig secondary antibodies were used at 1:400 dilution. During the immunolocalization procedures, solutions were changed using a pipetting robot (InsituPro, Intavis Bioanalytical Instruments AG).

IAA Quantification

Root tips were collected from 4 days post-germination (dpg) seedlings exposed to ethanol for the indicated time and frozen immediately in liquid nitrogen. The frozen samples were homogenized in 0.5 ml 50 mM sodium-phosphate buffer pH 7.0 containing 0.02% diethyldithiocarbamic acid (Sigma) and 500 pg $[^{13}C_6]$-IAA (Cambridge Isotope Laboratories, Andover, MA, USA) internal standard for 2 min at a frequency of 30 Hz, using a Retsch MM 301 vibration mill (Retsch GmbH, Haan, Germany) and a 3 mm tungsten carbide bead. The samples were then incubated for 15 min at + 4°C under continuous shaking. The pH was adjusted to 2.7 with 1 M HCl, and the samples were purified by solid phase extraction on a 500 mg Isolute C8 (EC) column (International Sorbent Technology), conditioned with 2 ml methanol and 2 ml 1% acetic acid. After sample application, the column was first washed with 2 ml 10% methanol in 1% acetic acid and then eluted with 2 ml 70% methanol in 1% acetic acid. The dried samples were dissolved in 0.2 ml 2-propanol and 1 ml dichloromethane and 5 µl 2 M trimethylsilyl-diazomethane in hexane (Aldrich) was added to methylate the samples. After methylation, the samples were trimethylsilylated and IAA was quantified by gas chromatography-selected reaction monitoring-mass spectrometry as described in Edlund et al., (1995). All samples were analyzed in triplicates from two biological repeats.

Real-Time RT-PCR Analysis

RNA was extracted from roots of 4dpg seedlings treated with 5 µM NAA for 30 min to 6 hr using an Qiagen RNeasy kit and digested with RNAse free DNAse on the column following the manufacturer's instructions (Qiagen S.A., Courtaboeuf, France). First-strand cDNAs were synthesized from 5 µg of total RNA using Superscript II reverse transcriptase according to the manufacturer's instructions. Quantitative RT-PCR analyses were performed using SYBR Green QPCR master mix (Roche) with specific primers as reported (Braun et al., 2008). Two biological repeats were analyzed in duplicates.

DISCUSSION

Our data show that ABP1 is implicated to various degrees in the control of auxin responses mediating root growth, especially cell division, cell elongation, and gene expression. The ABP1 is essential to maintain the mitotic activity of meristematic cells and stem cells (Figures 6 and 7). This critical control of cell division in roots confirms previous data obtained on BY2 cell suspension (David et al., 2007), embryo (Chen et al., 2001), and shoot tissues (Braun et al., 2008) and reveals the general nature of the control exerted by ABP1 on cell division. Interestingly, this effect is consistent with auxin's permissive role in cell division. Based on analysis of root stem cells, ABP1 affects the D-type Cyclin/RBR regulatory pathway (Figure 7). Although the molecular link between ABP1 and these cell cycle regulators may not be direct, it is clear that ABP1 is essential for the regulation of the G1/S transition. Within the Cyclin D/RBR pathway there are multiple potential targets, amongst which CDK inhibitors (KRPs) and E2F transcriptional factors are additional relevant targets. For example, E2Fc has been reported to negatively affect cell division (del Pozo et al., 2002) and we cannot exclude that the increased accumulation of E2Fc mRNAs in ABP1 inactivated seedlings also contributes to the inhibition of cell division. Overproduction of E2Fc was shown to inhibit cell proliferation (del Pozo et al., 2002, 2006). The E2Fc protein is however submitted to rapid degradation in an ubiquitin dependent manner and changes at the RNA level might not reflect the amount of protein. Protein turnover of E2Fc, as well as activator E2Fb which has been shown to be stabilized in response to auxin in BY2 cells (Magyar et al., 2005), will merit investigation in plants inactivated for ABP1.

The role of ABP1 is however not restricted to the auxin control of cell division, as our results clearly show that the protein is implicated in the auxin regulation of cell elongation and gene expression (Figure 4). By conferring on the root cells a relative sensitivity to auxin, ABP1 contributes to the regulation of cell behavior at the transition zone, likely contributing to deciphering the gradient of auxin (Blilou et al., 2005; Grieneisen et al., 2007; Ljung et al., 2005) and the cross-talk between auxin and cytokinin (Dello Ioio et al., 2007, 2008). It is tempting to hypothesize that the effect of ABP1 on PLT gradient leads to the switch between division and elongation that occurs in the transition zone. Interestingly, cell elongation in roots is inhibited by auxin and occurs as auxin content decreases, whereas cell expansion in shoot tissues is promoted by auxin. Importantly, the differential effect of ABP1 on cell elongation in roots and expansion in shoots (Braun et al., 2008; Jones et al., 1998) perfectly matches known differential effects of auxin on both tissues.

A differential auxin response is also observed on gene expression. The ABP1 is required for auxin induced expression of a subset of *Aux/IAA* genes in roots (Figure 4B) whereas we previously observed enhanced auxin responses for the same *Aux/IAA* genes and others in shoot tissues of ABP1 inactivated plants (Braun et al., 2008), indicating that the fine tuning exerted by ABP1 on the regulation of gene expression differs between root and shoot. Expression of *Aux/IAAs* in response to auxin is known to be mediated by the E3 ubiquitin ligase complex SCF^{TIR1}, where the F-box and auxin receptor TIR1 recruits the Aux/IAA repressor protein for ubiquitination and further degradation by the 26S proteasome (Mockaitis and Estelle, 2008). Little is known relative to a differential regulation of this mechanism to sustain differences of tissue sensitivity to auxin (Perez-Torres et al., 2008) and our data suggest that ABP1 is involved in this process. The ABP1 sits at the plasma membrane, where its role of auxin receptor transducing at least part of the auxin signal was initially demonstrated (Badescu and Napier, 2006; Napier et al., 2002) whereas TIR1 and Aux/IAA substrates are mainly located in the nucleus, thus physical interaction between these proteins is highly unlikely. Multiple signaling components such as MAP kinases (Kovtun et al., 1998; Mockaitis and Howell, 2000), IBR5 protein phosphatase (Monroe-Augustus et al., 2003; Strader et al., 2008), phospholipase A2 (Scherer, 2002; Scherer et al., 2007), and RAC GTPases (Tao et al., 2002, 2005) have been reported to be involved in auxin signaling yet have not been implicated in the short auxin signaling SCF^{TIR1} pathway. They are possible candidates mediating ABP1 action on downstream targets. How these regulators interact with each other to mediate ABP1 and auxin dependent gene expression remains to be determined. As only a subset of *Aux/IAA* genes is affected by ABP1 inactivation, we can hypothesize that either ABP1 acts on other transcription factors co-regulating expression of these genes (as Myb77) (Shin et al., 2007). independently of TIR1 or ABP1 somehow alters the relative affinity of Aux/IAA and ARF interaction which governs their expression thus interfering with the SCF^{TIR1} pathway. Elucidating the molecular basis of the cross-talk between ABP1-mediated responses and the SCF^{TIR1} pathway will be one of the important challenges for the coming years, in particular the relative positioning of ABP1 and TIR1 in the regulation of gene expression will shed new lighting on the complex network of auxin signaling.

In conclusion, it appears that ABP1 and TIR1 collectively contribute to mediate auxin responses. Based on the data presented here and previous reports (reviewed by (Badescu and Napier, 2006; Mockaitis and Estelle, 2008)) we propose as a working model that ABP1 shares auxin regulation of gene expression and control of cell elongation with the TIR1 pathway whereas ABP1 is the master regulator for the auxin control of cell division (Figure 4C).

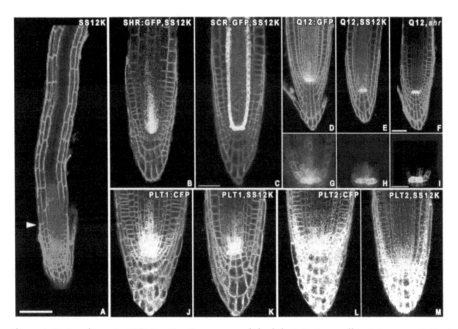

Figure 1. Post-embryonic ABP1 inactivation causes subtle defects in stem cell maintenance. A, Root phenotype of long-term ABP1 inactivated SS12K plant. Ethanol induction was maintained from germination to 14 dpg. Long cells are observed close to the QC, the elongation zone is absent and the meristem is restricted to five cells. The arrow points the end of the meristematic zone. Scale bar 80 μm. B–C, Unmodified patterning of pSHR:GFP (B) and pSCR:GFP (C) expression in induced ABP1AS at 4dpg. The GFP in green, FM4-64 staining in red. D–I, Expression pattern of pQ12: GFP (AT5G17800) in 4 dpg seedlings ethanol induced Col0 (D,G), ABP1AS (E,H), and *shr2* (F,I) showing strong reduction in the stele stem cells of both ABP1AS and *shr2*. J–M, Expression pattern of Plethora using full-size promoters driven CFP reporter (in blue) in 4 dpg seedlings ethanol induced since germination. *pPLT1: CFP* in Col0 (J) and SS12K (K); *pPLT2:CFP* in Col0 (L) and in SS12K (M). Inactivation of ABP1 provokes a reduction of the overall expression area of PLT.

To investigate whether arrest of cell division and reduction of meristem size result from a defect in stem cell activity, we studied expression of genes involved in stem cell specification. Two members of the GRAS family of transcription factors, short-root (SHR) and scarecrow (SCR) are required for QC identity and are also involved in root radial patterning during embryogenesis (Nakajima and Benfey, 2002; Sabatini et al., 2003). Expression patterns of SHR and SCR are not altered in SS12K ethanol induced plants (Figure 1B,C). Expression of Q12, a putative transcription factor specifically expressed in the QC and the stele stem cells (Lee et al., 2006), was maintained in the

QC but was consistently decreased or absent from stem cells (Figure 1D, E, G, H). Interestingly, similar results were performed when this marker was introduced into the shr-2 mutant (Figure 1F, I). The expression of SHR and SCR is however not modified in roots inactivated for ABP1 and no differentiation of the columella stem cells is observed, contrary to what was seen in the *SHR* mutant, suggesting that ABP1 does not act on SHR.

Genes of the Plethora family, encoding AP2-domain transcription factors, are also essential for root stem cell maintenance and are related to auxin action. PLTs were recently revealed to control distinct aspects of root development in a dose-dependent manner (Aida et al., 2004; Galinha et al., 2007). The general patterning of PLTs was unchanged in roots inactivated for ABP1 but the overall area of expression was consistently reduced compared to controls (Figure 1J–M) suggesting that the resulting PLT activity is limited to fewer cells, around the QC. To explore whether increased levels of PLT would be sufficient to either restore cell division within the meristem or to delay differentiation of meristematic cells, we overexpressed PLT2-GR in SS12K plants (Galinha et al., 2007). Dexamethasone-induced PLT2 expression in ABP1 inactivated roots did not reactivate cell division (Figure 2), coherent with previous data supporting that PLT overexpression sustains cell division only in cells that still have the capacity to divide (Galinha et al., 2007). The PLT2 overexpression at an early stage of ABP1 inactivation (before most meristematic cells have elongated) however, inhibits cell expansion of cells from the basal meristem and the transition zone, thus maintaining meristem size (not shown). This observation confirms the dose-dependent PLT requirement for the transition between the meristematic and the elongation zones and suggests that PLT2 is acting downstream of ABP1 to facilitate elongation. All these results suggest that ABP1 activity defines a zone of competence for PLT activity essential to control the transition between the meristem and the elongation zone.

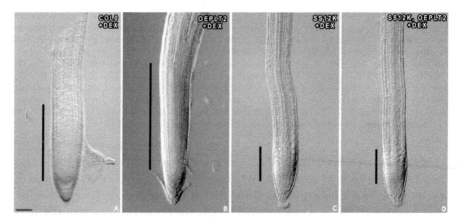

Figure 2. Overexpression of PLT2 does not reactivate cell division in roots inactivated for ABP1. Ethanol induced 35S-PLT2-GR roots in Col0 (A–B) and in SS12K (C–D) without dexamethasone (A,C) and after 24 h application of 10 µM dex (B,D). Overexpression of PLT2 cannot bypass cell cycle arrest mediated by ABP1 inactivation. Vertical bars represent meristem length. Scale bar 40 µm.

ABP1 Mediates Auxin Responsiveness

Observation of root growth defects resulting from ABP1 inactivation reinforces the correlation between the protein and auxin. These defects could, however, result from various initial defects including an alteration of auxin content, an alteration of auxin transport or a shift in auxin sensitivity.

To discriminate among these hypotheses, we first measured free IAA content in root samples after various times of ethanol induction. No significant differences were observed over a period of 48 hr of ABP1 inactivation (Figure 3A), suggesting that no global change in auxin content has occurred within this time frame. Second, to assess whether the observed root phenotype is due to more subtle changes in auxin distribution, we performed immunolocalization of PIN1 and PIN2 proteins. PIN proteins have been reported to control root meristem size through their tight control of auxin redistribution at the root apex (Blilou et al., 2005; Grieneisen et al., 2007). Two days after inactivation of ABP1, PIN protein localization was not significantly modified (Figures 3B–E). Upon prolonged inactivation, the signal decreases in correlation with cell differentiation especially for PIN2 (not shown). Thus, changes in PIN expression are likely to be a secondary consequence of ABP1 inactivation's effect on cell differentiation rather than a primary effect. As another indicator of auxin redistribution in induced SS12K roots, we monitored expression of the auxin responsive reporter DR5:GFP (Ottenschlager et al., 2003) in the absence and in the presence of the auxin transport inhibitor 1-N-naphthylphthalamic acid (NPA). We found that inactivation of ABP1 has no significant effect on the expression of the DR5:GFP reporter at the root apex (Figures 3F–G) and that NPA was able to disturb the auxin maxima independently of ABP1 activity (Figures 3H–I). This data indicates that auxin transport is still efficient in roots inactivated for ABP1 revealing that the carriers are functional. Taken together these results show that ABP1 is not essential to maintain PIN activities or the auxin gradient in the root apex.

We thus investigate the third hypothesis that ABP1 contributes to auxin responsiveness. We analyzed the effect of exogenous auxin on the major auxin responses: cell division, cell elongation and gene expression. We first attempted to reactivate cell division in the root meristem by treatment with various concentrations of exogenous auxin. Based on DboxCYCB1;1:GUS detection, we found that cell division cannot be reactivated in root meristem of ABP1 inactivated plants whatever the auxin concentration applied (not shown), thus revealing an insensitivity to auxin of the cells towards the response of division. This data confirms that ABP1 exerts a strict control on cell division and it suggests that ABP1 is required for the capacity of meristematic cells to divide in response to auxin.

Second, we examined the inhibitory effect of exogenous auxin on root elongation. As shown in Figure 4A, there was no effect of auxin on ABP1 inactivated root below 10^{-6}M IAA indicating that cell elongation was not reduced (also verified by microscopy). In the same conditions, Col0 plants exhibited 50% inhibition of growth corresponding to the inhibition of cell elongation. For IAA concentrations higher than 10^{-6}M, a significant inhibition of root growth was observed despite the initial reduced length of ABP1 inactivated roots. This effect resulted from a reduced elongation of the cells as observed for the control at lower concentrations. Roots inactivated for

ABP1 were less sensitive than control to primary root elongation inhibition caused by auxin, indicating that ABP1 is also required for this auxin response. Interestingly, it is not a complete insensitivity to auxin as observed for division but a shift in sensitivity suggesting that ABP1 is likely to contribute to this auxin response together with other regulators.

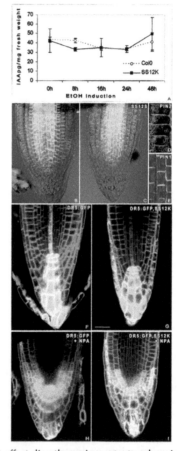

Figure 3. The ABP1 does not affect directly auxin content and auxin transport in roots. A, Analysis of IAA accumulation in Col0 and SS12K roots after various times of ethanol induction as indicated. Free IAA content was measured in root samples of 4 dpg control and SS12K lines. Inactivation of ABP1 does not affect the auxin content of root tissues. B–E, Immunolocalization of PIN1 (in green) and PIN2 (in red) in 4dpg Col0 (B) and SS12S (C–E) seedlings induced for 48 h with ethanol. D, E Inserts are enlargements showing apical and basal localization for PIN1 and PIN2, respectively. No significant difference of PIN localization is observed. PIN1 is located at the apical and lateral sides of stele cells whereas PIN2 is located at the basal side of epidermal cells and apically and laterally in cortical cells. F–I, DR5:GFP reporter expression pattern in 4dpg Col0 (F,H) and SS12K (G,I) ethanol induced seedlings grown in the absence (F–G) and in the presence of 10 µM NPA (H–I). Scale bars 40 µm. DR5:GFP visualizes the auxin response maximum and shows similar expression in roots inactivated for ABP1 and controls with a maximum in the QC, columella stem cells and differentiated cells and within the proximal meristem in some provascular cells. In the presence of 10 µM NPA, the DR5:GFP maximum was expanded laterally and shifted back in epidermis and cortical cells from the proximal meristem in both samples (H–I).

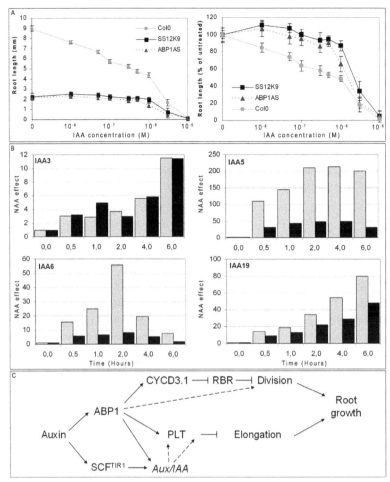

Figure 4. Inactivation of ABP1 impairs auxin responsiveness. (A) Auxin dependent root growth curve of ethanol induced Col0, ABP1AS, and SS12K. Seedlings were grown in the presence of various concentrations of IAA as indicated and induced with ethanol since germination. On the first panel, root length measurements at indicated IAA concentrations are plotted. On the second panel, data are expressed relative to the root length of each genotype grown without auxin. Standard deviations were calculated from samples >40. (B) Kinetic effects of 1 μM NAA treatment on transcript accumulation of *Aux/IAA* genes in roots of overnight ethanol induced control (open bars) and SS12K (grey bars). Data were normalized with *ACTIN2-8* then to the expression level at time zero of auxin application. (C) A model for ABP1 mediated auxin responses in roots. The permissive effect of auxin on cell division is dependent on ABP1. In root stem cells, the D-type cyclin/retinoblastoma (RBR) pathway acts downstream of ABP1 and controls the G1/S transition. In meristematic cells, ABP1 might also affect the D-type Cyclin/RBR pathway but other critical regulators of the G1/S transition phase are dependent on ABP1 activity. The ABP1 contributes to the auxin control of cell elongation by modulating a zone of competence for Plethora and by acting on the auxin-mediated regulation of Aux/IAA transcriptional repressors. It is worthwhile noticing that expression of PLTs was reported to be regulated downstream of ARF transcription factors (auxin response factors) (Aida et al., 2004) and consequently of Aux/IAAs. The ABP1 might act indirectly on PLT via Aux/IAAs regulation. It is well established that regulation of gene expression by auxin involves the TIR1 receptor which, within the SCFTIR1 E3 ligase, controls the degradation of Aux/IAA repressors (Mockaitis and Estelle, 2008). The ABP1 and TIR1 might collectively control gene regulation and elongation.

Third, we analyzed the kinetics of *Aux/IAA* mRNA accumulation in response to auxin in short-term ethanol induced control and SS12K lines. Auxin efficiently induced the expression of most *Aux/IAA* genes in root samples with minor changes in comparison to controls as illustrated for IAA3 (out of 14 genes tested) with the notable exception of *IAA5, IAA6,* and *IAA19* (Figure 4B), three genes belonging to the same clade of *Aux/IAAs* (Overvoorde et al., 2005). Auxin responsiveness for these *Aux/IAAs* was reduced by four to five fold indicating that, in roots, ABP1 is somehow required for optimal gene response to an auxin stimulus. The *Aux/IAAs* genes are submitted to various combinatorial regulations which are far to be elucidated for each member of the *Aux/IAA* gene family. The differential effect of ABP1 on the auxin responsiveness of various *Aux/IAAs* in roots illustrates the complexity of the regulatory pathway controlling their expression and indicates that ABP1 is involved.

RESULTS

ABP1 is Essential for Root Growth

To circumvent the embryo-lethality of ABP1 knock-out (Chen et al., 2001), we used ethanol inducible conditional knock-down plants generated via an antisense ABP1 construct to decrease its expression (ABP1AS lines) or via cellular immunization to inactivate ABP1 protein through its *in vivo* interaction with the recombinant antibody scFv12. The latter recognizes a conformational epitope of ABP1 overlapping the auxin binding site (SS12S and SS12K lines) thus impairing the capacity of the protein to bind and respond to auxin (Braun et al., 2008; Leblanc et al., 1999). The recombinant antibody was detected in enriched microsomal samples of ethanol induced SS12K (Figure 5A) and we showed by reciprocal co-immunoprecipitation experiments that the scFv12 produced in *Arabidopsis* interacts with AtABP1 *in vivo* (Figure 5B). The ABP1 was still detected in root samples expressing the scFv12 whereas the protein was not detected in induced antisense samples (Figure 5C). At 3 days post-germination (dpg), ethanol induced SS12S, SS12K, and ABP1AS plants exhibited similar phenotypes displaying drastic root growth reduction of 60–80% compared to ethanol induced control plants (Figures 5D–H). To determine which cellular alterations were responsible for such severe root growth defect, we performed a detailed analysis of SS12K and ABP1AS primary roots. The size of the meristem of ABP1 inactivated roots is about one third of that of controls which correlates with a reduced number of meristematic cells (Figures 6 A–H). Differentiated cortical cells reach a similar length as in control roots (Figure 6J), indicating that longitudinal elongation is not defective in ABP1 inactivated plants. The root diameter is, however, reduced by more than 40% due to decreased radial expansion but the radial tissue organization inherited from the embryonic root pattern is unaltered (Figures 6I, K, L). Introgression of a series of specific cell type GFP marker lines (Lee et al., 2006) confirmed maintenance of radial patterns (Figures 6Q–V). At the root apex, a cell layer is missing in both columella and lateral root cap in more than 80% of roots with repressed ABP1 activity (Figures 6M–N). Changes in the longitudinal gradient of root differentiation was confirmed by the use of the S17 GFP marker (Lee et al., 2006), which is expressed in phloem pole pericycle cells and is detected in the differentiation zone (Figure 6O). After ABP1 inactivation, S17 marker expression is observed at a more distal position, indicating that cells that have left the meristem rapidly begin differentiation (Figure 6P).

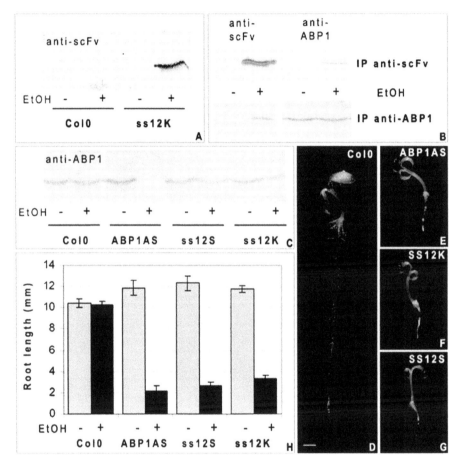

Figure 5. Inactivation of ABP1 affects post-embryonic root growth. (A) Immunodetection of scFv using anti-Etag antibody in Col0 and SS12K induced by ethanol or not. The recombinant antibody is detected in samples of ethanol induced SS12K. B, Western blot analysis of scFv and ABP1 on immunoprecipitates using anti-ABP1 mAb34 antibody, which recognizes an ABP1 epitope distinct from scFv12, or using anti-Etag scFv as indicated. The ABP1 was co-immunoprecipitated with scFv12/E-Tag and reciprocally scFv12 was co-immunoprecipitated with ABP1/mAb34 in seedling samples of ethanol induced SS12K. (C) Western blot analysis of ABP1 protein accumulation in Col0, ABP1-AS, SS12S, and the SS12K, induced or non induced using anti-ABP1 mAb34 antibody. The protein is not detected in root samples of ethanol induced ABP1AS whereas it is detected in controls and scFv12 producing lines. D–G, Severe inhibition of root growth after inactivation of ABP1 function. 3dpg seedlings treated by ethanol vapor since germination are shown. (D) Col0; E, ABP1AS; F, SS12K; G, ss12S. Scale bar 1 mm. H, Root length with or without ethanol induction of Col0, ABP1AS, SS12S, and SS12K lines at 3dpg. Error bars represent standard deviation (n > 40).

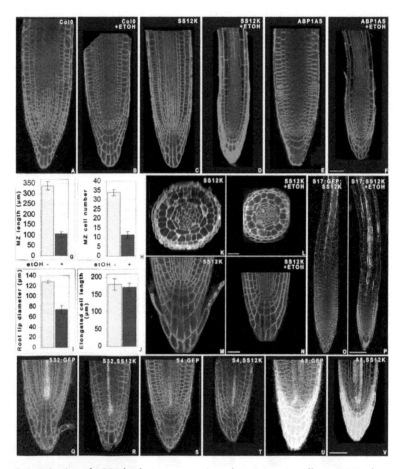

Figure 6. Inactivation of ABP1 leads to consumption of meristematic cells. A–F, Histology of root apex visualized on optical longitudinal sections of living roots stained with FM4-64 (in red). A,B, Col0; C,D, SS12K; E,F, ABP1-AS of 4-day-old seedlings grown in the absence (A,C,E) or in the presence of ethanol vapor (B,D,F). Scale bar 40 µm. G–J, Morphometric analysis comparing ethanol induced (blue bars) and non induced (grey bars) ABP1AS at 3dpg. G, meristematic zone length measured from the QC to the end of the lateral root cap; H, cortex cell number in the meristematic zone; I, root tip diameter measured at the end of the lateral root cap; J, epidermal and cortical cell length of differentiated cells, measured after the emergence of root hairs. Standard deviation were calculated from sample number >40. K–L, Optical radial sections of non induced (K) and ethanol induced (L) SS12K roots taken below the end of the lateral root cap. M–N, Close up of the root tip organization with a focus on the QC and the columella of SS12K not induced (M) or ethanol induced (N). Scale bar 40 µm. O–P, S17:GFP marker AT2G22850 of differentiated phloem pole pericycle (in green) in non induced (O) and induced (P) SS12K seedlings at 4 dpg. Scale bar 80 µm. Q–V, Patterning of cell type specific markers in induced control (Q, S, U) and ABP1-AS (R, T, V) roots. Protophloem S32:GFP AT2G18380 (Q, R); protoxylem S4:GFP AT3G25710 (S, T) and root cap and procambium A8:GFP AT3G48100 (U, V) (Lee et al., 2006). Scale bar 40 µm. GFP marker lines (Lee et al., 2006) were introgressed into SS12K and ABP1AS plants and double homozygous plants were selected from F3 progeny. No significant changes of GFP (in green) expression pattern were detected between ethanol induced or non induced ABP1AS plants or in comparison with reported expression in wild-type background.

The ABP1 Controls Meristem Size

The decrease in root meristem size could result from decreased activity of stem cells, a reduction of division of the daughter cells within the proximal meristem or accelerated consumption of meristematic cells towards elongation and differentiation. To discriminate among these possibilities we first made use of the G2/M phase pCYCB1::DboxCYCB1;1:UIDA reporter (Colon-Carmona et al., 1999) to evaluate the effect of ABP1 inactivation on the mitotic activity of meristematic cells. After short term ABP1 inactivation, expression of DboxCYCB1;1:GUS is severely reduced (Figures 7A–C), indicating that cells are no longer dividing and also that cells are not arrested in G2 or at the G2/M transition. Expression analysis of G1/S cell cycle markers revealed rapid changes in the mRNA accumulation of various markers after inactivation of ABP1, notably an increase in mRNA accumulation for the retinoblastoma-related protein (RBR) and the E2Fc transcriptional repressor, and a moderate to strong decrease for cyclin dependent kinase inhibitors (KRPs) and early D-type Cyclins, respectively (Figure 7D). The D-type cyclins have been proposed to integrate various signals and to be limiting for the G1 to S phase transition through their interaction with CDK and further phosphorylation of RBR (Menges et al., 2006). We hypothesized that the rapid decrease in D-type Cyclin expression and/or the increase in RBR following inactivation of ABP1 could disturb the Cyclin D/RBR regulatory pathway and thus contribute to the arrest of cell division. We then explored whether compensating for these expression changes by either overexpressing CYCD3.1 or reducing RBR expression would be sufficient to restore cell division (Figures 7E–L). In a wild-type background, increased CYCD3.1 (Menges et al., 2006) or decreased RBR (rRBr line) (Wildwater et al., 2005) do not change root growth rate and meristem size but supernumerary stem cells appear post-embryonically in the columella and lateral root cap (Figure 7E, G, I, K). We crossed these lines with SS12K plants and focused on root stem cells. In ethanol induced SS12K,CYCD3.1OE roots, supernumeray cells sometimes with aberrant cell plate formation are observed as in the CYCD3.1OE (Figure 7G, H), indicating that increased expression of CYCD3.1 in ABP1 inactivated cells is sufficient to restore cell division in stem cells. Similarly, in ethanol induced SS12K, rRBr roots, additional stem cell layers are generated as in the rRBr control line (Figure 7K, L). Therefore, both overexpression of CYCD3.1 and reduction of RBR bypass the cell cycle arrest mediated by inactivation of ABP1 in root cap stem cells, suggesting that the CYCLIN D/RBR pathway is operating downstream of ABP1. Future investigations will be necessary to identify the signaling pathway connecting ABP1 to the G1/S regulatory complex and to provide a comprehensive view of the role of ABP1 and auxin in cell division.

The roots of SS12K seedlings inactivated for ABP1 for up to 15 days exhibit elongated and differentiated cells next to a residual meristem made of five to six contiguous cells above the QC. These meristematic cells have been prevented from differentiation even if they have lost the capacity to divide suggesting that they were not competent for elongation and differentiation.

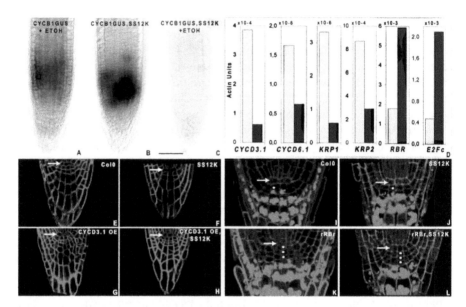

Figure 7. The ABP1 inactivation inhibits cell division in root meristem. A–C, Gus staining revealing *DboxCYCB1;1-GUS* pattern at the root tip of A, ethanol induced wild-type; B, non induced SS12K and C, 24 h ethanol induced SS12K. A strong decrease of DboxCYCB1;1-GUS expression follows inactivation of ABP1. D, Quantitative RT-PCR analysis of core cell cycle markers expression in Col0 (grey bars) and SS12K (blue bars) root seedlings treated for overnight with ethanol vapors. All data were normalized with respect to ACTIN2-8 and expressed in equivalent ACTIN units. E–H, Overexpression of CYCD3.1 restores additional divisions with altered division plane at the columella and lateral root cap in plants inactivated for ABP1. All seedlings were ethanol induced since germination. E, Wt; F, SS12K; G, CYCD3.1OE in Wt background; H, CYCD3.1OE in SS12K background. Living roots stained with FM4-64. I–L, Decreased expression of RBR at the root apex bypasses cell division arrest in the columella stem cells of plants inactivated for ABP1 and leads to the formation of additional stem cell layers. I, Col0 control; J, SS12K; K, pRCH1:RBR RNAi line (rRBr) in Wt background; L, rRBr in SS12K background. All seedlings were ethanol induced since germination. Fixed tissues stained with propidium iodide revealing statholiths in differentiated columella cells. Arrows point the QC and dots the columella stem cell layers.

KEYWORDS

- **Arabidopsis**
- **Auxin**
- **Auxin binding protein**
- **Aux/IAAs**
- **Indole 3 Acetic Acid**

ACKNOWLEDGMENTS

We acknowledge Syngenta and P. Doerner who kindly provided ethanol inducible constructs and DboxCYCB1;1-GUS plants, respectively; A. Eschstruth, M. Pierre and members of the IFR87 Cell imagery platform for technical assistance; L. Hobbie for helpful discussions and critical reading.

AUTHORS' CONTRIBUTIONS

Conceived and designed the experiments: Alexandre Tromas, Nils Braun, Ivan A. Paponov, Karin Ljung, James A. H. Murray, and Catherine Perrot-Rechenmann. Performed the experiments: Alexandre Tromas, Nils Braun, Philippe Muller, Tatyana Khodus, Ivan A. Paponov, and Klaus Palme. Analyzed the data: Alexandre Tromas, Nils Braun, Philippe Muller, Tatyana Khodus, Ivan A. Paponov, Klaus Palme, Karin Ljung, Ji-Young Lee, Philip Benfey, James A. H. Murray, Ben Scheres, and Catherine Perrot-Rechenmann. Contributed reagents/materials/analysis tools: Klaus Palme, Ji-Young Lee, Philip Benfey, and Ben Scheres. Wrote the chapter: Alexandre Tromas, Ben Scheres, and Catherine Perrot-Rechenmann.

Permissions

Chapter 1: Sampling Nucleotide Diversity in Cotton was originally published as "Sampling Nucleotide Diversity in Cotton" in *BioMed Central Ltd* 10:20, 2009. Reprinted with permission under the Creative Commons Attribution License or equivalent.

Chapter 2: Pitcher Plant *Sarraceniapurpurea* and the Inorganic Nitrogen Cycle was originally published as "The Pitcher Plant *Sarraceniapurpurea* Can Directly Acquire Organic Nitrogen and Short-Circuit the Inorganic Nitrogen Cycle" in *PloS ONE*7:7, 2009. Reprinted with permission under the Creative Commons Attribution License or equivalent.

Chapter 3: Analysis of Pathogen Resistance and Fruit Quality Traits in Melon (*Cucumismelo* L.) was originally published as "An oligo-based microarray offers novel transcriptomic approaches for the analysis of pathogen resistance and fruit quality traits in melon (*Cucumismelo* L.)" in *BioMed Central Ltd* 10:12, 2009. Reprinted with permission under the Creative Commons Attribution License or equivalent.

Chapter 4: *Arabidopsis* Gene Co-Expression Network was originally published as "*Arabidopsis* gene co-expression network and its functional modules" in *BioMed Central Ltd* 10:21, 2009. Reprinted with permission under the Creative Commons Attribution License or equivalent.

Chapter 5: Gene Expression and Physiological Responses in Mexican Maize Landraces under Drought Stress and Recovery Irrigation was originally published as "Analysis of Gene Expression and Physiological Responses in Three Mexican Maize Landraces under Drought Stress and Recovery Irrigation" in *PloS ONE* 10:30, 2009. Reprinted with permission under the Creative Commons Attribution License or equivalent.

Chapter 6: Regulators in Transgenic *Arabidopsis* Plants was originally published as "Soybean GmPHD-Type Transcription Regulators Improve Stress Tolerance in Transgenic Arabidopsis Plants" in *PloS ONE* 9:30, 2009. Reprinted with permission under the Creative Commons Attribution License or equivalent.

Chapter 7: Cell Wall Biogenesis of *Arabidopsis Thaliana* was originally published as "Cell wall biogenesis of *Arabidopsis thaliana* elongating cells: transcriptomics complements proteomics" in *BioMed Central Ltd* 10:31, 2009. Reprinted with permission under the Creative Commons Attribution License or equivalent.

Chapter 8: Spindle Assembly Checkpoint Protein Dynamics and Roles in Plant Cell Division was originally published as "Spindle Assembly Checkpoint Protein Dynamics Reveal Conserved and Unsuspected Roles in Plant Cell Division" in *PloS ONE*8:27, 2009. Reprinted with permission under the Creative Commons Attribution License or equivalent.

Chapter 9: Soybeans Iron Deficiency Response was originally published as "Integrating microarray analysis and the soybean genome to understand the soybeans iron deficiency response" in *BioMed Central Ltd* 13:8, 2009. Reprinted with permission under the Creative Commons Attribution License or equivalent.

Chapter 10: Global Characterization of *Artemisia annua* Glandular TrichomeTranscriptome was originally published as "Global characterization of Artemisia annua glandular trichometranscriptome using 454 pyrosequencing" in *BioMed Central Ltd* 10:9, 2009. Reprinted with permission under the Creative Commons Attribution License or equivalent.

Chapter 11: Tissue-Specific Expression of the Adjacent Genes in *Arabidopsis thaliana*was originally published as "A small intergenic region drives exclusive tissue-specific expression of the adjacent genes in *Arabidopsis thaliana*" in *BioMed Central Ltd* 10:16, 2009. Reprinted with permission under the Creative Commons Attribution License or equivalent.

Chapter 12: Plant Physiological Response to Elevated Night Temperature was originally published as "Instrumentation enabling study of plant physiological response to elevated night temperature" in *BioMed Central Ltd* 6:11, 2009. Reprinted with permission under the Creative Commons Attribution License or equivalent.

Chapter 13: Auxin Binding Protein 1 for Root Growth was originally published as "The AUXIN BINDING PROTEIN 1 is Required for Differential Auxin Responses Mediating Root Growth" in *BioMed Central Ltd* 9:24, 2009. Reprinted with permission under the Creative Commons Attribution License or equivalent.

References

1

An, C., Saha, S., Jenkins, J. N., Ma, D. P., Scheffler, B. E., Kohel, R. J., Yu, J. Z., and Stelly, D. M. (2008). Cotton (*Gossypium* spp.) R2R3-MYB transcription factors SNP identification, phylogenomic characterization, chromosome localization, and linkage mapping. *Theor. Appl. Genet.* **116**(7), 1015–1026.

An, C., Saha, S., Jenkins, J. N., Scheffler, B. E., Wilkins, T. A., and Stelly, D. M. (2007). Transcriptome profiling, sequence characterization, and SNP-based chromosomal assignment of the *EXPANSIN* genes in cotton. *Mol. Genet. Gen.* **278**(5), 539–553.

Arpat, A., Waugh, M., Sullivan, J., Gonzales, M., Frisch, D., Main, D., Wood, T., Leslie, A., Wing, R., and Wilkins, T. (2004). Functional genomics of cell elongation in developing cotton fibers. *Pla. Mol. Biol.* **54**(6), 911–929.

Blenda, A., Scheffler, J., Scheffler, B., Palmer, M., Lacape, J. M., Yu, J. Z., Jesudurai, C., Jung, S., Muthukumar, S., Yellambalase, P., Ficklin, S., Staton, M., Eshelman, R., Ulloa, M., Saha, S., Burr, B., Liu, S., Zhang, T., Fang, D., Pepper, A., Kumpatla, S., Jacobs, J., Tomkins, J., Cantrell, R., and Main, D. (2006). CMD: A Cotton Microsatellite Database resource for *Gossypium* genomics. *BMC Geno.* **7**, 132.

Blenda, A., Van Deynze, A. and Main, D. (2005). CMD: A Cotton Microsatellite Database Resource for *Gossypium* genomics. UC Davis SNP project. [http://www.cottonmarker.org/Downloads.shtml].

Bowman, D. T. (2000). Attributes of public and private cotton breeding programs. *J. Cotton Sci.* **4**, 130–136.

Caldwell, K. S., Dvorak, J., Lagudah, E. S., Akhunov, E., Luo, M. C., Wolters, P., and Powell, W. (2004). Sequence polymorphism in polyploid wheat and their D-genome diploid ancestor. *Genet.* **167**(2), 941–947.

Chee, P. W., Rong, J., Williams-Coplin, D., Schulze, S. R., and Paterson, A. H. (2004). EST derived PCR-based markers for functional gene homologues in cotton. *Genome/National Research Council Canada = Genome/Conseil national de recherches Canada* **47**(3), 449–462.

Chen, Z. J. (2007). Genetic and epigenetic mechanisms for gene expression and phenotypic variation in plant polyploids. *Annua. rev. plant boil.* **58**, 377–406.

Ching, A., Caldwell, K. S., Jung, M., Dolan, M., Smith, O. S., Tingey, S., Morgante, M., and Rafalski, A. J. (2002). SNP frequency, haplotype structure and linkage disequilibrium in elite maize inbred lines. *BMC Genet.* **3**(1), 19.

Eveland, A. L., McCarty, D. R., and Koch, K. E. (2008). Transcript profiling by 3'-untranslated region sequencing resolves expression of gene families. *Pla phy.* **146**(1), 32–44.

Ewing, B. and Green, P. (1998). Base-calling of automated sequencer traces using phred. II. Error probabilities. *Genome. Res.* **8**(3), 186–194.

Frary, A., Fulton, T. M., Zamir, D., and Tanksley, S. D. (2004). Advanced backcross QTL analysis of a *Lycopersicon esculentum* × *L. pennellii* cross and identification of possible orthologs in the Solanaceae. *Theor. Appl. Genet.* **108**(3), 485–496.

Frelichowski, J. E., Palmer, M. B., Main, D., Tomkins, J. P., Cantrell, R. G., Stelly, D. M., Yu, J., Kohel, R. J., and Ulloa, M. (2006). Cotton genome mapping with new microsatellites from Acala "Maxxa" BAC-ends. *Mol. Gent. Gen.* **275**(5), 479–491.

Gao, W., Chen, Z. J., Yu, J. Z., Raska, D., Kohel, R. J., Womack, J. E., and Stelly, D. M. (2004). Wide-cross whole-genome radiation hybrid mapping of cotton (*Gossypium hirsutum* L.). *Genet.* **167**(3), 1317–1329.

Guo, W., Cai, C., Wang, C., Han, Z., Song, X., Wang, K., Niu, X., Lu, K., Shi, B., and Zhang, T. (2007). A microsatellite-based, gene-rich linkage map reveals genome structure, function and evolution in *Gossypium. Genet.* **176**(1), 527–541.

Han, Z. G., Guo, W. Z., Song, X. L., and Zhang, T. Z. (2004). Genetic mapping of EST-derived microsatellites from the diploid *Gossypium arboreum* in allotetraploid cotton. *Mol. Genet. Gen.* **272**(3), 308–327.

He, D. H., Lin, Z. X., Zhang, X. L., Nie, Y. C., Guo, X. P., Zhang, Y. X., and Li, W. (2007). QTL mapping for economic traits based on a dense genetic map of cotton with PCR-based markers using the interspecific cross of *Gossypium hirsutum* × *Gossypium barbadense*. *Euphyt.* **153**(1–2), 181–197.

Hsu, C. Y., An, C., Saha, S., Ma, D. P., Jenkins, J. N., Scheffler, B., and Stelly, D. M. (2008). Molecular and SNP characterization of two genome specific transcription factor genes GhMyb8 and GhMyb10 in cotton species. *Euphyt.* **159**(1–2), 259–273.

Kosambi, D. D. (1944). The estimation of map distances from recombination values. *Ann. Eugen.* **12**, 172–175.

Lacape, J. M., Nguyen, T. B., Thibivilliers, S., Bojinov, B., Courtois, B., Cantrell, R. G., Burr, B., and Hau, B. (2003). A combined RFLP-SSR-AFLP map of tetraploid cotton based on a *Gossypium hirsutum* × *Gossypium barbadense* backcross population. Genome/National Research Council Canada = Genome/Conseil national de recherches Canada **46**(4), 612–626.

Liu, S., Cantrell, R. G., McCarty, J. C., and Stewart, J. M. (2000). Simple sequence repeat-based assessment of genetic diversity in cotton race stock accessions. *Crop. Sci.* **40**(5), 1459–1469.

Liu, S., Saha, S., Stelly, D., Burr, B., and Cantrell, R. G. (2000). Chromosomal assignment of microsatellite loci in cotton. *J. Hered.* **91**(4), 326–332.

Park, Y. H., Alabady, M. S., Ulloa, M., Sickler, B., Wilkins, T. A., Yu, J., Stelly, D. M., Kohel, R. J., el-Shihy, O. M., and Cantrell, R. G. (2005). Genetic mapping of new cotton fiber loci using EST-derived microsatellites in an interspecific recombinant inbred line cotton population. *Mol Genet. Gen.* **274**(4), 428–441.

Reinisch, A. J., Dong, J. M., Brubaker, C. L., Stelly, D. M., Wendel, J. F., and Paterson, A. H. (1994). A detailed RFLP map of cotton, *Gossypium hirsutum* × *Gossypium barbadense*: chromosome organization and evolution in a disomic polyploid genome. *Genet.* **138**(3), 829–847.

Robinson, A. F., Bell, A. A., Dighe, N. D., Menz, M. A., Nichols, P. L., and Stelly, D. M. (2007). Introgression of resistance to nematode Rotylenchulus reniformis into upland cotton (*Gossypium hirsutum*) from *Gossypium longicalyx*. *Crop. Sci.* **47**(5), 1865–1877.

Rong, J., Abbey, C., Bowers, J. E., Brubaker, C. L., Chang, C., Chee, P. W., Delmonte, T. A., Ding, X., Garza, J. J., Marler, B. S., Park, C. H., Pierce, G. J., Rainey, K. M., Rastogi, V. K., Schulze, S. R., Trolinder, N. L., Wendel, J. F., Wilkins, T. A., Williams-Coplin, T. D., Wing, R. A., Wright, R. J., Zhao, X., Zhu, L., and Paterson, A. H. (2004). A 3347-locus genetic recombination map of sequence-tagged sites reveals features of genome organization, transmission and evolution of cotton (*Gossypium*). *Genet.* **166**(1), 389–417.

Rong, J., Pierce, G. J., Waghmare, V. N., Rogers, C. J., Desai, A., Chee, P. W., May, O. L., Gannaway, J. R., Wendel, J. F., Wilkins, T. A., and Paterson, A. H. (2005). Genetic mapping and comparative analysis of seven mutants related to seed fiber development in cotton. *Theor. Appl. Genet.* 1–10.

Rozen, S. and Skaletsky, H. (2000). Primer3 on the WWW for general users and for biologist programmers. *Meth. mol. boil. (Clifton, NJ)* **132**, 365–386.

Rungis, D., Llewellyn, D., Dennis, E. S., and Lyon, B. R. (2005). Simple sequence repeat (SSR) markers reveal low levels of polymorphism between cotton (*Gossypium hirsutum* L.) cultivars. *Aust. J. Agr. Res.* **56**(3), 301–307.

Senchina, D. S., Alvarez, I., Cronn, R. C., Liu, B., Rong, J., Noyes, R. D., Paterson, A. H., Wing, R. A., Wilkins, T. A., and Wendel, J. F. (2003). Rate variation among nuclear genes and the age of polyploidy in *Gossypium*. *Mol. bio. evol.* **20**(4), 633–643.

Small, R. L., Ryburn, J. A., and Wendel, J. F. (1999). Low levels of nucleotide diversity at homoeologous *Adh* loci in allotetraploid cotton (*Gossypium* L.). *Mol. bio. evol.* **16**(4), 491–501.

Tearse, B. *Polyphred to Excel Converter.* [http://dendrome.ucdavis.edu/resources/].

Udall, J. A., Swanson, J. M., Haller, K., Rapp, R. A., Sparks, M. E., Hatfield, J., Yu, Y., Wu, Y., Dowd, C., Arpat, A. B., Sickler, B. A., Wilkins, T. A., Guo, J. Y., Chen, X. Y., Scheffler, J., Taliercio, E., Turley, R., McFadden, H., Payton, P., Klueva, N., Allen, R., Zhang, D., Haigler, C., Wilkerson, C., Suo, J., Schulze, S. R., Pierce, M. L., Essenberg, M., Kim, H., Llewellyn, D. J., Dennis, E. S., Kudrna, D., Wing, R., Paterson, A. H., Soderlund, C., and Wendel, J. F. (2006). A global assembly of cotton ESTs. *Genome. Res.* **16**(3), 441–450.

Van Becelaere, G., Lubbers, E. L., Paterson, A. H., and Chee, P. W. (2005). Pedigree- vs. DNA marker-based genetic similarity estimates in cotton. *Crop. Sci.* **45**(6), 2281–2287.

Van Deynze, A. E., Stoffel, K., Buell, R. C., Kozik, A., Liu, J., Knaap, E., and Francis, D. (2007). Diversity in conserved genes in tomato. *BMC Gen.* **8**(1), 465.

Van Ooijen, J. W. (2006). *JoinMap 4.0.* Software for the calculation of genetic linkage maps in experimental populations. B. V. Kyazma (Ed.). Wageningen, Netherlands.

Volker, B., Lushbough, C. and Lawrence, C. PlantGDB: *Resources for Plant Comparative Genomics.* [http://www.plantgdb.org/prj/EST-Cluster/progress.php].

Waghmare, V. N., Rong, J. K., Rogers, C. J., Pierce, G. J., Wendel, J. F., and Paterson, A. H. (2005). Genetic mapping of a cross between *Gossypium hirsutum* (cotton) and the *Hawaiian endemic, Gossypium tomentosum. Theor. Appl. Genet.* **111**(4), 665–676.

Wang, C. B., Guo, W. Z., Cai, C. P., and Zhang, T. Z. (2006). Characterization, development and exploitation of EST-derived microsatellites in *Gossypium raimondii* Ulbrich. *Chin. Sci. Bull.* **51**(5), 557–561.

Wendel, J. F. and Cronn, R. C. (2003). Polyploidy and the evolutionary history of cotton. *Adva. Ag.* **78**, 139–186.

Wilkins, T. A., Wan, C. Y., and Lu, C. C. (1994). Ancient origin of the vacuolar H+-Atpase 69-Kilodalton catalytic subunit superfamily. *Theo. Appl. Genet.* **89**(4), 514–524.

Wu, F. N., Mueller, L. A., Crouzillat, D., Petiard, V., and Tanksley, S. D. (2006). Combining bioinformatics and phylogenetics to identify large sets of single-copy orthologous genes (COSII) for comparative, evolutionary and systematic studies: A test case in the euasterid plant clade. *Genet.* **174**(3), 1407–1420.

Yang, S. S., Cheung, F., Lee, J. J., Ha, M., Wei, N. E., Sze, S. H., Stelly, D. M., Thaxton, P., Triplett, B., Town, C. D., and Jeffrey Chen, Z. (2006). Accumulation of genome-specific transcripts, transcription factors and phytohormonal regulators during early stages of fiber cell development in allotetraploid cotton. *Pla. J.* **47**(5), 761–775.

Zhang, J., Pang, M. X., Niu, C., Wang, W., Percy, R. G., Cantrell, R. G., and Stewart, J. M. (2007). AFLP-Based SNP Discovery. In *Cotton*. Plant & Animal Genomes XV Conference, San Diego, CA.

Zhang, T., Yuan, Y., Yu, J., Guo, W., and Kohel, R. J. (2003). Molecular tagging of a major QTL for fiber strength in Upland cotton and its marker-assisted selection. *Theor. Appl. Genet.* **106**(2), 262–268.

2

Addicott, J. F. (1974). Predation and prey community structure: An experimental study of the effect of mosquito larvae on the protozoan communities of pitcher plants. *Ecology* **55**, 475–492.

Atkin, O. K. (1996). Reassessing the nitrogen relations of Arctic plants: A mini-review. *Plant Cell Env.* **19**, 695–704.

Bedford, B. L., Walbridge, M. R., and Aldous, A. (1999). Patterns in nutrient availability and plant diversity of temperate North American wetlands. *Ecology* **80**, 2151–2169.

Błędzki, L. A. and Ellison, A. M. (1998). Population growth and production of *Habrotrocha rosa* Donner (Rotifera: Bdelloidea) and its contribution to the nutrient supply of its host, the northern pitcher plant, *Sarracenia purpurea* L. (*Sarraceniaceae*). *Hydrobiologia* **385**, 193–200.

Błędzki, L. A. and Ellison, A. M. (2002). Nutrient regeneration by rotifers in New England (USA) bogs. *Ver. Int. Ver. Limnol.* **28**, 1328–1331.

Bradshaw, W. E. and Creelman, R. A. (1984). Mutualism between the carnivorous purple pitcher plant *Sarracenia purpurea* and its inhabitants. *Am. Midl. Nat.* **112**, 294–304.

Bridgham, S. D., Pastor, J., Janssens, J. A., Chapin, C. T., and Malterer, T. J. (1996). Multiple limiting gradients in peatlands: A call for a new paradigm. *Wetlands* **16**, 45–65.

Butler, J. L. and Ellison, A. M. (2007). Nitrogen cycling dynamics in the carnivorous northern pitcher plant, *Sarracenia purpurea. Func. Ecol.* **21**, 835–843.

Butler, J. L., Gotelli, N. J., and Ellison, A. M. (2008). Linking the brown and green: Nutrient transformation and fate in the *Sarracenia microecosystem. Ecology* **89**, 898–904.

Chapin, C. T. and Pastor, J. (1995). Nutrient limitations in the northern pitcher plant *Sarracenia purpurea. Can. J. Bot.* **73**, 728–734.

Chapin, F. S., Moilanen, L., and Kielland, K. (1993). Preferential use of organic nitrogen for growth by a non-mycorrhizal arctic sedge. *Nature* **361**, 150–153.

Clarkson, D. T. (1985). Factors affecting mineral nutrient acquisition by plants. *Ann. Rev. Plant. Physiol.* **36**, 77–115.

Clemmensen, K. E., Sorensen, P. L., Michelsen, A., Jonasson, S., and Ström, L. (2008). Site-dependent N uptake from N-form mixtures by arctic plants, soil microbes and ectomycorrhizal fungi. *Oecologia* **155**, 771–738.

Cochran-Stafira, D. L. and von Ende, C. N. (1998). Integrating bacteria into food webs: Studies with *Sarracenia purpurea* inquilines. *Ecology* **79**, 880–898.

Damman, A. W. H. (1990). Nutrient status of ombrotrophic peat bogs. *Aquilo. Ser. Bot.* **28**, 5–14.

Darwin, C. (1875). *Insectivorous Plants.* Appleton and Company, New York.

Ellison, A. M. and Gotelli, N. J. (2001). Evolutionary ecology of carnivorous plants. *Trends Ecol. Evol.* **16**, 623–629.

Ellison, A. M. and Gotelli, N. J. (2002). Nitrogen availability alters the expression of carnivory in the northern pitcher plant *Sarracenia purpurea. Proc. Natl. Acad. Sci.* USA **99**, 4409–4412.

Ellison, A. M. and Gotelli, N. J. (2009). Energetics and the evolution of carnivorous plants––Darwin's "most wonderful plants in the world." *J. Exp. Bot.* **60**, 19–42.

Ellison, A. M., Gotelli, N. J., Brewer, J. S., Cochran-Stafira, D. L., Kneitel, J., et al. (2003). The evolutionary ecology of carnivorous plants. *Adv. Ecol. Res.* **33**, 1–74.

Environment Canada (2004). *Canadian Acid Deposition Science Assessment.* Meteorological Service of Canada, Toronto.

Finzi, A. D. and Berthrong, S. T. (2005) The uptake of amino acids by microbes and trees in three cold-temperate forests. *Ecology* **86**, 3345–3353.

Fish, D. and Hall, D. W. (1978). Succession and stratification of aquatic insects inhabiting the leaves of the insectivorous pitcher plant *Sarracenia purpurea. Am. Midl. Nat.* **99**, 172–183.

Gallie, D. R. and Chang, S. C. (1997). Signal transduction in the carnivorous plant *Sarracenia purpurea. Plant Physiol.* **115**, 1461–1471.

Givnish, T. J. (1989). Ecology and evolution of carnivorous plants. In *Plant-Animal Interactions.*

Gotelli, N. J. and Ellison, A. M. (2002). Nitrogen deposition and extinction risk in the northern pitcher plant *Sarracenia purpurea. Ecology* **83**, 2758–2765.

Gotelli, N. J., and Ellison, A. M. (2006). Food-web models predict species abundance in response to and habitat change. *PLoS. Biol.* **44**, e324.

Harrison, K. A., Bol, R., and Bardgett, R. D. (2007). Preferences for different nitrogen forms by coexisting plant species and soil microbes. *Ecology* **88**, 989–999.

Heard, S. (1998). Capture rates of invertebrate prey by the pitcher plant, *Sarracenia purpurea* L. *Am. Midl. Nat.* **139**, 79–89.

Heard, S. B. (1994). Pitcher plant midges and mosquitoes: A processing chain commensalism. *Ecology* **75**, 1647–1660.

Henry, H. A. L. and Jefferies, R. L. (2002). Free amino acid, ammonium and nitrate concentrations in soil solutions of a grazed coastal marsh in relation to plant growth. *Plant Cell Env.* **25**, 665–675.

Hepburn, J. S., Jones, F. M., and St. John, E. Q. (1927). The biochemistry of the American pitcher plants. *Trans. Wagner Free Inst. Sci.* **11**, 1–95.

Hodge, A., Robinson, D., and Fitter, A. (2000). Are microorganisms more effective than plants at competing for nitrogen? *Trends Plant Sci.* **5**, 304–307.

Jones, D. L., Farrar, J. F., and Newsham, K. K. (2004). Rapid amino acid cycling in Arctic and Antarctic soils. *Water Air Soil Pollut., Focus* **4**, 169–175.

Kielland, K. (1994). Amino acid absorption by arctic plants: Implications for plant nutrition and nitrogen cycling. *Ecology* **75**, 2373–2383.

Lipson, D. and Nasholm, T. (2001). The unexpected versatility of plants: Organic nitrogen use and availability in terrestrial ecosystems. *Oecologia* **128**, 305–316.

Lipson, D. A., Raab, T. K., Schmidt, S. K., and Monson, R. K. (1999). Variation in competitive abilities of plants and microbes for specific amino acids. *Biol. Fert. Soil* **29**, 257–261.

Lloyd, F. E. (1942). *The Carnivorous Plants.* Ronald Press, New York.

Lüttge, U. (1965). Studies on the physiology of carnivores-Druze. 2nd edition. On the absorption of various substances. *Planta* **66**, 331–334.

Resource-based niches provide a basis for plant species diversity and dominance in arctic tundra. *Nature* **415**, 68–71.

Melin, E. and Nilsson, H. (1953). Transfer of labelled nitrogen from glutamic acid to pine seedlings through the mycelium of *Boletus variegatus* (Sw.) Fr. *Nature* **171**, 134.

Miller, A. E. and Bowman, W. D. (2003). Alpine plants show species-level differences in the uptake of organic and inorganic nitrogen. *Plant Soil* **250**, 283–292.

Miller, A. E., Bowman, W. D., and Nash Suding, K. (2007). Plant uptake of inorganic and organic nitrogen: Neighbor identity matters. *Ecology* **88**, 1832–1840.

Miller, R. H. and Schmidt, E. L. (1965). Uptake and assimilation of amino acids supplied to the sterile soil: Root environment of the bean plant (*Phaseolus vulgaris*). *Soil Sci.* **100**, 323–330.

Nadelhoffer, K. and Fry, B. (1994). Nitrogen isotope studies in forested ecosystems. In *Methods in Ecology: Stable Isotopes in Ecology and Environmental Science*. R. H. Michener and K. Lajtha (Ed.). Blackwell Scientific Publications, London, pp. 22–44.

Naeem, S (1988). Resource heterogeneity fosters coexistence of a mite and a midge in pitcher plants. *Ecol. Mono.* **58**, 215–227.

Nasholm, T., Ekblad, A., Nordin, A., Giesler, R., Hogberg, M., et al. (1998) Boreal forest plants take up organic nitrogen. *Nature* **392**, 914–916.

National Atmospheric Deposition Program (NRSP-3) (2009). *Champaign*. NADP Program Office, Illinois State Water Survey, Illinois.

Neff, J. C., Chapin, F. S., and Vitousek, P. M. (2003). Breaks in the cycle: Dissolved organic nitrogen in terrestrial ecosystems. *Front. Ecol. Env.* **1**, 205–211.

Newell, S. J. and Nastase, A. J. (1998). Efficiency of insect capture by *Sarracenia purpurea* (Sarraceniaceae), the northern pitcher plant. *Am. J. Bot.* **85**, 88–91.

Nordin, A., Hogberg, P., and Nasholm, T. (2001). Soil nitrogen form and plant nitrogen uptake along a boreal forest productivity gradient. *Oecologia* **129**, 125–132.

Nordin, A., Schmidt, I., and Shaver, G. (2004). Nitrogen uptake by arctic soil microbes and plants in relation to soil nitrogen supply. *Ecology* **85**, 955–962.

Persson, J., Hogberg, P., Ekblad, A., Hogberg, M., Nordgren, A., et al. (2003). Nitrogen acquisition from inorganic and organic sources by boreal forest plants in the field. *Oecologia* **137**, 252–257.

Persson, J. and Nasholm, T. (2001). Amino acid uptake: A widespread ability among boreal forest plants. *Ecol Lett* **4**, 434–438.

Peterson, C. N., Day, S., Wolfe, B. E., Ellison, A. M., Kolter, R., et al. (2008). A keystone predator controls bacterial diversity in the pitcher plant

(*Sarracenia purpurea*) microecosystem. *Env. Microbiol.* **10**, 2257–2266.

Plummer, G. L. and Kethley, J. B. (1964). Foliar absorption of amino acids, peptides, and other nutrients by the pitcher plant, *Sarracenia flava. Bot. Gaz.* **125**, 245–260.

Porembski, S. and Barthlott, W. (2000) Inselbergs: *Biotic Diversity of Isolated Rock Outcrops in Tropical and Temperate Regions.* Springer-Verlag, Berlin.

Raab, T. K., Lipson, D. A., and Monson, R. K. (1999). Soil amino acid utilization among species of the Cyperaceae: Plant and soil processes. *Ecology* **80**, 2408–2419.

Schnell, D. E. (2002). *Carnivorous Plants of the United States and Canada,* 2nd edition. Timber Press, Portland.

Schulze, W., Schulze, E. -D., Pate, J. S., and Gillison, A. N. (1997). The nitrogen supply from soils and insects during growth of the pitcher plants *Nepenthes mirabilis, Cephalotus follicularis* and *Darlingtonia californica. Oecologia* **112**, 464–471.

Senwo, Z. N. and Tabatabai, M. A. (1998). Amino acid composition of soil organic matter. *Biol. Fert. Soil* **26**, 235–242.

Stark, J. M. (2000). Nutrient transformations. In *Methods in Ecosystem Science.* O. E Sala., R. B. Jackson, H. A. Mooney, and R. W. Howarth (Eds.).Springer-Verlag, New York, pp. 215–234.

Swan, J. M. and Gill, A. M. (1970). The origins, spread, and consolidation of a floating bog in Harvard Pond, Petersham, Massachusetts. *Ecology* **51**, 829–840.

Virtanen, A. I. and Linkola, H. (1946). Organic nitrogen compounds as nitrogen nutrition for higher plants. *Nature* **158**, 515.

Vitousek, P. M., Aber, J., Howarth, R. W., Likens, G. E., Matson, P. A., et al. (1997). Human alteration of the global nitrogen cycle: Sources and consequences. *Ecol. App.* **7**, 737–750.

von Felton, S., Buchmann, N., and Scherer-Lorenzen, M. (2008). Preferences for different nitrogen forms by coexisting plant species and soil microbes: Comment. *Ecology* **89**, 878–879.

Wiegelt, A., Bol, R., and Bardgett, R. D. (2005). Preferential uptake of soil nitrogen forms by grassland plant species. *Oecologia* **142**, 627–635.

3

Aharoni, A. and Vorst, O. (2002). DNA microarrays for functional plant genomics. *Plant. Mol. Biol.* **48**(1-2), 99-118.

Alba, R., Fei, Z., Payton, P., Liu, Y., Moore, S. L., Debbie, P., Cohn, J., D'Ascenzo, M., Gordon, J. S., Rose, J. K., et al. (2004). ESTs, cDNA microarrays, and gene expression profiling: tools for dissecting plant physiology and development. *Plant. J.* **39**(5), 697–714.

Albrecht, V., Ritz, O., Linder, S., Harter, K., and Kudla, J. (2001). The NAF domain defines a novel protein-protein interaction module conserved in Ca2+-regulated kinases. *EMBO J* **20**(5), 1051–1063.

Aranda, M. and Maule, A. (1998). Virus-induced host gene shutoff in animals and plants. *Virology.* **243**(2), 261–267.

Arumuganathan, K. E. E. (1991). Nuclear DNA content of some important plant species. Plant *Mol. Biol. Rep.* **9**, 208–218.

Ayub, R., Guis, M., Ben Amor, M., Giot, L., Roustan, J. P., Latche, A., Bouzayen, M., and Pech, J. C. (1996). Expression of ACC oxidase antisense gene inhibits ripening of cantaloupe melon fruits. *Nat Biotechnol.* **14**(7), 862–866.

Balague, C., Watson, C. F., Turner, A. J., Rouge, P., Picton, S., Pech, J. C., and Grierson, D. (1993). Isolation of a ripening and wound-induced cDNA from *Cucumis melo* L. encoding a protein with homology to the ethylene-forming enzyme. *Eur. J. Biochem.* **212**(1), 27–34.

Benjamini, Y. and Hochberg, Y. (1995). Controlling the false discovery rate: A practical and powerful approach to multiple testing. *J. Roy. Stat. Soc. Ser. B. (Methodological).* **57**, 289–300.

Bolstad, B. M., Irizarry, R. A., Astrand, M., and Speed, T. P. (2003). A comparison of normalization methods for high density oligonucleotide array data based on variance and bias. *Bioinformatics.* **19**(2), 185–193.

Boualem, A., Fergany, M., Fernandez, R., Troadec, C., Martin, A., Morin, H., Sari, M. A., Collin, F., Flowers, J. M., Pitrat, M., et al. (2008). A conserved mutation in an ethylene biosynthesis enzyme leads to andromonoecy in melons. *Science* **321**(5890), 836–838.

Camacho-Cristobal, J., Herrera-Rodriguez, B., Beato, V., Rexach, J., Navarro-Gochicoa, M., Maldonado, J., and Gonzalez-Fontes, A. (2008). The expression of several cell wall-related genes in *Arabidopsis* roots is down-regulated under boron deficiency. *Environ. Exp. Bot.* **63**, 351–358.

Carvalho, B., Bengtsson, H., Speed, T. P., and Irizarry, R. A. (2007). Exploration, normalization, and genotype calls of high-density oligonucleotide SNP array data. *Biostatistics* **8**(2), 485–499.

Chen, J. Q., Rahbe, Y., Delobel, B., Sauvion, N., Guillaud, J., and Febvay, G. (1997). Resistance to the aphid *Aphis gossypii*: Behavioral analysis and chemical correlations with nitrogenous compounds. *Entomol. Exp. Appl.* **85**, 33–44.

Clarke, J. D. and Zhu, T. (2006). Microarray analysis of the transcriptome as a stepping stone towards understanding biological systems: practical considerations and perspectives. *Plant. J.* **45**(4), 630–650.

Cope, L. M., Irizarry, R. A., Jaffee, H. A., Wu, Z., and Speed, T. P. (2004). A benchmark for Affymetrix GeneChip expression measures. *Bioinformatics.* **20**(3), 323–331.

Coppack, S. W. (1990). Limitations of the Pearson product-moment correlation. *Clin. Sci. (Lond).* **79**(3), 287.

Dias, R., De Cassia, R., Picó, B., Espinós, A., and Nuez, F. (2004). Resistance to melon vine decline derived from *C. melo* subsp. *agrestis*: Genetic analysis of root structure and root response to disease. *Plant. Breeding.* **123**, 1–7.

Eisen, M. B., Spellman, P. T., Brown, P. O., and Botstein, D. (1998). Cluster analysis and display of genome-wide expression patterns. *Proc Natl. Acad. Sci. USA.* **95**(25), 14863–14868.

Fernandez-Silva, I., Eduardo, I., Blanca, J., Esteras, C., Pico, B., Nuez, F., Arus, P., Garcia-Mas, J., and Monforte, A. J. (2008). Bin mapping of genomic and EST-derived SSRs in melon (*Cucumis melo* L.). *Theor. Appl. Genet.*, **118**(1), 139–150.

Garcia-Mas, J., Monforte, A. J., and Arus, P. (2004). Phylogenetic relationships among *Cucumis* species based on the ribosomal internal transcriber spacer sequence and microsatellite markers. *Plant. Syst. Evol.* **248**, 191–203.

Gelhaye, E., Rouhier, N., Navrot, N., and Jacquot, J. P. (2005). The plant thioredoxin system. *Cell. Mol. Life. Sci.* **62**(1), 24–35.

Gentleman, R. C., Carey, V. J., Bates, D. M., Bolstad, B., Dettling, M., Dudoit, S., Ellis, B., Gautier, L., Ge, Y., Gentry, J., et al. (2004). Bioconductor: Open software development for computational biology and bioinformatics. *Genome. Biol.* **5**(10), R80.

Goff, S. A., Ricke, D., Lan, T. H., Presting, G., Wang, R., Dunn, M., Glazebrook, J., Sessions A., Oeller, P., Varma, H., et al. (2002). A draft sequence of the rice genome (*Oryza sativa* L. sp. *japonica*). *Science* **296**(5565), 92–100.

Gomez, G., Torres, H., and Pallas, V. (2005). Identification of translocatable RNA-binding phloem proteins from melon, potential components of the long-distance RNA transport system. *Plant. J.* **41**(1), 107–116.

Gonzalez-Ibeas, D., Blanca, J., Roig, C., Gonzalez-To, M., Pico, B., Truniger, V., Gomez, P., Deleu, W., Cano-Delgado, A., Arus, P., et al. (2007). MELOGEN: An EST database for melon functional genomics. *BMC. Genomics.* **8**, 306.

Gonzalez-Lamothe, R., Tsitsigiannis, D. I., Ludwig, A. A., Panicot, M., Shirasu, K., and Jones, J. D. (2006). The U-box protein CMPG1 is required for efficient activation of defense mechanisms triggered by multiple resistance genes in tobacco and tomato. *Plant. Cell.* **18**(4), 1067–1083.

Gozzo, F. (2003). Systemic acquired resistance in crop protection: From nature to a chemical approach. *J Agric Food. Chem.* **51**(16), 4487–4503.

Haritatos, E., Keller, F., and Turgeon, R. (1996). Raffinose oligosaccharide concentrations measured in individual cell and tissue types in *Cucumis melo* L. leaves: Implications for phloem loading. *Planta.* **198**, 614–622.

He, Y. D., Dai, H., Schadt, E. E., Cavet, G., Edwards, S. W., Stepaniants, S. B., Duenwald, S., Kleinhanz, R., Jones, A. R., Shoemaker, D. D., et al. (2003). Microarray standard data set and figures of merit for comparing data processing methods and experiment designs. *Bioinformatics* **19**(8), 956–965.

Huala, E., Dickerman, A. W., Garcia-Hernandez, M., Weems, D., Reiser, L., LaFond, F., Hanley, D., Kiphart, D., Zhuang, M., Huang, W., et al. (2001). The *Arabidopsis* Information Resource (TAIR): A comprehensive database and web-based information retrieval, analysis, and visualization system for a model plant. *Nucleic. Acids. Res.* **29**(1), 102–105.

Irizarry, R. A., Bolstad, B. M., Collin, F., Cope, L. M., Hobbs, B., and Speed, T. P. (2003). Summaries of Affymetrix GeneChip probe level data. *Nucleic. Acids. Res.* **31**(4), e15.

Kirkbride, J. (1993). *Biosystematic Monograph of the Genus Cucumis (Cucurbitaceae)*. Parkway Publishers, Boone, North Carolina.

Kirsch, C., Logemann, E., Lippok, B., Schmelzer, E., and Hahlbrock, K. (2001). A highly specific pathogen-responsive promoter element from the immediate-early activated CMPG1 gene in *Petroselinum crispum*. *Plant. J.* **26**(2), 217–227.

Kotchoni, S. O. and Gachomo, E. W. (2006). The reactive oxygen species network pathways: An essential prerequisite for perception of pathogen attack and the acquired disease resistance in plants. *J. Biosci.* **31**(3), 389–404.

Larkin, J. E., Frank, B. C., Gavras, H., Sultana, R., and Quackenbush, J. (2005). Independence and reproducibility across microarray platforms. *Nat. Meth.* **2**(5), 337–344.

Libault, M., Wan, J., Czechowski, T., Udvardi, M., and Stacey, G. (2007). Identification of 118 *Arabidopsis* transcription factor and 30 ubiquitin-ligase genes responding to chitin, a plant-defense elicitor. *Mol. Plant. Microb. Interact.* **20**(8), 900–911.

Lionetti, V., Raiola, A., Camardella, L., Giovane, A., Obel, N., Pauly, M., Favaron, F., Cervone, F., and Bellincampi, D. (2007). Overexpression of pectin methylesterase inhibitors in *Arabidopsis* restricts fungal infection by *Botrytis cinerea*. *Plant. Physiol.* **143**(4), 1871–1880.

Liu, L., Kakihara, F., and Kato, M. (2004). Characterization of six varieties of *Cucumis melo* L. based on morphological and physiological characters, including shelf-life of fruit. *Euphytica* **135**, 305–313.

Liu, Y., Meng, Q., Chen, R., Wang, J., Jiang, S., and Hu, Y. (2004). A new method to evaluate the similarity of chromatographic fingerprints: Weighted Pearson product-moment correlation coefficient. *J. Chromatogr. Sci.* **42**(10), 545–550.

Lopez-Bigas, N., Kisiel, T. A., Dewaal, D. C., Holmes, K. B., Volkert, T. L., Gupta, S., Love, J., Murray, H. L., Young, R. A., and Benevolenskaya, E. V. (2008). Genome-wide analysis of the H3K4 histone demethylase RBP2 reveals a transcriptional program controlling differentiation. *Mol. Cell.* **31**(4), 520–530.

Louvet, R., Cavel, E., Gutierrez, L., Guenin, S., Roger, D., Gillet, F., Guerineau, F., and Pelloux, J. (2006). Comprehensive expression profiling of the pectin methylesterase gene family during silique development in *Arabidopsis thaliana*. *Planta.* **224**(4), 782–791.

Lu, Z. X., Gaudet, D. A., Frick, M., Puchalski, B., Genswein, B., and Laroche, A. (2005). Identification and characterization of genes differentially expressed in the resistance reaction in wheat infected with *Tilletia tritici*, the common bunt pathogen. *J. Biochem. Mol. Biol.* **38**(4), 420–431.

Martinez-Godoy, M. A., Mauri, N., Juarez, J., Marques, M. C., Santiago, J., Forment, J., and Gadea, J. (2008). A genome-wide 20 K citrus microarray for gene expression analysis. *BMC Gen.* **9**, 318.

Millenaar, F. F., Okyere, J., May, S. T., van Zanten, M., Voesenek, L. A., and, Peeters, A. J. (2006). How to decide? Different methods of calculating gene expression from short oligonucleotide array data will give different results. BMC *Bioinformatics.* **7**, 137.

Moreno, E., Obando, J. M., Dos-Santos, N., Fernandez-Trujillo, J. P., Monforte, A. J., and Garcia-Mas, J. (2008). Candidate genes and QTLs for fruit ripening and softening in melon. *Theor. Appl. Genet.* **116**(4), 589–602.

Morgante, M., Hanafey, M., and Powell, W. (2002). Microsatellites are preferentially associated with nonrepetitive DNA in plant genomes. *Nat. Genet.* **30**(2), 194–200.

Morgulis, A., Gertz, E. M., Schaffer, A. A., and Agarwala, R. (2006). WindowMasker: Window-based masker for sequenced genomes. *Bioinformatics* **22**(2), 134–141.

Nieto, C., Morales, M., Orjeda, G., Clepet, C., Monfort, A., Sturbois, B., Puigdomenech, P., Pitrat, M., Caboche, M., Dogimont, C., et al. (2006). An *eIF4E* allele confers resistance to an uncapped and non-polyadenylated RNA virus in melon. *Plant. J.* **48**(3), 452–462.

NimbleGen Arrays User's Guide v3.0. [http://www.nimblegen.com/products/].

Nunez-Palenius, H. G., Gomez-Lim, M., Ochoa-Alejo, N., Grumet, R., Lester, G., and Cantliffe, D. J. (2008). Melon fruits: Genetic diversity, physiology, and biotechnology features. *Crit. Rev. Biotechnol.* **28**(1), 13–55.

Oliveros, J. C. (2007). *VENNY*. An interactive tool for comparing lists with Venn Diagrams. [http://bioinfogp.cnb.csic.es/tools/venny/index.html].

Peirson, S. N., Butler, J. N., and Foster, R. G. (2003). Experimental validation of novel and conventional approaches to quantitative real-time PCR data analysis. *Nucleic. Acids. Res.* **31**(14), e73.

R statistical software (2008). A Language and Environment for Statistical Computing R Foundation for Statistical Computing RDC Team. Vienna, Austria.

Rafalski, J. A. (2002). Novel genetic mapping tools in plants: SNPs and LD-based approaches. *Plant. Sci.* **162**, 329–333.

Raychaudhuri, S., Stuart, J. M., and Altman, R. B. (2000). Principal components analysis to summarize microarray experiments: application to sporulation time series. *Pac. Symp. Biocomput.* 455–466.

Rensink, W. A. and Buell, C. R. (2005). Microarray expression profiling resources for plant genomics. *Trends. Plant. Sci.* **10**(12), 603–609.

Roche NimbleGen Probe Design Fundaments. [http://www.nimblegen.com/products/lit/probe_design_2008_06_04.pdf].

Rudd, S. (2003). Expressed sequence tags: Alternative or complement to whole genome sequences? *Trends. Plant. Sci.* **8**(7), 321–329.

Saeed, A. I., Bhagabati, N. K., Braisted, J. C., Liang, W., Sharov, V., Howe, E. A., Li, J.,

Thiagarajan, M., White, J. A., and Quackenbush, J. (2006). TM4 microarray software suite. *Methods Enzymol.* **411**, 134–193.

Schmelzer, E. (2002). Cell polarization, a crucial process in fungal defence. *Trends. Plant. Sci.* **7**(9), 411–415.

Schutz, I., Gus-Mayer, S., and Schmelzer, E. (2006). Profilin and Rop GTPases are localized at infection sites of plant cells. *Protoplasma.* **227**(2–4), 229–235.

Takemoto, D., Doke, N., and Kawakita, K. (2001). Characterization of Elicitor-inducible tobacco genes isolated by differential hybridization. *J. Gen. Plant. Pathol.* **67**, 89–96.

Tusher, V. G., Tibshirani, R., and Chu, G. (2001). Significance analysis of microarrays applied to the ionizing radiation response. *Proc Natl. Acad. Sci. USA.* **98**(9), 5116–5121.

Vidali, L., Augustine, R. C., Kleinman, K. P., and Bezanilla, M. (2007). Profilin is essential for tip growth in the moss Physcomitrella patens. *Plant Cell* **19**(11), 3705–3722.

Wechter, W. P., Levi, A., Harris, K. R., Davis, A. R., Fei, Z., Katzir, N., Giovannoni, J. J., Salman-Minkov, A., Hernandez, A., Thimmapuram, J., et al. (2008). Gene expression in developing watermelon fruit. *BMC Gen.* **9**, 275.

Whitham, S. A., Quan, S., Chang, H. S., Cooper, B., Estes, B., Zhu, T., Wang, X., and Hou, Y. M. (2003). Diverse RNA viruses elicit the expression of common sets of genes in susceptible *Arabidopsis thaliana* plants. *Plant. J.* **33**(2), 271–283.

Wong, C. E., Bhalla, P. L., Ottenhof, H., and Singh, M. B. (2008). Transcriptional profiling of the pea shoot apical meristem reveals processes underlying its function and maintenance. *BMC Plant. Biol.* **8**, 73.

Woo, Y., Affourtit, J., Daigle, S., Viale, A., Johnson, K., Naggert, J., and Churchill, G. (2004). A comparison of cDNA, oligonucleotide, and Affymetrix GeneChip gene expression microarray platforms. *J. Biomol. Tech.* **15**(4), 276–284.

Xiang, D., Datla, R., Li, F., Cutler, A., Malik, M. R., Krochko, J. E., Sharma, N., Fobert, P., Georges, F., Selvaraj, G., et al. (2008). Development of a Brassica seed cDNA microarray. *Genome* **51**(3), 236–242.

Yamamoto, Y., Sato, E., Shimizu, T., Nakamich, N., Sato, S., Kato, T., Tabata, S., Nagatani, A., Yamashino, T., and Mizuno, T. (2003). Comparative genetic studies on the *APRR5* and *APRR7* genes belonging to the APRR1/TOC1 quintet implicated in circadian rhythm, control of flowering time, and early photomorphogenesis. *Plant. Cell. Physiol.* **44**(11), 1119–1130.

Yu, J., Hu, S., Wang, J., Wong, G. K., Li, S., Liu, B., Deng, Y., Dai, L., Zhou, Y., Zhang, X., et al. (2002). A draft sequence of the rice genome (*Oryza sativa* L. sp. *indica*). *Science* **296**(5565), 79–92.

Zhou, W., Eudes, F., and Laroche, A. (2006). Identification of differentially regulated proteins in response to a compatible interaction between the pathogen *Fusarium graminearum* and its host, *Triticum aestivum*. *Proteomics* **6**(16), 4599-4609.

4

Aoki, K., Ogata, Y., and Shibata, D. (2007). Approaches for extracting practical information from gene coexpression networks in plant biology. *Plant Cell Physiol.* **48**(3), 381–390.

Barabasi, A. L. and Oltvai, Z. N. (2004). Network biology: Understanding the cell's functional organization. *Natl. Rev. Genet.* **5**(2), 101–113.

Barrero, J. M., Gonzalez-Bayon, R., del Pozo, J. C., Ponce, M. R., and Micol, J. L. (2007). INCURVATA2 Encodes the catalytic subunit of DNA polymerase α and interacts with genes Involved in chromatin-mediated cellular memory in *Arabidopsis thaliana*. *Plant Cell* **19**(9), 2822–2838.

Batada, N. N., Reguly, T., Breitkreutz, A., Boucher, L., Breitkreutz, B. J., Hurst, L. D., and Tyers, M. (2006). Stratus not altocumulus: A new view of the yeast protein interaction network. *PLoS Biology* **4**(10), e317.

Beemster, G. T., De Veylder, L., Vercruysse, S., West, G., Rombaut, D., Van Hummelen, P., Galichet, A., Gruissem, W., Inze, D., and Vuylsteke, M. (2005). Genome-wide analysis of gene expression profiles associated with cell cycle transitions in growing organs of *Arabidopsis*. *Plant Physiol.* **138**(2), 734–743.

Beemster, G. T., Fiorani, F., and Inze, D. (2003). Cell cycle: The key to plant growth control? *Trends Plant Sci.* **8**(4), 154–158.

Bergmann, S., Ihmels, J., and Barkai, N. (2004). Similarities and differences in genome-wide expression data of six organisms. *PLoS Biology* **2**(1), e9.

Binarova, P., Cenklova, V., Prochazkova, J., Doskocilova, A., Volc, J., Vrlik, M., and Bogre, L. (2006). γ-Tubulin is essential for acentrosomal microtubule nucleation and coordination of late mitotic events in *Arabidopsis*. *Plant Cell* **18**(5), 1199–1212.

Brohee, S. and van Helden, J. (2006). Evaluation of clustering algorithms for protein-protein interaction networks. *BMC. Bioinformatics* **7**(1), 488.

Carlson, M. R. J., Zhang, B., Fang, Z. X., Mischel, P. S., Horvath, S., and Nelson, S. F. (2006). Gene connectivity, function, and sequence conservation: Predictions from modular yeast coexpression networks. *BMC Genomics* **7**, 15.

Choi, K., Park, C., Lee, J., Oh, M., Noh, B., and Lee, I. (2007). *Arabidopsis* homologs of components of the SWR1 complex regulate flowering and plant development. *Development* **134**(10), 1931–1941.

Conklin, P. L., Saracco, S. A., Norris, S. R., and Last, R. L. (2000). Identification of ascorbic acid-deficient *Arabidopsis thaliana* mutants. *Genetics* **154**(2), 847–856.

Cook, D., Fowler, S., Fiehn, O., and Thomashow, M. F. (2004). A prominent role for the CBF cold response pathway in configuring the low-temperature metabolome of *Arabidopsis*. *Proc. Natl. Acad. Sci. USA* **101**(42), 15243–15248.

De Schutter, K., Joubes, J., Cools, T., Verkest, A., Corellou, F., Babiychuk, E., Schueren, E., Beeckman, T., Kushnir, S., Inze, D., et al. (2007). *Arabidopsis* WEE1 kinase controls cell cycle arrest in response to activation of the DNA integrity checkpoint. *Plant Cell* **19**(1), 211–225.

Diaz-Trivino, S., del Mar Castellano, M., de la Paz Sanchez, M., Ramirez-Parra, E., Desvoyes, B., and Gutierrez, C. (2005). The genes encoding *Arabidopsis* ORC subunits are E2F targets and the two *ORC1* genes are differently expressed in proliferating and endoreplicating cells. *Nucl. Acids Res.* **33**(17), 5404–5414.

Egelkrout, E. M., Mariconti, L., Settlage, S. B., Cella, R., Robertson, D., and Hanley-Bowdoin, L. (2002). Two E2F elements regulate the proliferating cell nuclear antigen promoter differently during leaf development. *Plant Cell* **14**(12), 3225–3236.

Enright, A. J., Van Dongen, S., and Ouzounis, C. A. (2002). An efficient algorithm for large-scale detection of protein families. *Nucl. Acids Res.* **30**(7), 1575–1584.

Fatma, K., Joachim, K., Dong Yul, S., Wei, Z., Mick, P., Ron, P., and Charles, L. G. (2007). Transcript and metabolite profiling during cold acclimation of *Arabidopsis* reveals an intricate relationship of coldregulated gene expression with modifications in metabolite content. *Plant J.* **50**(6), 967–981.

Finnegan, E. J., Genger, R. K., Kovac, K., Peacock, W. J., and Dennis, E. S. (1998). DNA methylation and the promotion of flowering by vernalization. *Proc. Natl. Acad. Sci.* **95**(10), 5824–5829.

Freeman, T. C., Goldovsky, L., Brosch, M., Van Dongen, S., Maziere, P., Grocock, R. J., Freilich, S., Thornton, J., and Enright, A. J. (2007). Construction, visualisation, and clustering of transcription networks from Microarray expression data. *PLoS Comput Biol.* **3**(10), 2032–2042.

Girvan, M. and Newman, M. E. J. (2002). Community structure in social and biological networks. *Proc. Natl. Acad. Sci.* **99**(12), 7821–7826.

Gonzali, S., Loreti, E., Solfanelli, C., Novi, G., Alpi, A., and Perata, P. (2006). Identification of sugar-modulated genes and evidence for *in vivo* sugar sensing in *Arabidopsis*. *J. Plant Res.* **119**(2), 115–123.

Havaux, M., Eymery, F., Porfirova, S., Rey, P., and Dormann, P. (2005). Vitamin E protects against photoinhibition and photooxidative stress in *Arabidopsis thaliana*. *Plant Cell* **17**(12), 3451–3469.

Horan, K., Jang, C., Bailey-Serres, J., Mittler, R., Shelton, C., Harper, J. F., Zhu, J. K., Cushman, J. C., Gollery, M., and Girke, T. (2008). Annotating genes of known and unknown function by large-scale coexpression analysis. *Plant Physiol.* **147**(1), 41–57.

Horvath, S. and Dong, J. (2008). Geometric interpretation of gene coexpression network analysis. *PLoS Comput Biol.* **4**(8), 27.

Jordan, I. K., Marino-Ramirez, L., Wolf, Y. I., and Koonin, E. V. (2004). Conservation and coevolution in the scale-free human gene coexpression network. *Mol. Biol. Evol.* **21**(11), 2058–2070.

Kilian, J., Whitehead, D., Horak, J., Wanke, D., Weinl, S., Batistic, O., D'Angelo C., Bornberg-Bauer, E., Kudla, J., and Harter, K. (2007). The AtGenExpress global stress expression data set: Protocols, evaluation and model data analysis of UV-B light, drought and cold stress responses. *Plant J.* **50**(2), 347–363.

Lee, B. H., Henderson, D. A., and Zhu, J. K. (2005). The *Arabidopsis* cold-responsive transcriptome and its regulation by ICE1. *Plant Cell* **17**(11), 3155–3175.

Lee, H. K., Hsu, A. K., Sajdak, J., Qin, J., and Pavlidis, P. (2004). Coexpression analysis of human genes across many microarray data sets. *Genom. Res.* **14**(6), 1085–1094.

Lee, Y. R., Li, Y., and Liu, B. (2007). Two *Arabidopsis* phragmoplast-associated kinesins play a critical role in cytokinesis during male gametogenesis. *Plant Cell* **19**(8), 2595–2605.

Li, C. and Wong, W. H. (2001). Model-based analysis of oligonucleotide arrays: Expression index computation and outlier detection. *Proc. Natl. Acad. Sci.* **98**(1), 31–36.

Ma, S., Gong, Q., and Bohnert, H. J. (2007). An *Arabidopsis* gene network based on the graphical Gaussian model. *Genome Res.* **17**(11), 1614–1625.

Maere, S., Heymans, K., and Kuiper, M. (2005). BiNGO: A Cytoscape plugin to assess overrepresentation of gene ontology categories in Biological Networks. *Bioinformatics* **21**(16), 3448–3449.

Manfield, I. W., Jen, C. H., Pinney, J. W., Michalopoulos, I., Bradford, J. R., Gilmartin, P. M., and Westhead, D. R. (2006). *Arabidopsis* Co-expression Tool (ACT): Web server tools for microarray-based gene expression analysis. *Nucl. Acids Res.* **34**(suppl_2), W504–509.

Mentzen, W. I. and Wurtele, E. S. (2008). Regulon organization of *Arabidopsis*. *BMC Plant Biol.* **8**, 22.

Meurer, J., Lezhneva, L., Amann, K., Godel, M., Bezhani, S., Sherameti, I., and Oelmuller, R. (2002). A peptide chain release factor 2 affects the stability of UGA-Containing transcripts in *Arabidopsis* chloroplasts. *Plant Cell* **14**(12), 3255–3269.

Mueller, L. A., Zhang, P., and Rhee, S. Y. (2003). AraCyc: A biochemical pathway database for *Arabidopsis*. *Plant Physiol.* **132**(2), 453–460.

Muller-Moule, P. Golan, T., and Niyogi, K. K. (2004). Ascorbate-Deficient mutants of *Arabidopsis* grow in high light despite chronic Photooxidative stress. *Plant Physiol.* **134**(3), 1163–1172.

Obayashi, T., Kinoshita, K., Nakai, K., Shibaoka, M., Hayashi, S., Saeki, M., Shibata, D., Saito, K., and Ohta, H. (2007). ATTED-II: A database of co-expressed genes and *cis* elements for identifying co-regulated gene groups in *Arabidopsis*. *Nucl. Acids Res.* **35**(suppl_1), D863–869.

Oswald, F., Dobner, T., and Lipp, M. (1996). The E2F transcription factor activates a replication-dependent human *H2A* gene in early S phase of the cell cycle. *Mol. Cell Biol.* **16**(5), 1889–1895.

Pu, S., Vlasblom, J., Emili, A., Greenblatt, J., and Wodak, S. J. (2007). Identifying functional modules in the physical interactome of *Saccharomyces cerevisiae*. *Proteomics* **7**(6), 944–960.

Ramirez-Parra, E. and Gutierrez, C. (2007). E2F regulates FASCIATA1, a Chromatin Assembly gene whose loss switches on the endocycle and activates gene expression by changing the epigenetic status. *Plant Physiol.* **144**(1), 105–120.

Raynaud, C., Sozzani, R., Glab, N., Domenichini, S., Perennes, C., Cella, R., Kondorosi, E., and Bergounioux, C. (2006). Two cell-cycle regulated SET-domain proteins interact with proliferating cell nuclear antigen (PCNA) in *Arabidopsis*. *Plant J.* **47**(3), 395–407.

Reinbothe, S., Quigley, F., Springer, A., Schemenewitz, A., and Reinbothe, C. (2004). The outer plastid envelope protein Oep16: Role as precursor translocase in import of protochlorophyllide oxidoreductase A. *Proc. Natl. Acad. Sci.* **101**(7), 2203–2208.

Rhee, S. Y., Beavis, W., Berardini, T. Z., Chen, G., Dixon, D., Doyle, A., Garcia-Hernandez, M., Huala, E., Lander, G., Montoya, M., et al. (2003). The *Arabidopsis* Information Resource (TAIR): A model organism database providing a centralized, curated gateway to *Arabidopsis* biology, research materials and community. *Nucl. Acids Res.* **31**(1), 224–228.

Ruan, J. and Zhang, W. (2007). Identification and evaluation of functional modules in gene co-expression networks. In *Systems Biology and Computational Proteomics*. Vol. **4532**. Berlin-Heidelberg, Springer, pp. 57–76.

Sane, A. P., Stein, B., and Westhoff, P. (2005). The nuclear gene *HCF107* encodes a membrane-associated R-TPR (RNA tetratricopeptide repeat)-containing protein involved in expression of the plastidial *psbH* gene in *Arabidopsis*. *Plant J.* **42**(5), 720–730.

Schadt, E. E. and Lum, P. Y. (2006). Thematic review series: Systems biology approaches to metabolic and cardiovascular disorders. Reverse engineering gene networks to identify key drivers of complex disease phenotypes. *J. Lipid. Res.* **47**(12), 2601–2613.

Schmid, M., Davison, T. S., Henz, S. R., Pape, U. J., Demar, M., Vingron, M., Scholkopf, B., Weigel, D., and Lohmann, J. U. (2005). A gene expression map of *Arabidopsis thaliana* development. *Natl. Genet.* **37**(5), 501–506.

Segal, E., Friedman, N., Koller, D., and Regev, A. (2004). A module map showing conditional activity of expression modules in cancer. *Natl. Genet.* **36**(10), 1090–1098.

Shannon, P., Markiel, A., Ozier, O., Baliga, N. S., Wang, J. T., Ramage, D., Amin, N., Schwikowski, B., and Ideker, T. (2003), Cytoscape: A software environment for integrated models of biomolecular interaction networks. *Genome Res.* **13**(11), 2498–2504.

Smith, S. M., Fulton, D. C., Chia, T., Thorneycroft, D., Chapple, A., Dunstan, H., Hylton, C., Zeeman, S. C., and Smith, A. M. (2004). Diurnal changes in the transcriptome encoding enzymes of starch metabolism provide evidence for both transcriptional and posttranscriptional regulation of starch metabolism in *Arabidopsis* leaves. *Plant Physiol.* **136**(1), 2687–2699.

Sozzani, R., Maggio, C., Varotto, S., Canova, S., Bergounioux, C., Albani, D., and Cella, R. (2006). Interplay between *Arabidopsis* activating factors E2Fb and E2Fa in cell cycle progression and development. *Plant Physiol* **140**(4), 1355–1366.

Stitt, M. and Hurry, V. (2002). A plant for all seasons: Alterations in photosynthetic carbon metabolism during cold acclimation in *Arabidopsis*. *Curr. Opin. Plant Biol.* **5**(3), 199–206.

Stuart, J. M., Segal, E., Koller, D., and Kim, S. K. (2003). A gene-coexpression network for global discovery of conserved genetic modules. *Science* **302**(5643), 249–255.

Thompson, M. D., Jacks, C. M., Lenvik, T. R., and Gantt, J. S. (1992). Characterization of *rps17*, *rpl9*, and *rpl15*: Three nucleus-encoded plastid ribosomal protein genes. *Plant Mol. Biol.* **18**(5), 931–944.

Titiz, O., Tambasco-Studart, M., Warzych, E., Apel, K., Amrhein, N., Laloi, C., and Fitzpatrick, T. B. (2006). PDX1 is essential for vitamin B6 biosynthesis, development and stress tolerance in *Arabidopsis*. *Plant J.* **48**(6), 933–946.

Van Dongen, S. (2000a). Graph Clustering by Flow Simulation. PhD thesis. University of Utrecht.

Van Dongen, S. (2000b). *Performance Criteria for Graph Clustering and Markov Cluster Experiments*. Technical Report INS-R0012. National Research Institute for Mathematics and Computer Science in the Netherlands, Amsterdam.

Vanstraelen, M., Inze, D., and Geelen, D. (2006). Mitosis-specific kinesins in *Arabidopsis*. *Trends Plant Sci.* **11**(4), 167–175.

Wang, C. and Liu, Z. (2006). *Arabidopsis* ribonucleotide reductases are critical for cell cycle progression, DNA damage repair, and plant development. *Plant Cell* **18**(2), 350–365.

Wei, H., Persson, S., Mehta, T., Srinivasasainagendra, V., Chen, L., Page, G. P., Somerville C., and Loraine, A. (2006). Transcriptional coordination of the metabolic network in *Arabidopsis*. *Plant Physiol* **142**(2), 762–774.

Wilson, C. L. and Miller, C. J. (2005). Simpleaffy: A BioConductor package for affymetrix quality control and data analysis. *Bioinformatics* **21**(18), 3683–3685.

Woo, H. R., Goh, C. H., Park, J. H., de la Serve, B. T., Kim, J. H., Park, Y. I., and Nam, H. G. (2002). Extended leaf longevity in the *ore4-1* mutant of *Arabidopsis* with a reduced expression of a plastid ribosomal protein gene. *Plant J.* **31**(3), 331–340.

Wurtele, Eve, Li, J., Diao, L., Zhang, H., Foster, C., Fatland, B., Dickerson, J., Brown, A., Cox, Z., Cook, D., Lee, E. K., and Hofmann, H. (2003). MetNet: software to build and model the biogenetic lattice of *Arabidopsis*. *Comp. Funct. Genom.* **4**(2), 239–245.

Yabuta, Y., Mieda, T., Rapolu, M., Nakamura, A., Motoki, T., Maruta, T., Yoshimura, K., Ishikawa, T., and Shigeoka, S. (2007). Light regulation of ascorbate biosynthesis is dependent on the photosynthetic electron transport chain but independent of sugars in *Arabidopsis*. *J. Exp. Bot.* **58**(10), 2661–2671.

Yu, M., Yuan, M., and Ren, H. (2006). Visualization of actin cytoskeletal dynamics during the cell cycle in tobacco (*Nicotiana tabacum* L. cv Bright Yellow) cells. *Biol. Cell* **98**(5), 295–306.

Zhang Shihua, J. G. and Zhang X-S., C. L. (2007). Discovering functions and revealing mechanisms at molecular level from biological networks. *Proteomics* **7**(16), 2856–2869.

5

Aquino, P., Carrion, F., Calvo, R., and Flores, D. (2001). Selected maize statistics. Part 4. In *CIMMYT 1999–2000 World Maize Facts and Trends Meeting World Maize Needs Technological Opportunities and Priorities for the Public Sector*. P. L. Pingali (Ed.). CIMMYT, Mexico, D.F. pp. 45–57.

Bartels, D. and Sunkar, R. (2005). Drought and salt tolerance in plants. *Crit. Rev. Plant Sci.* **24**. 23–58.

Benjamini, Y. and Hochberg, Y (1995). Controlling the false discovery rate: A practical and powerful approach to multiple testing. *J. Roy. Statist. Soc. B.* **57**, 289–300.

Bhatnagar-Mathur, P., Vadez, V., and Sharma, K. (2008). Transgenic approaches for abiotic stress tolerance in plants:retrospect and prospects. *Plant Cell Rep.* **27**, 411–424.

Brady, S. M., Sarkar, S. F., Bonetta, D., and Mc-Court, P. (2003). The Abscisic Acid Insensitive 3 (*ABI3*) gene is modulated by farnesylation and is involved in auxin signaling and lateral root development in *Arabidopsis*. *Plant J.* **34**, 67–75.

Bray, E. A. (1997). Plant responses to water deficit. *Trend. Plant Sci.* **2**, 48–54.

Bruce, W. B., Edmeades, G. O., and Barker, T. (2002). Molecular and physiological approaches to maize improvement for drougth tolerance. *J. Exp. Bot.* **53**, 13–25.

Brugiére, N., Jiao, S., Hantke, S., Zinselmeier, C., Roessler, J. A., et al. (2003). Cytokinin oxidase genes expression in maize is localized to the vasculature, and is induced by cytokinins, abscisic acid, and abiotic stress. *Plant Phys.* **132**, 1228–1240.

Calderón-Vázquez, C., Ibarra-Laclette, E., Caballero-Perez, J., and Herrera-Estrella, L. (2008). Transcript profiling of *Zea mays* roots reveals gene responses to phosphate deficiency at the plant- and species-specific levels. *J. Exp. Bot.* **59**, 2479–2497.

Chaves, M. M., Maroco, J. P., and Pereira, J. S. (2003). Understanding plant responses to drought-from genes to the whole plant. *Fun. Plant Biol.* **30**, 239–264.

Degenkolbe, T., Do, P., Zuther, E., Repsilber, D., Walther, D., et al. (2009). Expression profiling of rice cultivars differing in their tolerance to long-term drought stress. *Plant Mol. Biol.* **69**, 133–153.

Edgar, R., Domrachev, M., and Lash, A. E. (2002). Gene Expression Omnibus: *NCBI* gene expression and hybridization array data repository. *Nucleic Acids Res.* **30**, 207–210.

Fischer, K. S., Johnson, E. C., and Edmeades, G. O. (1983). *Breeding and Selection for Drought Resistance in Tropical Maize*. CIMMYT, Mexico City (Mexico), P. 20.

Foyer, C., Valadier, M., Migge, A., and Becker, T. (1998). Drought-Induced effects on nitrate reductase activity and mRNA and on the coordination of nitrogen and carbon metabolism in maize leaves. *Plant Phys.* 283–292.

Garcion, C., Applimath, F., and Metraux, J. (2006). FiRe and microarrays: A fast answer to burning questions. *Trends Plant Sci.* **11**, 320–322.

Gibson, G. and Wolfinger, R. (2004). Gene expression profiling using mixed models. In *Genetic analysis of complex traits using SAS*. A. Saxton (ed.). SAS Press, Cary, NC. USA, pp. 251–278.

Gutiérrez, R. A., Gifford, M. L., Poultney, C., Wang, R., Shasha, D. E., et al. (2007). Insights into the genomic nitrate response using genetics and the Sungear Software System. *J. Exp. Bot.* **58**, 2359–2367.

Hegeman, C., Good, L., and Grabau, E. (2001). Expression of D-myo-Inositol-3-Phosphate Synthase in Soybean. Implications for Phytic Acid Biosynthesis. *Plant Phys.* **125**, 1941–1948.

Ingram, J. and Bartels, D. (1996). The molecular basis of dehydration tolerance in plants. *Annu. Rev. Plant Physiol. Plant Mol. Biol.* **47**, 377–403.

Jia, J., Fu, J., Zheng, J., Zhou, X., Huai, J., et al. (2006). Annotation and expression profile analysis of 2073 full-length cDNAs from stress-induced maize (*Zea mays* L.) seedlings. *Plant J.* **48**, 710–727.

Kawaguchi, R., Girke, T., Bray, E. A., and Bailey-Serres, J. (2004). Differential mRNA translation contributes to gene regulation under non-stress and dehydration stress conditions in *Arabidopsis thaliana*. *Plant J.* **38**, 823–839.

Kishor, P. B. K., Sangam, S., Amrutha, R. N., Laxmi, P. S., Naidu, K. R., et al. (2005). Regulation of proline biosynthesis, degradation, uptake and transport in higher plants: Its implications in plant growth and abiotic stress tolerance. *Curr. Sci.* **88**, 424–438.

Koch, K. E. (1996). Carbohydrate-modulated gene expresion in plants. *Annu. Rev. Plant Physiol. Plant Mol. Biol* **47**, 509–540.

Li, W. E., Oono, Y., Zhu, J., He, X. J., Wu, J. M., et al. (2008). The *Arabidopsis* NFYA5 Transcription Factor is regulated transcriptionally and posttranscriptionally to promote drought resistance. *Plant Cell* **20**, 2238–2251.

Mahajan, S. and Tuteja, N. (2005). Cold, salinity and drought stresses: An overview. *Arch. Biochem.Biophys.* **444**, 139–158.

Massonneau, A., Houba-Hérin, N., Pethe, C., Madzak, C., Falque, M., et al. (2004). Maize cytokinin oxidase genes: Differential expression and cloning of the two cDNAs. *J. Exp. Bot.* **55**, 2549–2557.

Mohammadi, M., Kav, N. N. V., and Deyholos, M. K. (2007). Transcriptional profiling of hexaploid wheat (*Triticum aestivum* L.) roots identifies novel, dehydration-responsive genes. *Plant, Cell Env.* **30**, 630–645.

Morcuende, R., Bari, R., Gibon, Y., Zheng, W., Pant, B., et al. (2007) Genome-wide reprogramming of metabolism and regulatory networks of *Arabidopsis* in response to phosphorus. *Plant, Cell Env.* **30**, 85–112.

Muñoz, O. A., Stevenson, K. R., Ortiz, C. J., Thurtell, G. W., and Carballo, C. A. (1983). Transpiration, photosynthesis, water use efficiency and water potential in maize drought and frost resistant. *Agrociencia* **51**, 115–153.

Nelson, D., Repetti, P. P., Adams, T. R., Creelman, R. A., Wu, J., et al. (2007). Plant nuclear factor Y (NF-Y) B subunits confer drought tolerance and lead to improved corn yields on water-limited acres. *Proc. Natl. Acad. Sci.* **104**, 16450–16455.

Oono, Y., Seki, M., Nanjo, T., Narusaka, M., Fujita, M., et al. (2003). Monitoring expression profiles of *Arabidopsis* gene expression during rehydration process after dehydration using ca. 7000 full-length cDNA microarray. *Plant J.* **34**, 868–887.

Ozturk, Z. N., Talame, V., Deyholos, M., Michalowski, C. B., Galbraith, D. W., et al. (2002). Monitoring large-scale changes in transcript abundance in drought- and salt-stressed barley. *Plant Mol. Biol.* **48**, 551–573.

Park, J. E., Park, J. Y., Kim, Y. S., Staswick, P. E., Jeon, J., et al. (2007). GH3-mediated Auxin Homeostasis links growth regulation with stress adaptation response in *Arabidopsis*. *J. Biol. Chem.* **282**, 10036–10046.

Parry, M. A. J., Flexas, J., and Medrano, H. (2005). Prospects for crop production under drought: Research priorities and future directions. *Ann. Appl. Biol.* **147**, 211–226.

Pelleschi, S., Rocher, J., and Prioul, J. (1997). Effect of water restriction on carbohydrate metabolism and photosynthesis in mature maize leaves. *Plant, Cell Env.* **20**, 493–503.

Pérez, J. G. (1979). *Behavior of Caramel Corn Under Different Moisture Levels.* Texcoco, Estado de Mexico, Colegio de Postgraduados, Chapingo, Mexico, p. 193.

Phillips, R. D. and Jennings, D. H. (1976). Succulence, cations and organic acids in leaves of Kalanchoe daigremontiana grown in long and short days in soil and water culture. *New Phytol.* **77**, 599–611.

Pingali, P. L. and Pandey, S. (2001). Meeting world maize needs: Technological opportunities and priorities for the public sector. In *CIMMYT 1999–2000 World Maize Facts and Trends Meeting World Maize Needs Technological Opportunities and Priorities for the Public Sector.*

Poroyko, V., Spollen, W. G., Hernandez, A. G., LeNoble, M. E., Davis, G., et al. (2007). Comparing regional transcript profiles from maize primary roots under well-watered and low water potential conditions. *J. Exp. Bot.* **2**, 279–289.

Rabbani, M. A., Maruyama, K., Abe, H., Khan, M. A., Katsura, K., et al. (2003). Monitoring expression profiles of rice genes under cold, drought, and high-salinity stresses and Abscisic acid application using cDNA microarray and RNA gel-blot analyses. *Plant Phys.* **133**, 1755–1767.

Reynolds, M., Dreccer, F., and Trethowan, R. (2007). Drought-adaptive traits derived from wheat wild relatives and landraces. *J. Exp. Bot.* **58**, 177–186.

Ribaut, J. M. and Ragot, M. (2007). Marker-assisted selection to improve drought adaptation in maize: the backcross approach, perspectives, limitations, and alternatives. *J. Exp. Bot.* **58**, 351–360.

Schafleitner, R., Gutierrez-Rosales, R. O., Gaudin, A., Alvarado-Aliaga, C. A., Nombeo-Martinez, G, et al. (2007). Capturing candidate drought tolerance traits in two native Andean potato clones by transcription profiling of field grown plants under water stress. *Plant Phys. Biochem.* **45**, 673–690.

Seki, M., Narusaka, M., Ishida, J., Nanjo, T., Fujita, M., et al. (2002). Monitoring the expression profiles of 7000 *Arabidopsis* genes under drought, cold and high-salinity stresses using a full-length cDNA microarray. *Plant J.* **31**, 279–292.

Seki, M., Umezawa, T., Urano, K., and Shinozaki, K. (2007). Regulatory metabolic networks in drought stress responses. *Curr. Opin. Biotechnol* **10**, 296–302.

Shen, Q. C., Chen, C. N., Brands, A., Pan, S. M., and Ho, T. H. D. (2001). The stress- and abscisic acid-induced barley gene HVA22: developmental regulation and homologues in diverse organisms. *Plant Mol. Biol.* **45**, 327–340.

Shinozaki, K. and Yamaguchi-Shinozaki, K. (1997). Gene expression and signal transduction in water-stress response. *Plant Physiol.* **115**, 327–334.

Shinozaki, K. and Yamaguchi-Shinozaki, K. (2007). Gene networks involved in drought stress response and tolerance. *J. Exp. Bot.* **58**, 221–227.

Smyth, G., Thorne, N., and Wettenhall, J. (2003). Limma: Linear Models for Microarray Data User's Guide.Software manual available from http://www.bioconductor.org.

Spollen, W. G., Tao, W., Valliyodan, B., Chen, K., Hejlek, L. G., et al. (2008). Spatial distribution of transcript changes in the maize primary root elongation zone at low water potential. *BMC Plant Biol.* **3**, 8–32.

Sun, W., Van, Montagu, M., and Verbruggen, N. (2002). Small heat shock proteins and stress tolerance in plants Biochimica et Biophysica Acta (BBA). *Gene Struct. Exp.* **1577**, 1–9.

Talamé, V., Ozturk, Z. N., Bohnert, H. J., and Tuberosa, R. (2007). Barley transcript profiles under dehydration shock and drought stress treatments: a comparative analysis. *J. Exp. Bot.* **58**, 229–240.

Tamminen, I., Mäkelä, P., Heino, P., and Palva, E. T. (2001). Ectopic expression of ABI3 gene enhances freezing tolerance in response to abscisic acid and low temperature in *Arabidopsis* thaliana. *Plant J.* **25**, 1–8.

Thimm, O., Blasing, O., Gibon, Y., Nagel, A., Meyer, S., et al. (2004). MAPMAN:a user-driven tool to display genomics data sets onto diagrams of metabolic pathways and other biological processes. *Plant J.* **37**, 914–939.

Tuberosa, R., Salvi, S., Giuliani, S., Sanguineti, M. C., Bellotti, M., et al. (2007). Genome-wide Approaches to Investigate and Improve Maize Response to Drought. *Crop. Sci.* **47** (S3), S120–141.

Umezawa, T., Fujita, M., Fujita, Y., Yamaguchi-Shinozaki, K., and Shinozaki, K. (2006). Engineering drought tolerance in plants: discovering and tailoring genes to unlock the future. *Curr. Opin. Biotechnol.* **17**, 113–122.

Vendruscolo, E. C. G., Schuster, I., Pileggi, M., Scapim, C. A., Correa, H. B., et al. (2007). Stress-induced synthesis of proline confers tolerance to water deficit in transgenic wheat. *J. Plant Phys.* **10**, 1367–1376.

Vinocur, B. and Altman, A. (2005). Recent advances in engineering plant tolerance to abiotic stress: achievements and limitations. *Curr. Opin. Biotechnol.* **16**, 123–132.

Wang, W., Vinocur, B., and Altman, A. (2003). Plant responses to drought, salinity and extreme temperatures: towards genetic engineering for stress tolerance. *Planta* **218**, 1–14.

Wasilewska, A., Vlad, F., Sirichandra, C., Redko, Y., Jammes, F., et al. (2008). An update on Abscisic acid signaling in plants and more. *Mol. Plant* **1**, 198–217.

Wolfinger, R., Gibson, G., and Wolfinger, E. (2001). Assessing gene significance from cDNA microarray expression data via mixed models. *J. Comput. Biol.* **8**, 625–637.

Xiao, W., Sheen, J., and Jang, J. C. (2000). The role of hexokinase in plant sugar signal transduction and growth and development. *Plant Mol. Biol.* **44**, 451–461.

Xiong, L., Schumaker, K. S., and Zhu, J. K. (2002). Cell signaling during cold, drought, and salt stress. *Plant Cell* **14**, S165–S183.

Xue, P., McIntyre, C., Glassop, D., and Shorter, R. (2008). Use of expression analysis to dissect alterations in carbohydrate metabolism in wheat leaves during drought stress. *Plant Mol. Biol.* **67**, 197–214.

Yamaguchi-Shinozaki, K. and Shinozaki, K. (2006). Transcriptional regulatory networks in cellular responses and tolerance to dehydration and cold stresses. *Annu. Rev. Plant Biol.* **57**, 781–803.

Yang, Y., Dudoit, S., Luu, P., Lin, D., Peng, V., et al. (2002). Normalization for cDNA microarray data: A robust composite method addressing single and multiple slide systematic variation. *Nucleic Acids Res.* **30**, e15.

Yu, L. and Setter, T. L. (2003). Comparative transcriptional profiling of placenta and endosperm in developing maize kernels in response to water deficit. *Plant Phys.* **131**, 568–582.

Zheng, J., Zhao, J., Wang, J., Liu, Y., Fu, J., et al. (2004). Isolation and analysis of water stress induced genes in maize seedlings by subtractive PCR and cDNA macroarray. *Plant Mol. Bio.* **55**, 807–823.

6

Aasland, R., Gibson, T. J., and Stewart, A. F. (1995). The PHD finger: implications for chromatin-mediated transcriptional regulation. *Trends Biochem. Sci.* **20**, s56–59.

Abe, H., Urao, T., Ito, T., Seki, M., Shinozaki, K., et al. (2003). *Arabidopsis* AtMYC2 (bHLH) and AtMYB2 (MYB) function as transcriptional activators in abscisic acid signaling. *Plant Cell* **15**, 63–78.

Anderson, M., Fair, K., Amero, S., Nelson, S., Harte, P. J., et al. (2002) A new family of cyclophilins with an RNA recognition motif that interact with members of the trx/MLL protein family in Drosophila and human cells. *Dev. Gen. Evol.* **212**, 107–113.

Apel, K. and Hirt, H. (2004). Reactive oxygen species: metabolism, oxidative stress, and signal transduction. *Annu Rev. Plant Biol.* **55**, 373–399.

Aravind, L., Iyer, L. M., and Koonin, E. V. (2003). Scores of RINGS but no PHDs in ubiquitin signaling. *Cell Cycle* **2**, 123–126.

Bastola, D. R., Pethe, V. V., and Winicov, I. (1998). Alfin1, a novel zinc-finger protein in alfalfa roots that binds to promoter elements in the salt-inducible *MsPRP2* gene. *Plant Mol. Biol.* **38**, 1123–1135.

Borden, K. L. and Freemont, P. S. (1996). The RING finger domain: A recent example of a sequence-structure family. *Curr. Opin. Struct. Biol.* **6**, 395–401.

Bottomley, M. J., Stier, G., Pennacchini, D., Legube, G., Simon, B., et al. (2005). NMR structure of the first PHD finger of autoimmune regulator protein (AIRE1). Insights into autoimmune polendocrinopathy-candidiasis-ectodermal dystrophy (APECED) disease. *J. Biol. Chem.* **280**, 11505–11512.

Cao, W. H., Liu, J., He, X. J., Mu, R. L., Zhou, H. L., et al. (2007). Modulation of ethylene responses affects plant salt-stress responses. *Plant Physiol.* **143**, 707–719.

Capili, A. D., Schultz, D. C., Rauscher, F. J. III, and Borden, K. L. (2001). Solution structure of the PHD domain from the KAP-1 corepressor: Structural determinants for PHD, RING, and LIM zinc-binding domains. *E.M.B.O. J.* **20**, 165–177.

Coscoy, L. and Ganem, D. (2003). PHD domains and E3 ubiquitin ligases: Viruses make the connection. *Trends Cell Biol.* **13**, 7–12.

Dawid, I. B., Breen, J. J., and Toyama, R. (1998). LIM domains: multiple roles as adapters and functional modifiers in protein interactions. *Trends Genet.* **14**, 156–162.

De la Paz Sanchez, M., and Gutierrez, C. (2009). *Arabidopsis* ORC1 is a PHD-containing H3K4me3 effector that regulates transcription. *Proc. Natl. Acad. Sci. USA* **106**, 2065–2070.

Eulgem, T., Tsuchiya, T., Wang, X. J., Beasley, B., Cuzick A., et al. (2007). EDM2 is required for RPP7-dependent disease resistance in *Arabidopsis* and affects RPP7 transcript levels. *Plant J.* **49**, 829–839.

Fair, K., Anderson, M., Bulanova, E., Mi, H., Tropschug, M., et al. (2001) Protein interactions of the MLL PHD fingers modulate MLL target gene regulation in human cells. *Mol. Cell. Biol.* **21**, 3589–3597.

Gao, L. and Xiang, C. B. (2008). The genetic locus *At1g73660* encodes a putative MAPKKK and negatively regulates salt tolerance in *Arabidopsis*. *Plant Mol. Biol.* **67**, 125–134.

Goto, E., Ishido, S., Sato, Y., Ohgimoto, S., Ohgimoto, K., et al. (2003). c-MIR, a human E3 ubiquitin ligase, is a functional homolog of herpesvirus proteins MIR1 and MIR2 and has similar activity. *J. Biol. Chem.* **278**, 14657–14658.

Greb, T., Mylne, J. S., Crevillen, P., Geraldo, N., An, H., et al. (2007) The PHD finger protein VRN5 functions in the epigenetic silencing of *Arabidopsis* FLC. *Curr. Biol.* **17**, 73–78.

Halbach, T., Scheer, N., and Werr, W. (2000). Transcriptional activation by the PHD finger is inhibited through an adjacent leucine zipper that binds 14-3-3 proteins. *Nucleic Acids Res.* **28**, 3542–3550.

Hasegawa, P. M., Bressan, R. A., Zhu, J. K., and Bohnert, H. J. (2000). Plant cellular and molecular responses to high salinity. *Annu. Rev. Plant Physiol. Plant Mol. Biol.* **51**, 463–499.

He, X. J., Mu, R. L., Cao, W. H., Zhang, Z. G., and Zhang, J. S., et al. (2005). AtNAC2, a transcription factor downstream of ethylene and auxin signaling pathways, is involved in salt stress response and lateral root development. *Plant J.* **44**, 903–916.

Kalkhoven, E., Teunissen, H., Houweling, A., Verrijzer, C. P., and Zantema, A. (2002). The PHD type zinc finger is an integral part of the CBP acetyltransferase domain. *Mol. Cell. Biol.* **22**, 1961–1970.

Kant, P., Kant, S., Gordon, M., Shaked, R., and Barak, S. (2007). Stress Response Suppressor1 And Stress Response Suppressor2, two DEAD-Box RNA Helicases that attenuate *Arabidopsis* responses to multiple abiotic stresses. *Plant Physiol.* **145**, 814–830.

Kasuga, M., Liu, Q., Miura, S., Yamaguchi-Shinozaki, K., and Shinozaki, K. (1999). Improving plant drought, salt, and freezing tolerance by gene transfer of a single stress-induced transcription factor. *Nat. Biotechnol.* **17**, 287–291.

Kuzmichev, A., Zhang, Y., Erdjument-Bromage, H., Tempst, P., and Reinberg, D. (2002). Role of the Sin3-histone deacetylase complex in growth regulation by the candidate tumor suppressor p33(ING1). *Mol. Cell Biol.* **22**, 835–848.

Lee, W. Y., Lee, D., Chung, W. I., and Kwon, C. S. (2009). *Arabidopsis* ING and Alfin1-like protein families localize to the nucleus and bind to H3K4me3/2 via plant homeodomain fingers. *Plant J.* **58**, 511–524.

Liao, Y., Zou, H. F., Wang, H. W., Zhang, W. K., Ma, B., et al. (2008b). Soybean *GmMYB76*, *GmMYB92* and *GmMYB177* genes confer stress tolerance in transgenic *Arabidopsis* plants. *Cell Res.* **18**, 1047–1060.

Liao, Y., Zou, H. F., Wei, W., Hao, Y. J., Tian, A. G., et al. (2008) Soybean GmbZIP44, *GmbZIP62* and *GmbZIP78* genes function as negative regulator of ABA signaling and confer salt and freezing tolerance in transgenic *Arabidopsis*. *Planta* **228**, 225–240.

Linder, B., Newman, R., Jones, L. K., Debernardi, S., Young, B. D., et al. (2000). Biochemical analyses of the AF10 protein: the extended LAP/PHD-finger mediates oligomerisation. *J. Mol. Biol.* **299**, 369–378.

Lu, Z., Xu, S., Joazeiro, C., Cobb, M. H., and Hunter, T. (2002). The PHD domain of MEKK1 acts as an E3 ubiquitin ligase and mediates ubiquitination and degradation of ERK1/2. *Mol. Cell* **9**, 945–956.

Maehly, A. C. and Chance, B. (1954). The assay of catalases and peroxidases. *Methods of biochemical analysis* **1**, 357–425.

Mansouri, M., Bartee, E., Gouveia, K., Hovey Nerenberg, B. T., Barrett J., et al. (2003). The PHD/LAP-domain protein M153R of myxomavirus is a ubiquitin ligase that induces the rapid internalization and lysosomal destruction of CD4. *J. Virol.* **77**, 1427–1440.

Mittler, R., Vanderauwera, S., Gollery, M., and Van Breusegem, F. (2004). Reactive oxygen gene network of plants. *Trends Plant Sci.* **9**, 490–498.

Novillo, F., Alonso, J. M., Ecker, J. R., and Salinas, J. (2004). CBF2/DREB1C is a negative regulator of CBF1/DREB1B and CBF3/DREB1A expression and plays a central role in stress tolerance in *Arabidopsis. Proc. Natl. Acad. Sci. USA* **101**, 3985–3990.

Ohta, M., Matsui, K., Hiratsu, K., Shinshi, H., and Ohme-Takagi, M. (2001). Repression domains of class II ERF transcriptional repressors share an essential motif for active repression. *Plant Cell* **13**, 1959–1968.

Pena, P. V., Davrazou, F., Shi, X., Walter, K. L., Verkhusha, V. V., et al. (2006). Molecular mechanism of histone H3K4me3 recognition by plant homeodomain of ING2. *Nature* **442**, 100–103.

Piñeiro, M., Gomez-Mena, C., Schaffer, R., Martinez-Zapater, J., and Coupland G. (2003). EARLY BOLTING IN SHORT DAYS is related to chromatin remodeling factors and regulates flowering in *Aribidopsis* by repressing FT. *Plant Cell* 15, 1552–1562.

Reddy, T. V., Kaur, J., Agashe, B., Sundaresan, V., and Siddiqi, I. (2003). The *DUET* gene is necessary for chromosome organization and progression during male meiosis in *Arabidopsis* and encodes a PHD finger protein. *Development* 130, 5975–5987.

Saiga, S., Furumizu, C., Yokoyama, R., Kurata, T., Sato, S., et al. (2008). The *Arabidopsis* *OBERON1* and *OBERON2* genes encode plant homeodomain finger proteins and are required for apical meristem maintenance. *Development* 135, 1751–1759.

Scheel, H. and Hofmann, K. (2003). No evidence for PHD fingers as ubiquitin ligases. *Trends Cell Biol.* 13, 285–287.

Schindler, U., Beckmann, H., and Cashmore, A. R. (1993). HAT3.1, a novel *Arabidopsis* homeodomain protein containing a conserved cysteine-rich region. *Plant J.* 4, 137–150.

Schultz, D. C., Friedman, J. R., and Rauscher, F. J. 3rd (2001). Targeting histone deacetylase complexes via KRAB-zinc finger proteins: The PHD and bromodomains of KAP-1 form a cooperative unit that recruits a novel isoform of the Mi-2alpha subunit of NuRD. *Genes. Dev.* 15, 428–443.

Sebastian, J., Ravi, M., Andreuzza, S., Panoli, A. P., Marimuthu, M. P., et al. (2009). The plant adherin AtSCC2 is required for embryogenesis and sister-chromatid cohesion during meiosis in *Arabidopsis. Plant J.* 59, 1–13.

Seki, M., Kamei, A., Yamaguchi-Shinozaki, K., and Shinozaki, K. (2003). Molecular responses to drought, salinity, and frost: common and different paths for plant protection. *Curr. Opin. Biotechnol* 14, 194–199.

Shi, X., Hong, T., Walter, K. L., Ewalt, M., Michishita, E., et al. (2006). ING2 PHD domain links histone H3 lysine 4 methylation to active gene repression. *Nature* 442, 96–99.

Skowyra, D., Zeremski, M., Neznanov, N., Li, M.Y., Choi, Y.M., et al. (2001) Differential as-sociation of products of alternative transcripts of the candidate tumor suppressor ING1 with the mSin3/HDAC1 transcriptional corepressor complex. *J. Biol. Chem.* 276, 8734–8739.

Sung, S. and Amasino, R. M. (2004). Vernalization in *Arabidopsis thaliana* is mediated by the PHD finger protein VIN3. *Nature* 427, 159–164.

Sung, S., Schmitz, R. J., and Amasino, R. M. (2006). A PHD finger protein involved in both the vernalization and photoperiod pathways in *Arabidopsis. Genes. Dev.* 20, 3244–3248.

Thomas, C. L., Schmidt, D., Bayer, Em., Dreos, R., and Maule, A. J. (2009). *Arabidopsis* plant homeodomain finger proteins operate downstream of auxin accumulation in specifying the vasculature and primary root meristem. *Plant J.* 59, 426–436.

Townsley, F. M., Thompson, B., and Bienz, M. (2004). Pygopus residues required for its binding to Legless are critical for transciption and development. *J. Biol. Chem.* 278, 5177–5183.

Tripoulas, N., Lajeunesse, D., Gildea, J., and Shearn, A. (1996). The Drosophila *ash1* gene product, which is localized at specific sites on polytene chromosomes, contains a SET domain and a PHD finger. *Genetics.* 143, 913–928.

Venturini, L., You, J., Stadler, M., Galien, R. Lallemand, V., et al. (1999). TIF1, a novel member of the transcriptional intermediary factor 1 family. *Oncogene.* 18, 1209–1217.

Wang, Y. J., Li, Y. D., Luo, G. Z., Tian, A. G., Wang, H. W., et al. (2005). Cloning and characterization of an HDZip I gene GmHZ1 from soybean. *Planta* 221, 831–843.

Wang, Y. J., Zhang, Z. G., He, X. J., Zhou, H. L., Wen, Y. X., et al. (2003). A rice transcription factor OsbHLH1 is involved in cold stress response. *Theor. Appl. Genet.* 107, 1402–1409.

Wilson, Z. A., Morroll, S. M., Dawson, J., Swarup, R., and Tighe, P. J. (2001). The *Arabidopsis* Male Sterility1 (*MS1*) gene is a transcriptional regulator of male gametogenesis, with homology to the PHD-finger family of transcription factors. *Plant J.* 28, 27–39.

Winicov, I. (2000). Alfin1 transcription factor overexpression enhances plant root growth under normal and saline conditions and improves salt tolerance in alfalfa. *Planta* **210**, 416–22.

Winicov, I., Valliyodan, B., Xue, L., and Hoober, J. K. (2004). The MsPRP2 promoter enables strong heterologous gene expression in a root-specific manner and is enhanced by overexpression of Alfin 1. *Planta* **219**, 925–935.

Winicov, II and Bastola, D. R. (1999). Transgenic overexpression of the transcription factor alfin1 enhances expression of the endogenous *MsPRP2* gene in alfalfa and improves salinity tolerance of the plants. *Plant Physiol.* **120**, 473–480.

Wysocka, J., Swigut, T., Xiao, H., Milne, T. A., Kwon, S. Y., et al. (2006). A PHD finger of NURF couples histone H3 lysine 4 trimethylation with chromatin remodelling. *Nature* **442**, 86–90.

Xie, C., Zhang, J. S., Zhou, H. L., Li, J., Zhang, Z. G., et al. (2003). Serine/threonine kinase activity in the putative histidine kinase-like ethylene receptor NTHK1 from tobacco. *Plant J.* **33**, 385–393.

Yang, X., Makaroff, C. A., and Ma, H. (2003). The *Arabidopsis* MALE MEIOCYTE DEATH1 gene encodes a PHD-finger protein that is required for male meiosis. *Plant Cell* **15**, 1281–1295.

Yochum, G. S. and Ayer, D. E. (2001). Pf1, a novel PHD zinc finger protein that links the TLE corepressor to the mSin3A-histone deacetylase complex. *Mol. Cell Biol.* **21**, 4110–4118.

Zhou, Q. Y., Tian, A. G., Zou, H. F., Xie, Z. M., Lei, G., et al. (2008). Soybean WRKY-type transcription factor genes, *GmWRKY13*, *Gm-WRKY21*, and *GmWRKY54*, confer differential tolerance to abiotic stresses in transgenic *Arabidopsis* plants. *Plant Biotechnol. J.* **6**, 486–603.

Zhu. J. K. (2002). Salt and drought stress signal transduction in plants. *Annu. Rev. Plant Biol.* **53**, 247–273.

7

Allemeersch, J., Durinck, S., Vanderhaeghen, R., Alard, P., Maes, R., Seeuws, K., Bogaert, T., Coddens, K., Deschouwer, K., van Hummelen, P., Vuylsteke, M., Moreau, Y., Kwekkeboom, J., Wijfjes, A., May, S., Beynon, J., Hilson, P., and Kuiper, M. (2005). Benchmarking the CATMA Microarray. A novel tool for *Arabidopsis* transcriptome analysis. *Plant Physiol.* **137**, 588–601.

Altschul, S. F., Gish, W., Miller, W., Myers, E. W., and Lipman, D. (1990). Basic local alignement search tool. *J. Mol. Biol.* **215**, 403–410.

Bakalovic, N., Passardi, F., Ioannidis, V., Cosio, C., Penel, C., Falquet, L., and Dunand, C. (2006). PeroxiBase: A class III plant peroxidase database. *Phytochemistry* **67**, 534–539.

Baumberger, N., Doesseger, B., Guyot, R., Diet, A., Parsons, R. L., Clark, M. A., Simmons, M. P., Bedinger, P., Goff, S. A., Ringli, C., and Keller, B. (2003). Whole-genome comparison of leucine-rich repeat extensins in *Arabidopsis* and rice. A conserved family of cell wall proteins form a vegetative and a reproductive clade. *Plant Physiol.* **131**, 1313–1326.

Berger, D. and Altmann, T. (2000). A subtilisin-like serine protease involved in the regulation of stomatal density and distribution in *Arabidopsis thaliana*. *Genes Dev.* **14**, 1119–1131.

Boudart, G., Minic, Z., Albenne, C., Canut, H., Jamet, E., and Pont-Lezica, R. (2007). Cell wall proteome. In *Plant Proteomics*. S. Samaj and J. Thelen (Ed.). Springer, pp. 169–185.

Brady, J. D., Sadler, I. H., and Fry, S. C. (1996). Di-isodityrosine, a novel tetrametric derivative of tyrosine in plant cell wall proteins: A new potential cross-link. *Biochem J.* **315**, 323–327.

Collett, C. E., Harberd, N. P., and Leyser, O. (2000). Hormonal interactions in the control of *Arabidopsis* hypocotyl elongation. *Plant Physiol.* **124**, 553–562.

Cosgrove, D. J. (2005). Growth of the plant cell wall. *Nat. Rev. Mol. Cell. Biol.* **6**, 850–861.

Coutinho, P. M. and Henrissat, B. (1999). Carbohydrate-active enzymes: An integrated database approach. In *Recent Advances in Carbohydrate Bioengineering*. H. J. Gilbert, B. Davies, B. Henrissat, and B. Svensson (Eds.). The Royal Society of Chemistry, Cambridge, pp. 3–12.

Crowe, M. L., Serizet, C., Thareau, V., Aubourg, S., Rouze, P., Hilson, P., Beynon, J., Weisbeek, P., van Hummelen, P., Reymond, P., Paz-Ares, J., Nietfeld, W., and Trick, M. (2003). CATMA: A complete *Arabidopsis* GST database. *Nucleic Acids Res.* **31**, 156–158.

Derbyshire, P., Findlay, K., McCann, M. C., and Roberts, K. (2007). Cell elongation in *Arabidopsis* hypocotyls involves dynamic changes in cell wall thickness. *J. Exp. Bot.* **58**, 2079–2089.

Derbyshire, P., McCann, M. C., and Roberts, K. (2007). Restricted cell elongation in *Arabidopsis* hypocotyls is associated with a reduced average pectin esterification level. *BMC Plant Biol.* **7**, 31.

Desnos, T., Orbovic, V., Bellini, C., Kronenberger, J., Caboche, M., Traas, J., and Hofte, H. (1996). Procuste1 mutants identify two distinct genetic pathways controlling hypocotyl cell elongation, respectively in dark- and light-grown *Arabidopsis* seedlings. *Development* **122**, 683–693.

Egelund, J., Skjot, M., Geshi, N., Ulvskov, P., and Petersen, B. L. (2004). A complementary bioinformatics approach to identify potential plant cell wall glycosyltransferase-encoding genes. *Plant Physiol.* **136**, 2609–2620.

Emanuelsson, O., Nielsen, H., Brunak, S., and von Heijne, G. (2000). Predicting subcellular localization of proteins based on their N-terminal amino acid sequence. *J. Mol. Biol.* **300**, 1005–1016.

Fowler, T. J., Bernhardt, C., and Tierney, M. L. (1999). Characterization and expression of four proline-rich cell wall protein genes in *Arabidopsis* encoding two distinct subsets of multiple domain proteins. *Plant Physiol.* **121**, 1081–1092.

Gagnot, S., Tamby, J. P., Martin-Magniette, M. L., Bitton, F., Taconnat, L., Balzergue, S., Aubourg, S., Renou, J. -P., Lecharny, A., and Brunaud, V. (2007). CATdb: A public access to *Arabidopsis* transcriptome data from the URGV-CATMA platform. *Nucleic Acids Res.* **36** (database issue), D986–990.

Ge, Y., Dudoit, S., and Speed, T. P. (2003). Resampling-based multiple testing for microarray data analysis. *Test* **12**, 1–77.

Gendreau, E., Traas, J., Desnos, T., Grandjean, O., Caboche, M., and Hofte, H. (1997). Cellular basis of hypocotyl growth in *Arabidopsis thaliana*. *Plant Physiol.* **114**, 295–305.

Gygi, S. P., Rochon, Y., Franza, B. R., and Aebersold, R. (1999). Correlation between protein and mRNA abundance in yeast. *Mol. Cell Biol.* **19**, 1720–1730.

Hilson, P., Allemeersch, J., Altmann, T., Aubourg, S., Avon, A., Beynon, J., Bhalerao, R. P., Bitton, F., Caboche, M., Cannoot, B., Chardakov, V., Cognet-Holliger, C., Colot, V., Crowe, M., Darimont, C., Durinck, S., Eickhoff, H., de Longevialle, A. F., Farmer, E. E., Grant, M., Kuiper, M. T., Lehrach, H., Leon, C., Leyva, A., Lundeberg, J., Lurin, C., Moreau, Y., Nietfeld, W., Paz-Ares, J., Reymond, P., Rouze, P., Sandberg, G., Segura, M. D., Serizet, C., Tabrett, A., Taconnat, L., Thareau, V., Van Hummelen, P., Vercruysse, S., Vuylsteke, M., Weingartner, M., Weisbeek, P. J., Wirta, V., Wittink, F. R., Zabeau, M., and Small, I. (2004). Versatile gene-specific sequence tags for *Arabidopsis* functional genomics: transcript profiling and reverse genetics applications. *Genome Res.* **14**, 2176–2189.

Irshad, M., Canut, H., Borderies, G., Pont-Lezica, R., and Jamet, E. (2008). A new picture of cell wall protein dynamics in elongating cells of *Arabidopsis thaliana*: Confirmed actors and newcomers. *BMC Plant Biol.* **8**, 94.

Jacobs, J. and Roe, J. L. (2005). SKS6, a multicopper oxidase-like gene, participates in cotyledon vascular patterning during *Arabidopsis thaliana* development. *Planta* **222**, 652–666.

Jamet, E., Albenne, C., Boudart, G., Irshad, M., Canut, H., and Pont-Lezica, R. (2008). Recent advances in plant cell wall proteomics. *Proteomics* **8**, 893–908.

Jamet, E., Canut, H., Boudart, G., and Pont-Lezica, R. F. (2006). Cell wall proteins: A new insight through proteomics. *Trends Plant Sci.* **11**, 33–39.

Johnson, K. L., Jones, B. J., Schultz, C. J., and Bacic, A. (2003). Non enzymic cell wall (glyco) proteins. In *The Plant Cell Wall*. J. K. Rose (Ed.). Blackwell Publishing, CRC Press, Boca Raton, pp. 111–154.

Jones, A. M., Thomas, V., Truman, B., Lilley, K., Mansfield, J., and Grant, M. (2004). Specific changes in the *Arabidopsis* proteome in response to bacterial challenge: Differentiating basal and R-gene mediated resistance. *Phytochemistry* **65**, 1805–1816.

Kolkman, A., Daran-Lapujade, P., Fullaondo, A., Olsthoorn, M. M., Pronk, J. T., Slijper, M., and Heck, A. J. (2006). Proteome analysis of yeast response to various nutrient limitations. *Mol. Syst. Biol.* **2**, 26.

Lafarguette, F., Leple, J.C., Dejardin, A., Laurans, F., Costa, G., Lesage-Descauses, M. C., and Pilate, G. (2004). Poplar genes encoding fasciclin-like arabinogalactan proteins are highly expressed in tension wood. *New Phytol.* **164**, 107–121.

Liu, C. and Mehdy, M. (2007). A nonclassical arabinogalactan protein gene highly expressed in vascular tissues, AGP31, is transcriptionally repressed by methyl jasmonic acid in *Arabidopsis. Plant Physiol.* **145**, 863–874.

Lurin, C., Andres, C., Aubourg, S., Bellaoui, M., Bitton, F., Bruyere, C., Caboche, M., Debast, C., Gualberto, J., Hoffmann, B., Lecharny, A., Le Ret, M., Martin-Magniette, M. L., Mireau, H., Peeters, N., Renou, J. P., Szurek, B., Taconnat, L., and Small, I. (2004). Genome-wide analysis of *Arabidopsis* pentatricopeptide repeat proteins reveals their essential role in organelle biogenesis. *Plant Cell* **16**, 2089–2103.

Matsubayashi, Y. and Sakagami, Y. (2006). Peptide hormones in plants. *Annu. Rev. Plant Biol.* **57**, 649–674.

McCabe, P. F., Valentine, T. A., Forsberg, L. S., and Pennell, R. I. (1997). Soluble signals from cells identified at the cell wall establish a developmental pathway in carrot. *Plant Cell* **9**, 2225–2241.

McCaig, B. C., Meagher, R. B., and Dean, J. F. (2005). Gene structure and molecular analysis of the laccase-like multicopper oxidase (LMCO) gene family in *Arabidopsis thaliana. Planta* **221**, 619–636.

Micheli, F. (2001). Pectin methylesterases: cell wall enzymes with important roles in plant physiology. *Trends Plant Sci.* **6**, 414–419.

Murashige, T. and Skoog, F. (1962). A revised medium for rapid growth and bioassays with tobacco tissue culture. *Physiol. Plant* **15**, 473–497.

Narsai, R., Howell, K., Millar, A., O'Toole, N., Small, I., and Whealan, J. (2007). Genome-wide analysis of mRNA decay rates and their determinants in *Arabidopsis thaliana. Plant Cell* **19**, 3418–3436.

Nersissian, A. M. and Shipp, E. L. (2002). Blue copper-binding domains. *Adv. Protein Chem.* **60**, 71–340.

Nielsen, H., Engelbrecht, J., Brunak, S., and von Heijne, G. (1997). Identification of prokaryotic and eukaryotic signal peptides and prediction of their cleavage sites. *Prot. Eng.* **10**, 1–6.

Passardi, F., Penel, C., and Dunand, C. (2004). Performing the paradoxical: how plant peroxidases modify the cell wall. *Trends Plant Sci.* **9**, 534–540.

Pourcel, L., Routaboul, J. M., Kerhoas, L., Caboche, M., Lepiniec, L., and Debeaujon, I. (2005). TRANSPARENT TESTA10 encodes a laccase-like enzyme involved in oxidative polymerization of flavonoids in *Arabidopsis* seed coat. *Plant Cell* **17**, 2966–2980.

Quail, P. H., Boylan, M. T., Parks, B. M., Short, T. W., Xu, Y., and Wagner, D. (1995). Phytochromes: Photosensory perception and signal transduction. *Science* **268**, 675–680.

Quevillon, E., Silventoinen, V., Pillai, S., Harte, N., Mulder, N., Apweiler, R., and Lopez, R. (2005). InterProScan: protein domains identifier. *Nucleic Acids Res.* **33**, W116–120.

Raes, J., Rohde, A., Christensen, J. H., Van de Peer, Y., and Boerjan, W. (2003). Genome-wide characterization of the lignification toolbox in *Arabidopsis. Plant Physiol.* **133**, 1051–1071.

Rautengarten, C., Usadel, B., Neumetzler, L., Hartmann, J., Büssis, D., and Altmann, T. (2008). A subtilisin-like serine protease essential for mucilage release from *Arabidopsis* seed coats. *Plant J.* **54**, 466–480.

Raz, V. and Koornneef, M. (2001). Cell division activity during apical hook development. *Plant Physiol.* **125**, 219–226.

Refrégier, G., Pelletier, S., Jaillard, D., and Hofte, H. (2004). Interaction between wall deposition and cell elongation in dark-grown hypocotyl cells in *Arabidopsis. Plant Physiol.* **135**, 959–968.

Ringli, C., Keller, B., and Ryser, U. (2001). Glycine-rich proteins as structural components of plant cell walls. *Cell Mol. Life Sci.* **58**, 1430–1441.

Roudier, F., Fernandez, A. G., Fujita, M., Himmelspach, R., Borner, G. H., Schindelman, G., Song, S., Baskin, T. I., Dupree, P., Wasteneys, G. O., and Benfey, P. N. (2005). COBRA, an *Arabidopsis* extracellular glycosyl-phosphatidyl inositol-anchored protein, specifically controls highly anisotropic expansion through its involvement in cellulose microfibril orientation. *Plant Cell* **17**, 1749–1763.

Roxrud, I., Lid, S., Fletcher, J., Schmidt, E., and Opsahl-Sorteberg H, -G. (2007). GASA4, one of the 14-member *Arabidopsis* GASA family of small polypeptides, regulates flowering and seed development. *Plant Cell Physiol.* **48**, 471–483.

Saibo, N. J., Vriezen, W. H., Beemster, G. T., and Van Der Straeten, D. (2003). Growth and stomata development of *Arabidopsis* hypocotyls are controlled by gibberellins and modulated by ethylene and auxins. *Plant J.* **33**, 989–1000.

San Clemente, H., Pont-Lezica, R., and Jamet, E. (2009). Bioinformatics as a tool for assessing the quality of sub-cellular proteomic strategies and inferring functions of proteins: Plant cell wall proteomics as a test case. *Bioinform. Biol. Insights* **3**, 15–28.

Schnabelrauch, L. S., Kieliszewski, M. J., Upham, B. L., and Alizedeh, H. (1996). Lamport DTA: Isolation of pI 4.6 extensin peroxidase from tomato cell suspension cultures and identification of Val-Tyr-Lys as putative intermolecular cross-link site. *Plant J.* **9**, 477–489.

Schultz, C. J., Ferguson, K. L., Lahnstein, J., and Bacic, A. (2004). Post-translational modifications of arabinogalactan-peptides of *Arabidopsis thaliana*. *J. Biol. Chem.* **279**, 455103–445511.

Schultz, C. J., Rumsewicz, M. P., Johnson, K. L., Jones, B. J., Gaspar, Y. G., and Antony Bacic, A. (2002). Using genomic resources to guide research directions. The arabinogalactan protein gene family as a test case. *Plant Physiol.* **129**, 1448–1463.

Schwacke, R., Schneider, A., van der Graaff, E., Fischer, K., Catoni, E., Desimone, M., Frommer, W. B., Flugge, U. I., and Kunze, R. (2003). ARAMEMNON, a novel database for *Arabidopsis* integral membrane proteins. *Plant Physiol.* **131**, 16–26.

Shah, K., Penel, C., Gagnon, J., and Dunand, C. (2004). Purification and identification of a Ca2+-pectate binding peroxidase from *Arabidopsis* leaves. *Phytochemistry* **65**, 307–312.

Tanaka, H., Onouchi, H., Kondo, M., Hara-Nishimura, I., Nishimura, M., Machida, C., and Machida, Y. (2001). A subtilisin-like serine protease is required for epidermal surface formation in *Arabidopsis* embryos and juvenile plants. *Development* **128**, 4681–4689.

van Hengel, A. J. and Roberts, K. (2003). AtAGP30, an arabinogalactan-protein in the cell walls of the primary root, plays a role in root regeneration and seed germination. *Plant J.* **36**, 256–270.

Ye, Z. H., Song, Y. R., Marcus, A., and Varner, J. E. (1991). Comparative localization of three classes of cell wall proteins. *Plant J.* **1**, 175–183.

Zhang, S., Yang, C., Peng, J., Sun, S., and Wang, X. (2009). GASA5, a regulator of flowering time and stem growth in *Arabidopsis thaliana*. *Plant Mol. Biol.* **69**, 745–759.

8

Abad, P., Gouzy, J., Aury, J. M., Castagnone-Sereno, P., Danchin, E. G., et al. (2008). Genome sequence of the metazoan plant-parasitic nematode *Meloidogyne incognita*. *Nat. Biotechnol.* **26**, 909–915.

Austin, J. R. 2nd, Segui-Simarro, J. M., and Staehelin, L. A. (2005). Quantitative analysis of changes in spatial distribution and plus-end geometry of microtubules involved in plant-cell cytokinesis. *J. Cell Sci.* **118**, 3895–3903.

Basu, J., Logarinho, E., Herrmann, S., Bousbaa, H., Li, Z., et al. (1998). Localization of the *Drosophila* checkpoint control protein BUB3 to the kinetochore requires BUB1 but not Zw10 or Rod. *Chromosoma* **107**, 376–385.

Caillaud, M. C., Lecomte, P., Jammes, F., Quentin, M., Pagnotta, S., et al. (2008). MAP65-3 microtubule-associated protein is essential for nematode-induced giant cell ontogenesis in *Arabidopsis*. *Plant Cell* **20**, 423–437.

Chan, G. K., Jablonski, S. A., Sudakin, V., Hittle, J. C., and Yen, T. J. (1999). Human BUBR1 is a mitotic checkpoint kinase that monitors CENP-E functions at kinetochores and binds the cyclosome/APC. *J. Cell Biol.* **146**, 941–954.

Chen, R. H. (2002). BUBR1 is essential for kinetochore localization of other spindle checkpoint proteins and its phosphorylation requires MAD1. *J. Cell. Biol.* **158**, 487–496.

Chen, R. H., Waters, J. C., Salmon, E. D., and Murray, A. W. (1996). Association of spindle assembly checkpoint component XMAD2 with unattached kinetochores. *Science* **274**, 242–246.

Criqui, M. C., Parmentier, Y., Derevier, A., Shen, W. H., Dong, A., et al. (2000). Cell cycle-dependent proteolysis and ectopic overexpression of cyclin B1 in tobacco BY2 cells. *Plant J.* **24**, 763–773.

Deslandes, L., Olivier, J., Peeters, N., Feng, D. X., Khounlotham, M., et al. (2003). Physical interaction between RRS1-R, a protein conferring resistance to bacterial wilt, and PopP2, a type III effector targeted to the plant nucleus. *Proc. Natl. Acad Sci. USA* **100**, 8024–8029.

Fang, Y. and Spector, D. L. (2005) Centromere positioning and dynamics in living *Arabidopsis* plants. *Mol. Biol. Cell.* **16**, 5710–5718.

Fang, G., Yu, H., and Kirschner, M. W. (1998). The checkpoint protein MAD2 and the mitotic regulator CDC20 form a ternary complex with the anaphase-promoting complex to control anaphase initiation. *Genes Dev.* **12**, 1871–1883.

Fraschini, R., Beretta, A., Sironi, L., Musacchio, A., Lucchini, G., et al. (2001). BUB3 interaction with MAD2, MAD3 and CDC20 is mediated by WD40 repeats and does not require intact kinetochores. *Embo. J.* **20**, 6648–6659.

Genschik, P., Criqui, M. C., Parmentier, Y., Derevier, A., and Fleck, J. (1998). Cell cycle -dependent proteolysis in plants. Identification Of the destruction box pathway and metaphase arrest produced by the proteasome inhibitor mg132. *Plant Cell* **10**, 2063–2076.

Hardwick, K. G., Johnston, R. C., Smith, D. L., and Murray, A. W. (2000). MAD3 encodes a novel component of the spindle checkpoint which interacts with BUB3p, CDC20p, and MAD2p. *J Cell Biol.* **148**, 871–882.

Houben, A. and Schubert, I. (2003). DNA and proteins of plant centromeres. *Curr. Opin. Plant. Biol.* **6**, 554–560.

Howell, B. J., Hoffman, D. B., Fang, G., Murray, A. W., and Salmon, E. D. (2000). Visualization of MAD2 dynamics at kinetochores, along spindle fibers, and at spindle poles in living cells. *J. Cell Biol.* **150**, 1233–1250.

Howell, B. J., Moree, B., Farrar, E. M., Stewart, S., Fang, G., et al. (2004). Spindle checkpoint protein dynamics at kinetochores in living cells. *Curr. Biol.* **14**, 953–964.

Hoyt, M. A., Totis, L., and Roberts, B. T. (1991). *S. cerevisiae* genes required for cell cycle arrest in response to loss of microtubule function. *Cell* **66**, 507–517.

Hu, C. D., Chinenov, Y., and Kerppola, T. K. (2002). Visualization of interactions among bZIP and Rel family proteins in living cells using bimolecular fluorescence complementation. *Mol. Cell* **9**, 789–798.

Kimbara, J., Endo, T. R., and Nasuda, S. (2004). Characterization of the genes encoding for MAD2 homologues in wheat. *Chromosome Res.* **12**, 703–714.

Lee, Y. R., Giang, H. M., and Liu, B. (2001). A novel plant kinesin-related protein specifically associates with the phragmoplast organelles. *Plant Cell* **13**, 2427–2439.

Lermontova, I., Fuchs, J., and Schubert, I. (2008). The *Arabidopsis* checkpoint protein Bub3.1 is essential for gametophyte development. *Front. Biosci.* **13**, 5202–5211.

Lermontova, I., Schubert, V., Fuchs, J., Klatte, S., Macas, J., et al. (2006). Loading of *Arabidopsis* centromeric histone CENH3 occurs mainly during G2 and requires the presence of the histone fold domain. *Plant Cell* **18**, 2443–2451.

Li, Y. and Benezra, R. (1996). Identification of a human mitotic checkpoint gene: hsMAD2. *Science* **274**, 246–248.

Li, R. and Murray, A. W. (1991). Feedback control of mitosis in budding yeast. *Cell* **66**, 519–531.

Li, L., Stoeckert, C. J. Jr., and Roos, D. S. (2003). OrthoMCL: Identification of ortholog groups for eukaryotic genomes. *Genome. Res.* **13**, 2178–2189.

Martinez-Exposito, M. J., Kaplan, K. B., Copeland, J., and Sorger, P. K. (1999). Retention of the BUB3 checkpoint protein on lagging chromosomes. *Proc. Natl. Acad. Sci. USA* **96**, 8493–8498.

Menges, M., de Jager, S. M., Gruissem, W., and Murray, J. A. (2005). Global analysis of the core cell cycle regulators of *Arabidopsis* identifies novel genes, reveals multiple and highly specific profiles of expression and provides a coherent model for plant cell cycle control. *Plant J.* **41**, 546–566.

Musacchio, A. and Hardwick, K. G. (2002). The spindle checkpoint: Structural insights into dynamic signalling. *Nat. Rev. Mol. Cell. Biol.* **3**, 731–741.

Musacchio, A. and Salmon, E. D. (2007). The spindle-assembly checkpoint in space and time. *Nat. Rev. Mol. Cell. Biol.* **8**, 379–393.

Perez-Perez, J. M., Serralbo, O., Vanstraelen, M., Gonzalez, C., Criqui, M. C., et al. (2008). Specialization of CDC27 function in the *Arabidopsis thaliana* anaphase-promoting complex (APC/C). *Plant J.* **53**, 78–89.

Pfleger, C. M. and Kirschner, M. W. (2000). The KEN box: An APC recognition signal distinct from the D box targeted by Cdh1. *Genes Dev.* **14**, 655–665.

Ritzenthaler, C., Nebenfuhr, A., Movafeghi, A., Stussi-Garaud, C., Behnia, L., et al. (2002). Re-evaluation of the effects of brefeldin A on plant cells using tobacco Bright Yellow 2 cells expressing Golgi-targeted green fluorescent protein and COPI antisera. *Plant Cell* **14**, 237–261.

Sczaniecka, M., Feoktistova, A., May, K. M., Chen, J. S., Blyth, J., et al. (2008). The spindle checkpoint functions of MAD3 and MAD2 depend on a MAD3 KEN box-mediated interaction with CDC20-anaphase-promoting complex (APC/C). *J. Biol. Chem.* **283**, 23039–23047.

Stagljar, I., Korostensky, C., Johnsson, N., and te Heesen, S. (1998). A genetic system based on split-ubiquitin for the analysis of interactions between membrane proteins *in vivo*. *Proc. Natl. Acad. Sci. USA* **95**, 5187–5192.

Sudakin, V., Chan, G. K., and Yen, T. J. (2001). Checkpoint inhibition of the APC/C in HeLa cells is mediated by a complex of BUBR1, BUB3, CDC20, and MAD2. *J. Cell. Biol.* **154**, 925–936.

Talbert, P. B., Masuelli, R., Tyagi, A. P., Comai, L., and Henikoff, S. (2002). Centromeric localization and adaptive evolution of an *Arabidopsis* histone H3 variant. *Plant Cell* **14**, 1053–1066.

Tang, Z., Bharadwaj, R., Li, B., and Yu, H. (2001). MAD2-Independent inhibition of APC CDC20 by the mitotic checkpoint protein BUBR1. *Dev. Cell.* **1**, 227–237.

Taylor, S. S., Ha, E., and McKeon, F. (1998). The human homologue of BUB3 is required for kinetochore localization of BUB1 and a MAD3/BUB1-related protein kinase. *J. Cell Biol.* **142**, 1–11.

Van Damme, D., Bouget, F. Y., Van Poucke, K., Inze, D., and Geelen, D. (2004) Molecular dissection of plant cytokinesis and phragmoplast structure: a survey of GFP-tagged proteins. *Plant J.* **40**, 386–398.

Voinnet, O., Rivas, S., Mestre, P., and Baulcombe, D. (2003). An enhanced transient expression system in plants based on suppression of gene silencing by the p19 protein of tomato bushy stunt virus. *Plant J.* **33**, 949–956.

Vos, J. W., Pieuchot, L., Evrard, J. L., Janski, N., Bergdoll, M., et al. (2008). The plant TPX2 protein regulates prospindle assembly before nuclear envelope breakdown. *Plant Cell* **20**, 2783–2797.

Yu, H. G., Muszynski, M. G., and Kelly Dawe, R. (1999). The maize homologue of the cell cycle checkpoint protein MAD2 reveals kinetochore substructure and contrasting mitotic and meiotic localization patterns. *J. Cell. Biol.* **145**, 425–435.

9

Abadia, J., Lopez-Millan, A. F., Rombola, A., and Abadia, A. (2002). Organic acids and Fe deficiency: A review. *Plant and Soil* **241**, (75–86).

Affymetrix. [http://www.affymetrix.com].

Altschul, S. F., Madden, T. L., Schaffer, A. A., Zhang, J., Zhang, Z., Miller, W., and Lipman, D. J. (1997). Gapped BLAST and PSI-BLAST: A new generation of protein database search programs. *Nucl. Acids Res.* **25**, 3389–3402.

Bailey, T. L., Wiliams, N., Misleh, C., and Li, W. W. (1996). MEME: Discovering and analyzing

DNA and protein sequence motifs. *Nucl. Acids Res.* 34 Web Server, W369–373.

Bonferroni, C. E. (1935). Ill. Calculation of the insurance groups of heads. *Studies in Honour of Professor Salvatore Ortu Carboni* 13–60.

Borevitz, J. O., Hazen, S. P., Michael, T. P., Morris, G. P., Baxter, I. R., Hu, T. T., Chen, H., Werner, J. D., Nordborg, M., Salt, D. E., et al. (2007). Genome-wide patterns of single-feature polymorphism in *Arabidopsis thaliana*. *Proc. Natl. Acad. Sci.* **104**(29), 12057–12062.

Charlson, D. V., Grant, D., Bailey, T. B., Cianzio, S. R., and Shoemaker, R. C. (2005). Molecular marker Satt481 is sssociated with iron-deficiency chlorosis resistance in a soybean breeding population. *Crop. Sci.* 2394–2399.

Connolly, E. L., Campbell, N. H., Grotz, N., Prichard, C. L., and Guerinot, M. L. (2003). Overexpression of the FRO2 ferric chelate reductase confers tolerance to growth on low iron and uncovers posttranscriptional control. *Plant Phys.* **133**(3), 1102–1110.

Diers, B. W., Cianzio, S. R., and Shoemaker, R. C. (1992). Possible identification of quantitative trait loci affecting iron efficiency in soybean. *J. Plant Nut.* **15**(10), 2127–2136.

Espen, L., Dell'Orto, M., De Nissi, P., and Zocchi, G. (2000). Metabolic responses incucumber (*Cucumis sativus* L.) roots under Fe-deficiency: A 31P-nuclear magnetic resonance in-vivo study. *Planta* **210**, 985–992.

Fisher, R. (1949). A preliminary linkage test with Agouti and Undulated mice: The fifth linkage-group. *Heredity* **3**, 229–241.

Ghandilyan, A., Vreugdenhil, D. and Aarts, M. G. M. (2006). Progress in the genetic understanding of palnt iron and zinc nutrition. *Physiologia Plantarum* **126**, 407–417.

Grant, D., Cregan, P., and Shoemaker, R. C. (2000). Genome organization in dicots: Genome duplication in *Arabidopsis* and synteny between soybean and *Arabidopsis*. *Proc. Natl. Acad. Sci.* **97**(8), 4168–4173.

Guo, A. Y., Chen, X., Gao, G., Zhang, H., Zhu, Q. H., Liu, X. C., Zhong, Y. F., Gu, X., He, K., and Lou, J. (2008). PlantTFDB: A comprehensive plant transcription factor database. *Nucl. Acids Res.* **36**, 966–969.

Guo, A. Y., He, K., Liu, D., Bai, S., Gu, X., Wei, L., and Luo, J. (2005). DATF: A Database of *Arabidopsis* Transcription Factors. *Bioinformatics* **21**, 2568–2569.

Hansen, N. C., Schmitt, M. A., Anderson, J. E., and Strock, J. S. (2003). Soybean: Iron deficiency of soybean in the upper Midwest and associated soil properties. *Agr. J.* **95**, 1595–1601.

Hernandez, G., Ramirez, M., Valdes-Lopez, O., Tesfaye, M., Graham, M. A., Czechowski, T., Schlereth, A., Wandrey, M., Erban, A., Cheung, F., et al. (2007). Phosphorus stress in common bean: Root transcript and metabolic responses. *Plant Phys.* **144**(2), 752–767.

Hurst, L. D., Williams, E. J. B., and Pal, C. (2002). Natural selection promotes the conservation of linkage of co-expressed genes. *TRENDS in Genet.* **18**(12), 604–606.

Inskeep, W. P. and Bloom, P. R. (1987). Soil chemical factors associated with soybean chlorosis in calciaquolls of western minnesota. *Agr. J.* **79**, 779–786.

Irizarry, R. A., Hobbs, B., Collin, F., Beazer-Barclay, Y. D., Antonellis, K. J., Scherf, U., and Speed, T. P. (2003). Exploration, normalization and summaries of high density oligonucleotide array probe level data. *Biostatistics* **4**, 249–264.

Kosak, S. T., Scalzo, D., Alworth, S. V., Li, F., Palmer, S., Enver, T., Lee, J. S. J., and Groudine, M. (2007). Coordinate gene regulation during hematopoiesis is related to genomic organization. *PLoS Bio.* **5**(11), 2602–2613.

Kumar, R., Qiu, J., Joshi, T., Valliyodan, B., Xu, D., and Nguyen, H. T. (2007). Single feature polymorphism discovery in rice. *PLoS One* **2**(3), e284.

Li, C. and Wong, W. H. (2001). Model-based analysis of oligonucleotide arrays: Model validation, design issues and standard error application. *Gen. Biol.* **2**(8), 1–11.

Lin, S., Baumer, J. S., Ivers, D., Cianzio, S. R., and Shoemaker, R. C. (1998). Field and nutrient solution tests measure similar mechanisms controlling iron deficiency chlorosis in soybean. *Crop. Sci.* **38**, 254–259.

Lin, S., Cianzio, S. R., and Shoemaker, R. C. (1997). Mapping genetic loci for iron deficiency chlorosis in soybean. *Mol. Breed.* **3**, 219–229.

Matys, V. E., Fricke, R., Geffers, E., Gobling, M., Haubrock, R., Hehl, K., Hornischer, D., Karas, A. E., Kel, O. V., Kel-Margoulis, et al. (2003). TRANSFAC: Transcriptional regulation, from patterns to profiles. *Nucl. Acids Res.* **31**(374–378).

Mengel, K., Kirby, E. A., Kosegarten, H., and Appel, T. (2001). *Principles of Plant Nutrition.* 5th edition. Kluwer Academic Publishers, Boston.

Michalak, P. (2008). Coexpression, coregulation, and cofunctionality of neighboring genes in eukaryotic genomes. *Genomics* **91**(3), 243–248.

Mortel, M., Recknor, J. C., Graham, M. A., Nettleton, D., Dittman, J. D., Nelson, R. T., Godoy, C. V., Abdelnoor, R. V., Almeida, A. M., Baum, T. J., et al. (2007). Distinct biphasic mRNA changes in response to Asian soybean rust infection. *Mol. Plant Micro.Interac.* **20**(8), 887–899.

O'Rourke, J. A., Charlson, D. V., Gonzalez, D. O., Vodkin, L. O., Graham, M. A., Cianzio, S. R., Grusak, M. A., and Shoemaker, R. C. (2007b). Microarray analysis of iron deficiency chlorosis in near-isogenic soybean lines. *BMC Gen.* **8**, 476–489.

O'Rourke, J. A., Graham, M. A., Vodkin, L., Gonzalez, D. O., Cianzio, S. R., and Shoemaker, R. C. (2007a). Recovering from iron deficiency chlorosis in near-isogenic soybeans: A microarray study. *Plant Phys. Biochem.* **45**(5), 287–292.

Oliver, B., Parisi, M., and Clark, D. (2002). Gene expession neighborhoods. *J. Biol.* **1**(1), 1–4.

Ren, X., Stickema, W., and Nap, J. (2005). Local coexpression domains of two to four genes in the genome of *Arabidopsis*. *Plant Phys.* **138**, 923–934.

Ren, X. Y., Stickema, W., and Nap, J. P. (2007). Local coexpression domains in the genome of rice show no microsynteny with *Arabidopsis* domains. *Plant Mol. Biol.* **65**, 205–217.

Romheld, V. (1987). Different strategies for iron acquisition in higher plants. *Phys. Plant* **70**, 231–234.

Romheld, V. and Marschner, H. (1986). Evidence for a Specific uptake system for iron phytosiderophores in roots of grasses. *Plant Phys.* **80**(1), 175–180.

Schmittgen, T. D., Zakrajsek, B. A., Mills, A. G., Gorn, V., Singer, M. J., and Reed, M. W. (2000). Quantitative reverse transcription-polymerase chain reaction to study mRNA decay: Comparison of endpoint and real-time methods. *Anal. Biochem.* **285**, 194–204.

Singh, R. J. and Hymowitz, T. (1988). The genomic relationship between *Glycine max* (L.) Merr. and *G. soja* Sieb. and Zucc. as revealed by pachytene chromosome analysis. *Theor. Appl. Genet.* **76**, 705–711.

Sun, B., Jing, Y., Chen, K., Song, L., Chen, F., and Zhang, L. (2007). Protective effect of nitric oxide on iron deficiency-induced oxidative stress in maize (*Zea mays*). *J. plant phys.* **164**(5), 536–543.

Theil, E. C. (2004). Iron, ferritin, and nutrition. *Annu. Rev. Nutr.* **24**, 327–343.

Thimm, O., Essigmann, B., Kloska, S., Altmann, T., and Buckhout, T. J. (2001). Response of *Arabidopsis* to iron deficiency stress as revealed by microarray analysis. *Plant Phys.* **127**, 1030–1043.

USDA. (2004). *A National Genetic Resources Program: Germplasm Resource Information Network—GRIN.* National Germplasm Resources Laboratory, Beltsville, Maryland.

Vogel, J. H., von Heydebrek, A., Purmmann, A., and Sperling, S. (2005). Chromosomal clustering of a human transcriptome reveals regulatory Background. *BMC Bioinform.* **6**, 230.

Wang, D. and Portis, A. R. (2007). A novel nucleus-encoded chloroplast protein, PIFI, is involved in NAD(P)H dehydrogenase complex-mediated chlororespiratory electron transport in *Arabidopsis*. *Plant Phys.* **144**(4), 1742–1752.

Wang, H. Y., Klatte, M., Jakoby, M., Baumlein, H., Weisshaar,B., and Bauer, P. (2007). Iron deficiency-mediated stress regulation of four subgroup Ib BHLH genes in *Arabidopsis thaliana*. *Planta* **226**, 897–908.

Wang, J., McLean, P. E., Lee, R., Goos, R. J., and Helms, T. (2008). Association mapping of iron deficiency chlorosis loci in soybean (*Glycine max* L. Merr.) advanced breeding lines. *Theor. App. Gen.* **116**(6), 777–787.

West, M. A., van Leeuwen, H., Kozik, A., Kliebenstein, D. J., Doerge, R. W., St Clair, D. A., and Michelmore, R. W. (2006). High-density haplotyping with microarray-based expression and single feature polymorphism markers in *Arabidopsis*. *Gen. Res.* **16**(6), 787–795.

Williams, E. J. B. and Bowles, D. J. (2004). Co-expression of neighboring genes in the genome of *Arabidopsis thaliana*. *Gen. Res.* **14**, 1060–1067.

Youxi, Y., Wu, H., Wang, N., Li, J., Zhao, W., Du, J., Wang, D., and Ling, H. (2008). FIT interacts ith AtbHLH38 and AtbHLH39 in regulating iron uptake gene expression for iron homeostasis in *Arabidopsis*. *Cell Res.* **18**, 385–397.

Zhan, S., Horrocks, J., and Lukens, L. N. (2006). Islands of co-expressed neighbouring genes in *Arabidopsis thaliana* suggest higher-order chromosome domains. *Plant J.* **45**, 347–357.

Zocchi, G., De Nisi, P., Dell'Orto, M., Espen, L., and Gallina, P. M. (2007). Iron deficiency differently affects metabolic responses in soybean roots. *J. exp. bot.* **58**(5), 993–1000.

10

Arigoni, D., Sagner, S., Latzel, C., Eisenreich, W., Bacher, A., and Zenk, M. H. (1997). Terpenoid biosynthesis from 1-deoxy-D-xylulose in higher plants by intramolecular skeletal rearrangement. *Proc. Natl. Acad. Sci. USA* **94**(20), 10600–10605.

Ashburner, M., Ball, C. A., Blake, J. A., Botstein, D., Butler, H., Cherry, J. M., Davis, A. P., Dolinski, K., Dwight, S. S., and Eppig, J. T. (2000). Gene ontology: Tool for the unification of biology. The Gene Ontology Consortium. *Nat. Genet.* **25**(1), 25–29.

Bainbridge, M. N., Warren, R. L., Hirst, M., Romanuik, T., Zeng, T., Go, A., Delaney, A., Griffith, M., Hickenbotham, M., and Magrini, V. (2006). Analysis of the prostate cancer cell line LNCaP transcriptome using a sequencing-by-synthesis approach. *BMC Genomics* **7**, 246.

Bao, X. M., Katz, S., Pollard, M., and Ohlrogge, J. (2002). Carbocyclic fatty acids in plants: Biochemical and molecular genetic characterization of cyclopropane fatty acid synthesis of *Sterculia foetida*. *Proc. Natl. Acad. Sci. USA* **99**(10). 7172–7177.

Bertea, C. M., Freije, J. R, Woude, H., Verstappen, F. W. A., Perk, L., Marquez, V., De Kraker, J. W., Posthumus, M. A., Jansen, B. J. M., and de Groot, A. (2005). Identification of intermediates and enzymes involved in the early steps of artemisinin biosynthesis in *Artemisia annua*. *Planta Medica* **71**(1), 40–47.

Bertea, C. M., Voster, A., Verstappen, F. W., Maffei, M., Beekwilder, J., and Bouwmeester H. J. (2006). Isoprenoid biosynthesis in *Artemisia annua*: Cloning and heterologous expression of a germacrene A synthase from a glandular trichome cDNA library. *Arch. Biochem. Biophys.* **448**(1–2), 3–12.

Bouwmeester, H. J., Wallaart, T. E., Janssen, M. H., van Loo, B., Jansen, B. J., Posthumus, M. A., Schmidt, C. O., De Kraker, J. W., Konig, W. A., and Franssen, M. C. (1999). Amorpha-4,11-diene synthase catalyses the first probable step in artemisinin biosynthesis. *Phytochemistry* **52**(5), 843–854.

Cai, Y., Jia, J.W., Crock, J., Lin, Z. X., Chen, X. Y., and Croteau, R. (2002). A cDNA clone for beta-caryophyllene synthase from *Artemisia annua*. *Phytochemistry* **61**(5), 523–529.

Chang, Y. J., Song, S. H, Park, S. H., and Kim, S. U. (2000). Amorpha-4,11-diene synthase of *Artemisia annua*: cDNA isolation and bacterial expression of a terpene synthase involved in artemisinin biosynthesis. *Arch. Biochem. Biophys.* **383**(2), 178–184.

Cheung, F., Haas, B. J., Goldberg, S. M. D., May, G. D., Xiao, Y. L., and Town, C. D. (2006). Sequencing *Medicago truncatula* expressed sequenced tags using 454 Life Sciences technology. *BMC Genomics* **7**, 272.

Choi, Y. E., Harada, E., Wada, M., Tsuboi, H., Morita, Y., Kusano, T., and Sano, H. (2001). Detoxification of cadmium in tobacco plants: formation and active excretion of crystals containing cadmium and calcium through trichomes. *Planta* **213**(1), 45–50.

Disch, A., Hemmerlin, A., Bach, T.J., and Rohmer, M. (1998). Mevalonate-derived isopentenyl diphosphate is the biosynthetic precursor of ubiquinone prenyl side chain in tobacco BY-2 cells. *Biochem. J.* **331**, 615–621.

Dudareva, N., Andersson, S., Orlova, I., Gatto, N., Reichelt, M., Rhodes, D., Boland, W., and

Gershenzon, J. (2005). The nonmevalonate pathway supports both monoterpene and sesquiterpene formation in snapdragon flowers. *Proc. Natl. Acad. Sci. USA* **102**(3), 933–938.

Duke, S. O. and Paul, R. N. (1993). Development and fine-structure of the glandular trichomes of *Artemisia annua* L. *Int. J. Plant Sci.* **154**(1), 107–118.

Emrich, S. J., Barbazuk, W. B., Li, L., and Schnable, P. S. (2007). Gene discovery and annotation using LCM-454 transcriptome sequencing. *Gen. Res.* **17**(1), 69–73.

Hirai, N., Yoshida, R., Todoroki, Y., and Ohigashi, H. (2000). Biosynthesis of abscisic acid by the non-mevalonate pathway in plants, and by the mevalonate pathway in fungi. *Biosci. Biotechnol. Biochem.* **64**(7), 1448–1458.

Hua, L. and Matsuda, S. P. (1999). The molecular cloning of 8-epicedrol synthase from *Artemisia annua*. *Arch. Biochem. Biophys.* **369**(2), 208–212.

Huang, X. and Madan, A. (1999). CAP3: A DNA sequence assembly program. *Geno. Res.* **9**(9), 868–877.

Huse, S. M., Huber, J. A., Morrison, H. G., Sogin, M. L., and Welch, D. M. (2007). Accuracy and quality of massively parallel DNA pyrosequencing. *Genome Biol.* **8**(7), R143.

Jia, J. W., Crock, J., Lu, S., Croteau, R., and Chen, X. Y. (1999). (3R)-Linalool synthase from *Artemisia annua* L.: cDNA isolation, characterization, and wound induction. *Arch. Biochem. Biophys.* **372**(1), 143–149.

Kupper, H., Lombi, E., Zhao, F. J., and McGrath, S. P. (2000). Cellular compartmentation of cadmium and zinc in relation to other elements in the hyperaccumulator *Arabidopsis halleri*. *Planta* **212**(1), 75–84.

Laule, O., Furholz, A., Chang, H. S., Zhu, T., Wang, X., Heifetz, P. B., Gruissem, W., and Lange, B. M. (2003). Crosstalk between cytosolic and plastidial pathways of isoprenoid biosynthesis in *Arabidopsis thaliana*. *Proc. Natl Acad. Sci. USA* **100**(11), 6866–6871.

Lichtenthaler, H. K., Schwender, J., Disch, A., and Rohmer, M. (1997). Biosynthesis of isoprenoids in higher plant chloroplasts proceeds via amevalonate-independent pathway. *FEBS Lett.* **400**(3), 271–274.

Ma, C. F., Wang, H. H., Lu, X., Li, H. F., Liu, B. Y., and Xu, G. W. (2007). Analysis of *Artemisia annua* L. volatile oil by comprehensive two-dimensional gas chromatography time-of-flight mass spectrometry. *J. Chromatogr. A* **1150**(1–2), 50–53.

Ma, C. F., Wang, H. H., Lu, X., Xu, G. W., and Liu, B. Y. (2008). Metabolic fingerprinting investigation of *Artemisia annua* L. in different stages of development by gas chromatography and gas chromatography mass spectrometry. *J. Chromatogr. A* **1186**(1–2), 412–419.

Margulies, M., Egholm, M., Altman, W. E., Attiya, S., Bader, J. S., Bemben, L. A., Berka, J., Braverman, M. S., Chen, Y. J., and Chen, Z. (2005). Genome sequencing in microfabricated high-density picolitre reactors. *Nature* **437**(7057), 376–380.

Matsushita, Y., Kang, W., and Charlwood, B. V. (1996). Cloning and analysis of a cDNA encoding farnesyl diphosphate synthase from *Artemisia annua*. *Gene* **172**(2), 207–209.

Mercke, P., Bengtsson, M., Bouwmeester, H. J., Posthumus, M. A., and Brodelius, P. E. (2000). Molecular cloning, expression, and characterization of amorpha-4,11-diene synthase, a key enzyme of artemisinin biosynthesis in *Artemisia annua* L. *Arch. Biochem. and Biophys.* **381**(2), 173–180.

Mercke, P., Crock, J., Croteau, R., and Brodelius, P. E. (1999). Cloning, expression, and characterization of epi-cedrol synthase, a sesquiterpene cyclase from *Artemisia annua* L. *Arch. Biochem. Biophys.* **369**(2), 213–222.

Milborrow, B. V. and Lee, H. S. (1998). Endogenous biosynthetic precursors of (+)- abscisic acid. VI - Carotenoids and ABA are formed by the "nonmevalonate" triose-pyruvate pathway in chloroplasts. *Australian J. Plant Physiol.* **25**(5), 507–512.

Moore, M. J., Dhingra, A., Soltis, P. S., Shaw, R., Farmerie, W. G., Folta, K. M, and Soltis, D. E. (2006). Rapid and accurate pyrosequencing of angiosperm plastid genomes. *BMC Plant Biol.* **6**, 17.

Pertea, G., Huang, X., Liang, F., Antonescu, V., Sultana, R., Karamycheva, S., Lee, Y., White, J., Cheung, F., and Parvizi, B. (2003). TIGR Gene Indices clustering tools (TGICL): A software system for fast clustering of large EST datasets. *Bioinformatics* **19**(5), 651–652.

Picaud, S., Brodelius, M., and Brodelius, P. E. (2005). Expression, purification and characterization of recombinant (E)-beta-farnesene synthase from *Artemisia annua*. *Phytochemistry* **66**(9), 961–967.

Piel, J., Donath, J., Bandemer, K., and Boland, W. (1998). Mevalonate-independent biosynthesis of terpenoid volatiles in plants: Induced and constitutive emission of volatiles. *Angew. Chem. Int. Ed.* **37**(18), 2478–2481.

Poinar, H. N., Schwarz, C., Qi, J., Shapiro, B., Macphee, R. D., Buigues, B., Tikhonov, A., Huson, D. H., Tomsho, L. P., and Auch, A. (2006). Metagenomics to paleogenomics: Large-scale sequencing of mammoth DNA. *Science* **311**(5759), 392–394.

Ranger, C. M., and Hower, A. A. (2001). Role of the glandular trichomes in resistance of perennial alfalfa to the potato leafhopper (Homoptera: Cicadellidae). *J. Econ. Entomol.* **94**(4), 950–957.

Ro, D. K. P. E., Ouellet, M., Fisher, K. J., Newman, K. L., Ndungu, J. M., Ho, K. A., Eachus, R. A., Ham, T. S., and Kirby, J.(2006). Production of the antimalarial drug precursor artemisinic acid in engineered yeast. *Nature* **440**(7086), 940–943.

Schwender, J., Zeidler, J., Groner, R., Muller, C., Focke, M., Braun, S., Lichtenthaler, F. W., and Lichtenthaler, H. K. (1997). Incorporation of 1-deoxy-Dxylulose into isoprene and phytol by higher plants and algae. *FEBS Letters* **414**(1), 129–134.

Teoh, K. H., Polichuk, D. R., Reed, D. W., Nowak, G., and Covello, P. S. (2006). *Artemisia annua* L. (Asteraceae) trichome-specific cDNAs reveal CYP71AV1, a cytochrome P450 with a key role in the biosynthesis of the antimalarial sesquiterpene lactone artemisinin. *FEBS Lett.* **580**(5), 1411–1416.

Wagner, G. J., Wang, E., and Shepherd, R. W. (2004). New approaches for studying and exploiting an old protuberance, the plant trichome. *Ann. Bot. (Lond)* **93**(1), 3–11.

Wang, G. D., Tian, L., Aziz, N., Broun, P., Dai, X. B., He, J., King, A., Zhao, P. X., and Dixon, R. A. (2008). Terpene Biosynthesis in Glandular Trichomes of Hop. *Plant Physiol.* **148**(3), 1254–1266.

Weber, A. P., Weber, K. L., Carr, K., Wilkerson, C., and Ohlrogge, J. B. (2007). Sampling the *Arabidopsis* transcriptome with massively parallel pyrosequencing. *Plant Physiol.* **144**(1), 32–42.

Wicker, T., Schlagenhauf, E., Graner, A., Close, T. J., Keller, B., and Stein, N. (2006). 454 sequencing put to the test using the complex genome of barley. *BMC Genomics* **7**:275.

Zhang, Y., Teoh, K. H., Reed, D. W., Maes, L., Goossens, A, Olson, D. J. H., Ross, A. R. S., and Covello, P. S. (2008). The molecular cloning of artemisinic aldehyde Delta 11(13) reductase and its role in glandular trichomedependent biosynthesis of artemisinin in *Artemisia annua*. *J. Biol. Chem.* **283**(31), 21501–21508.

11

Aoyama, T., Dong, C. H., Wu, Y., Carabelli, M., Sessa, G., Ruberti, I., Morelli, G., and Chua, N. H. (1995). Ectopic expression of the *Arabidopsis* transcriptional activator Athb-1 alters leaf cell fate in tobacco. *Plant Cell* **7**, 1773–1785.

Apel, K. and Hirt, H. (2004). Reactive oxygen species: Metabolism, oxidative stress, and signal transduction. *Annu. Rev. Plant Biol.* **55**, 373–399.

Baier, M. and Dietz, K. J. (1999). Protective function of chloroplast 2-cysteine peroxiredoxin in photosynthesis. Evidence from transgenic *Arabidopsis*. *Plant Physiol.* **119**, 1407–1414.

Beck, C. F. and Warren, R. A. (1988). Divergent promoters, a common form of gene organization. *Microbiol. Rev.* **52**, 318–326.

Boyes, D. C., Zayed, A. M., Ascenzi, R., McCaskill, A. J., Hoffman, N. E., Davis, K. R., and Gorlach, J. (2001). Growth stage-based phenotypic analysis of *Arabidopsis*: Amodel for high throughput functional genomics in plants. *Plant Cell* **13**, 1499–1510.

Bradford, M. M. (1976). A rapid and sensitive method for the quantitation of microgram quantities of protein utilizing the principle of protein-dye binding. *Anal. Biochem.* **72**, 248–254.

Caporaletti, D., D'Alessio, A. C., Rodriguez-Suarez, R. J., Senn, A. M., Duek, P. D., and Wolosiuk, R. A. (2007). Non-reductive modulation of chloroplast fructose-1,6-bisphosphatase by 2-Cys peroxiredoxin. *Biochem. Biophys. Res. Commun.* **355**, 722–727.

Carabelli, M., Morelli, G., Whitelam, G., and Ruberti, I. (1996). Twilight-zone and canopy shade induction of the Athb-2 homeobox gene in green plants. *Proc. Natl. Acad. Sci. USA* **93**, 3530–3535.

Chaturvedi, C. P., Sawant, S. V., Kiran, K., Mehrotra, R., Lodhi, N., Ansari, S. A., and Tuli, R. (2006). Analysis of polarity in the expression from a multifactorial bidirectional promoter designed for high-level expression of transgenes in plants. *J. Biotechnol.* **123**, 1–12.

Clough, S. J. and Bent, A. F. (1998). Floral dip: a simplified method for Agrobacterium-mediated transformation of *Arabidopsis thaliana*. *Plant J.* **16**, 735–743.

Colinas, J., Schmidler, S. C., Bohrer, G., Iordanov, B., and Benfey, P. N. (2008). Intergenic and genic sequence lengths have opposite relationships with respect to gene expression. *PLoS ONE* **3**, e3670.

Core, L. J., Waterfall, J. J., and Lis, J. T. (2008). Nascent RNA sequencing reveals widespread pausing and divergent initiation at human promoters. *Science* **322**, 1845–1848.

Dhadi, S. R., Krom, N., and Ramakrishna, W. (2009). Genome-wide comparative analysis of putative bidirectional promoters from rice, *Arabidopsis and Populus*. *Gene* **429**, 65–73.

Dietz, K. J. (2007). The dual function of plant peroxiredoxins in antioxidant defence and redox signaling. *Subcell. Biochem.* **44**, 267–294.

Dietz, K. J., Horling, F., Konig, J., and Baier, M. (2002). The function of the chloroplast 2-cysteine peroxiredoxin in peroxide detoxification and its regulation. *J. Exp. Bot.* **53**, 1321–1329.

Dietz, K. J., Jacob, S., Oelze, M. L., Laxa, M., Tognetti, V., de Miranda, S. M., Baier, M., and Finkemeier I. (2006). The function of peroxiredoxins in plant organelle redox metabolism. *J. Exp. Bot.* **57**, 1697–1709.

Emanuelsson, O., Nielsen, H., and von, H. G. (1999). ChloroP, a neural network-based method for predicting chloroplast transit peptides and their cleavage sites. *Protein. Sci.* **8**, 978–984.

Goulas, E., Schubert, M., Kieselbach, T., Kleczkowski, L. A., Gardestrom, P., Schroder, W., and Hurry, V. (2006). The chloroplast lumen and stromal proteomes of *Arabidopsis thaliana* show differential sensitivity to short- and long-term exposure to low temperature. *Plant J.* **47**, 720–734.

Henriksson, E., Olsson, A. S., Johannesson, H., Johansson, H., Hanson, J., Engstrom, P., and Soderman, E. (2005). Homeodomain leucine zipper class I genes in *Arabidopsis*. Expression patterns and phylogenetic relationships. *Plant Physiol.* **139**, 509–518.

Higo, K., Ugawa, Y., Iwamoto, M., and Korenaga, T. (1999). Plant *cis*-acting regulatory DNA elements (PLACE) database. *Nucl. Acids Res.* **27**, 297–300.

Horling, F., Baier, M., and Dietz, K. J. (2001). Redox-regulation of the expression of the peroxide-detoxifying chloroplast 2-cys peroxiredoxin in the liverwort *Riccia fluitans*. *Planta* **214**, 304–313.

Jang, H. H., Lee, K. O., Chi, Y. H., Jung, B. G., Park, S. K., Park, J. H., Lee, J. R., Lee, S. S., Moon, J. C., Yun, J. W., Choi, Y. O., Kim, W. Y., Kang, J. S., Cheong, G. W., Yun, D. J., Rhee, S. G., Cho, M. J., and Lee, S. Y. (2004). Two enzymes in one; two yeast peroxiredoxins display oxidative stress-dependent switching from a peroxidase to a molecular chaperone function. *Cell* **117**, 625–635.

Keddie, J. S., Tsiantis, M., Piffanelli, P., Cella, R., Hatzopoulos, P., and Murphy, D. J. (1994). A seed-specific Brassica napus oleosin promoter interacts with a G-box-specific protein and may be bi-directional. *Plant Mol. Biol.* **24**, 327–340.

Kirch, T., Bitter, S., Kisters-Woike, B., and Werr, W. (1998). The two homeodomains of the Zm-Hox2a gene from maize originated as an internal gene duplication and have evolved different target site specificities. *Nucleic Acids Res.* **26**, 4714–4720.

Krom, N. and Ramakrishna, W. (2008). Comparative analysis of divergent and convergent gene pairs and their expression patterns in rice,

Arabidopsis, and populus. *Plant Physiol.* **147**, 1763–1773.

Laemmli, U. K. (1970). Cleavage of Structural Proteins during the Assembly of the Head of Bacteriophage T4. *Nature* **227**, 680–685.

Li, Y. Y., Yu, H., Guo, Z. M., Guo, T. Q., Tu, K., and Li, Y. X. (2006). Systematic analysis of head-to-head gene organization: evolutionary conservation and potential biological relevance. *PLoS Comput. Biol.* **2**, e74.

Mitra, A., Han, J., Zhang, Z. J., and Mitra, A. (2009). The intergenic region of *Arabidopsis thaliana* cab1 and cab2 divergent genes functions as a bidirectional promoter. *Planta.* **229**, 1015–1022.

Mittler, R., Vanderauwera, S., Gollery, M., and Van, B. F. (2004). Reactive oxygen gene network of plants. *Trends Plant Sci.* **9**, 490–498.

Novina, C. D. and Roy, A. L. (1996). Core promoters and transcriptional control. *Trends Genet* **12**, 351–355.

Rieping, M. and Schoffl, F. (1992). Synergistic effect of upstream sequences, CCAAT box elements, and HSE sequences for enhanced expression of chimaeric heat shock genes in transgenic tobacco. [http://www.springerlink.com/content/w1631u5475028246/] *Mol. Gen. Genet.* **231**, 226–232.

Rombauts, S., Florquin, K., Lescot, M., Marchal, K., Rouze, P., and Peer, Y. (2003). Computational approaches to identify promoters and *cis*-regulatory elements in plant genomes. *Plant Physiology* **132**, 1162–1176.

Scarpella, E., Simons, E. J., and Meijer, A. H. (2005). Multiple regulatory elements contribute to the vascular-specific expression of the rice HD-Zip gene Oshox1 in *Arabidopsis. Plant Cell Physiol.* **46**, 1400–1410.

Seila, A. C., Core, L. J., Lis, J. T., and Sharp, P. A. (2009). Divergent transcription: A new feature of active promoters. *Cell Cycle* **8**, 2557–2564.

Shin, R., Kim, M. J., and Paek, K. H. (2003). The CaTin1 (*Capsicum annuum* TMV-induced Clone 1) and *CaTin1-2* Genes are linked head-to-head and share a bidirectional promoter. *Plant Cell Physiol* **44**, 549–554.

Singh, A., Sahi, C., and Grover, A. (2009). Chymotrypsin protease inhibitor gene family in rice: Genomic organization and evidence for the presence of a bidirectional promoter shared between two chymotrypsin protease inhibitor genes. *Gene.* **428**, 9–19.

Smale, S. T. and Kadonaga, J. T. (2003). The RNA polymerase II core promoter. *Annu. Rev. Biochem.* **72**, 449–479.

Steffens, N. O., Galuschka, C., Schindler, M., Bulow, L., and Hehl, R. (2004). AthaMap: An online resource for *in silico* transcription factor binding sites in the *Arabidopsis thaliana* genome. *Nucleic Acids Res.* **32**, D368–372.

Stewart, C. N. Jr. and Via, L. E. (1993). A rapid CTAB DNA isolation technique useful for RAPD fingerprinting and other PCR applications. *Biotechniques* **14**, 748–750.

Trinklein, N. D., Aldred, S. F., Hartman, S. J., Schroeder, D. I., Otillar, R. P., and Myers, R. M. (2004). An abundance of bidirectional promoters in the human genome. *Genome. Res.* **14**, 62–66.

Venter, M. (2007). Synthetic promoters: Genetic control through *cis* engineering. *Trends Plant Sci.* **12**, 118–124.

Walther, D., Brunnemann, R., and Selbig, J. (2007). The regulatory code for transcriptional response diversity and its relation to genome structural properties in *A. thaliana. PLoS Genet.* **3**, e11.

Weigel, D. and Glazebrook, J. (2002). *Arabidopsis: A laboratory manual.* Cold Spring Harbor Lab. Press, Plainview, New York.

Williams, E. J. and Bowles, D. J. (2004). Coexpression of neighboring genes in the genome of *Arabidopsis thaliana. Genome. Res.* **14**, 1060–1067.

Yamamoto, A., Mizukami, Y., and Sakurai, H. (2005). Identification of a novel class of target genes and a novel type of binding sequence of heat shock transcription factor in *Saccharomyces cerevisiae. J. Biol. Chem.* **280**, 11911–11919.

Yamamoto, Y. Y. and Obokata, J. (2008). ppdb: A plant promoter database. *Nucleic Acids Res.* **36**, D977–981.

Yamamoto, Y. Y., Ichida, H., Matsui, M., Obokata, J., Sakurai, T., Satou, M., Seki, M., Shinozaki, K., and Abe, T. (2007). Identification of plant promoter constituents by analysis of local distribution of short sequences. *BMC Genomics.* **8**, 67.

Zimmermann, P., Hennig, L., and Gruissem, W. (2005). Gene-expression analysis and network discovery using Genevestigator. *Trends Plant Sci.* **10**, 407–409.

12

Adros, G., Weigel, H. J., and Jager, H. J. (1989). Environment in open-top chambers and its effect on growth and yield of plants II. Plant responses. *Eur. J. Hortic. Sci.* **54**, 252–256.

Allen, L. H., Drake, B. G., Rogers, H. H., and Shinn, J. H. (1992). Field techniques for exposure of plants and ecosystem to elevated CO_2 and other trace gases. *Crit. Rev. Plant Sci.* **11**, 85–119.

Chen, C. C., Turner, F. T., and Dixon, J. B. (1989). Ammonium fixation by charge smectite in selected Texas gulf coast soils. *Soil Sci. Soc. America J.* **53**, 1035–1040.

Fangmeier, A., Gnittke, J., and Steubing, L. (1986). Transportable open-tops for discontinuous fumigations. In *Microclimate and Plant Growth in Open-Top Chambers.* Air Pollution Research Report 5. Freiburg, Commission of the European Communities, Directorate-general for Science, Research and Development, Environment Research Programme, Freiburg, pp. 102–112.

Harte, J. and Shaw, R. (1995). Shifting dominance within a montane vegetation community: Results of a climate-warming experiment. *Science* **267**, 876–880.

Harte, J., Torn, M. S., Chang, F. R., Feifarek, B, Kinzig, A., Shaw, R., and Shen, K. (1995). Global warming and soil microclimate results from a meadow-warming experiment. *Ecol. Appl.* **5**, 132–150.

Houghton, J. T., Ding, Y., Griggs, D. J., Noguer, M., Linden, P. J., Dai, X., Maskell, K., and Johnson, C. A. (2001). In *Climate Change: The Scientific Basis.* Contribution of Working Group I of the Third Assessment Report of the Intergovernmental Panel on Climate Change. Change IPCC (Ed.). Cambridge University Press, New York, p. 555.

Kimball, B. A. (2005). Theory and performance of an infrared heater for ecosystem warming. *Glob. Chang. Biol.* **11**, 2041–2056.

Kimball, B. A., Conley, M. M., Wang, S., Lin, X., Luo, C., Morgan, J., and Smith, D. (2008). Infrared heater arrays for warming ecosystem field plots. *Glob. Chang. Biol.* **14**, 309–320.

Luo, Y., Wan, S., Hui, D., and Wallace, L. L. (2001). Acclimatization of soil respiration to warming in tallgrass prairie. *Nature* **413**, 622–625.

Mohammed, A. R., Rounds, E. W., and Tarpley, L. (2007). Response of rice (*Oryza sativa* L.) tillering to sub-ambient levels of ultraviolet-B radiation. *J. Agron. Crop Sci.* **193**, 324–335.

Mohammed, A. R. and Tarpley, L. (2009). Impact of high nighttime temperature on respiration, membrane stability, antioxidant capacity, and yield of rice plants. *Crop Sci.* **49**, 313–322.

Nijs, I., Ferris, R., Blum, H., Hendrey, G., and Impens, I. (1997). Stomatal regulation in a changing climate: A field study using Free Air Temperature Increase (FATI) and Free Air CO_2 Enrichment (FACE). *Plant Cell Environ.* **20**, 1041–1050.

Nijs, I., Kockelbergh, F., Teughels, H., Blum, H., Hendrey, G., and Impens, I. (1996). Free air temperature increase (FATI): A new tool to study global warming effects on plants in the field. *Plant Cell Environ.* **19**, 495–502.

Noormets, A., Chen, J., Bridgham, S. D., Weltzin, J. F, Pastor, J., Dewey, B., and LeMoine, J. (2004). The effects of infrared loading and water table on soil energy fluxes in northern peatlands. *Ecosystems* **7**, 573–582.

Omega: Radiant process heaters (2008). In *The Electric Heaters Handbook.* 21st edition. Omega Engineering Inc., Omega Publishers, Stamford, Connecticut, USA, pp. 72–77.

OPC Foundation. [http://opcfoundation.org].

Park, J., Mackay, S., and Wright, E. (2003). *Practical Data Communications for Instrumentation and Control.* Newnes-Elsevier, Oxford, UK.

Reddy, K. R., Hodges, H. F., Read, J. J., McKinion, J. M., Baker, J. T., Tarpley, L., and Reddy, V. R. (2001). Soil-Plant-Atmosphere-Research

(SPAR) facility: A tool for plant research and modeling. *Biotronics* **30**, 27–50.

Shaw, M. R., Zavaleta, E. S., Chiariello, N. R., Cleland, E. E., Mooney, H. A., and Field, C. B. (2002). Grassland responses to global environmental changes. *Science* **298**, 1987–1990.

Sun, W., Van Montagu, M., and Verbruggen, N. (2002). Small heat shock proteins and stress tolerance in plants. *Biochim. Biophys. Acta.* **1577**(1), 1–9.

Thomas, R. B. and Strain, B. R. (1991). Root restriction as a factor in photosynthetic acclimation of cotton seedlings growing in elevated carbon dioxide. *Plant Physiol.* **96**, 627–634.

Tingey, D. T., McVeety, B. D., Waschmann, R., Johnson, M. G., Phillips, D. L., Rygiewicz, P. T., and Oiszyk, D. M. (1996). A versatile sunlit controlled-environment facility for studying plant and soil processes. *J. Environ. Qual.* 614–625.

Wan, S., Luo, Y., and Wallace, L. L. (2002). Changes in microclimate induced by experimental warming and clipping in tallgrass prairie. *Glob. Chang. Biol.* **8**, 754–768.

13

Aida, M., Beis, D., Heidstra, R., Willemsen, V., Blilou, I., et al. (2004). The *Plethora* genes mediate patterning of the *Arabidopsis* root stem cell niche. *Cell* **119**, 109–120.

Badescu G.O. and Napier R.M. (2006). Receptors for auxin: Will it all end in TIRs? *Trends Plant Sci.* **11**, 217–223.

Blilou, I., Xu, J., Wildwater, M., Willemsen, V., Paponov, I., et al. (2005). The PIN auxin efflux facilitator network controls growth and patterning in *Arabidopsis* roots. *Nature* **433**, 39–44.

Braun, N., Wyrzykowska, J., Muller, P., David, K., Couch, D., et al. (2008). Conditional Repression of auxin binding protein1 reveals that it coordinates cell division and cell expansion during postembryonic shoot development in *Arabidopsis* and tobacco. *Plant Cell* **10**, 2746–2762.

Chen, J. G., Ullah, H., Young, J. C., Sussman, M. R., and Jones, A. M. (2001). ABP1 is required for organized cell elongation and division in *Arabidopsis* embryogenesis. *Genes Dev.* **15**, 902–911.

Colon-Carmona, A., You, R., Haimovitch-Gal, T., and Doerner, P. (1999). Technical advance: Spatio-temporal analysis of mitotic activity with a labile cyclin-GUS fusion protein. *Plant J.* **20**, 503–508.

David, K. M., Couch, D., Braun, N., Brown, S., Grosclaude, J., et al. (2007). The auxin-binding protein 1 is essential for the control of cell cycle. *Plant J.* **50**, 197–206.

del Pozo, J. C., Boniotti, M. B., and Gutierrez, C. (2002). *Arabidopsis* E2Fc functions in cell division and is degraded by the ubiquitin-SCF (AtSKP2) pathway in response to light. *Plant Cell* **14**, 3057–3071.

del Pozo, J. C., Diaz-Trivino, S., Cisneros, N., and Gutierrez, C. (2006). The balance between cell division and endoreplication depends on E2FC-DPB, transcription factors regulated by the ubiquitin-SCFSKP2A pathway in *Arabidopsis*. *Plant Cell* **18**, 2224–2235.

Dello Ioio, R., Linhares, F. S., Scacchi, E., Casamitjana-Martinez, E., Heidstra, R., et al. (2007). Cytokinins determine *Arabidopsis* root-meristem size by controlling cell differentiation. *Curr. Biol.* **17**, 678–682.

Dello Ioio, R., Nakamura, K., Moubayidin, L., Perilli, S., Taniguchi, M., et al. (2008). A genetic framework for the control of cell division and differentiation in the root meristem. *Science* **322**, 1380–1384.

Dewitte, W., Riou-Khamlichi, C., Scofield, S., Healy, J. M., Jacqmard, A., et al. (2003). Altered cell cycle distribution, hyperplasia, and inhibited differentiation in *Arabidopsis* caused by the d-type cyclin CYCD3. *Plant Cell* **15**, 79–92.

Dharmasiri, N., Dharmasiri, S., and Estelle, M. (2005). The F-box protein TIR1 is an auxin receptor. *Nature* **435**, 441–445.

Edlund, A., Eklof, S., Sundberg, B., Moritz, T., and Sandberg, G. (1995). A microscale technique for gas chromatography-mass spectrometry measurements of picogram amounts of indole-3-acetic acid in plant tissues. *Plant Physiol.* **108**, 1043–1047.

Fukaki, H., Wysockadiller, J., Kato, T., Fujisawa, H., Benfey, P.N., et al. (1998). Genetic evidence that the endodermis is essential for shoot gravitropism in *Arabidopsis thaliana*. *Plant J.* **14**, 425–430.

Galinha, C., Hofhuis, H., Luijten, M., Willemsen, V., Blilou, I., et al. (2007). Plethora proteins as dose-dependent master regulators of *Arabidopsis* root development. *Nature* **449**, 1053–1057.

Grieneisen, V. A., Xu, J., Maree, A. F., Hogeweg, P., and Scheres, B. (2007). Auxin transport is sufficient to generate a maximum and gradient guiding root growth. *Nature* **449**, 1008–1013.

Helariutta, Y., Fukaki, H., Wysocka-Diller, J., Nakajima, K., Jung, J., et al. (2000). The short-root gene controls radial patterning of the *Arabidopsis* root through radial signaling. *Cell* **101**, 555–567.

Jones, A. M., Im, K. H., Savka, M. A., Wu, M. J., Dewitt, N. G., et al. (1998). Auxin-dependent cell expansion mediated by overexpressed auxin-binding protein 1. *Science* **282**, 1114–1117.

Kepinski, S. and Leyser, O. (2005). The *Arabidopsis* F-box protein TIR1 is an auxin receptor. *Nature* **435**, 446–451.

Kovtun, Y., Chiu, W. L., Zeng, W., and Sheen, J. (1998). Suppression of auxin signal transduction by a MAPK cascade in higher plants. *Nature* **395**, 716–720.

Leblanc, N., David, K., Grosclaude, J., Pradier, J. M., Barbier-Brygoo, H., et al. (1999). A novel immunological approach establishes that the auxin-binding protein, Nt-abp1, is an element involved in auxin signaling at the plasma membrane. *J. Biol. Chem.* **274**, 28314–28320.

Lee, J. Y., Colinas, J., Wang, J. Y., Mace, D., Ohler, U., et al. (2006). Transcriptional and post-transcriptional regulation of transcription factor expression in *Arabidopsis* roots. *Proc. Natl. Acad. Sci. USA* **103**, 6055–6060.

Leyser, O. (2005). Auxin distribution and plant pattern formation: How many angels can dance on the point of PIN? *Cell* **121**, 819–822.

Ljung, K., Hull, A. K., Celenza, J., Yamada, M., Estelle, M., et al. (2005). Sites and regulation of auxin biosynthesis in *Arabidopsis* roots. *Plant Cell* **17**, 1090–1100.

Magyar, Z., De Veylder, L., Atanassova, A., Bako, L., Inze, D., et al. (2005). The role of the *Arabidopsis* E2FB transcription factor in regulating auxin-dependent cell division. *Plant Cell* **17**, 2527–2541.

Malamy, J. E. and Benfey, P. N. (1997). Organization and cell differentiation in lateral roots of *Arabidopsis thaliana*. *Development* **124**, 33–44.

Menges, M., Samland, A. K., Planchais, S., and Murray, J. A. (2006). The D-type cyclin CYCD3;1 is limiting for the G1-to-S-phase transition in *Arabidopsis*. *Plant Cell* **18**, 893–906.

Mockaitis, K. and Estelle, M. (2008). Auxin receptors and plant development: A new signaling paradigm. *Annu. Rev. Cell Dev. Biol.* **24**, 55–80.

Mockaitis, K. and Howell, S. H. (2000). Auxin induces mitogenic activated protein kinase (MAPK) activation in roots of *Arabidopsis* seedlings. *Plant J.* **24**, 785–796.

Monroe-Augustus, M., Zolman, B. K., and Bartel, B. (2003). IBR5, a dual-specificity phosphatase-like protein modulating auxin and abscisic acid responsiveness in *Arabidopsis*. *Plant Cell* **15**, 2979–2991.

Nakajima, K. and Benfey, P. N. (2002). Signaling in and out: Control of cell division and differentiation in the shoot and root. *Plant Cell* **14**, Suppl., S265–276.

Napier, R. M., David, K. M., and Perrot-Rechenmann, C. (2002). A short history of auxin-binding proteins. *Plant Mol. Biol.* **49**, 339–348.

Ottenschlager, I., Wolff, P., Wolverton, C., Bhalerao, R.P., Sandberg, G., et al. (2003). Gravity-regulated differential auxin transport from columella to lateral root cap cells. *Proc. Natl. Acad. Sci. USA* **100**, 2987–2991.

Overvoorde, P. J., Okushima, Y., Alonso, J. M., Chan, A., Chang, C., et al. (2005). Functional genomic analysis of the auxin/indole-3-acetic acid gene family members in *Arabidopsis thaliana*. *Plant Cell* **17**, 3282–3300.

Perez-Torres, C. A., Lopez-Bucio, J., Cruz-Ramirez, A., Ibarra-Laclette, E., Dharmasiri, S., et al. (2008). Phosphate availability alters lateral root development in *Arabidopsis* by modulating auxin sensitivity via a mechanism involving the TIR1 auxin receptor. *Plant Cell* **20**, 3258–3272.

Sabatini, S., Beis, D., Wolkenfelt, H., Murfett, J., Guilfoyle, T., et al. (1999). An auxin-dependent distal organizer of pattern and polarity in the *Arabidopsis* root. *Cell* **99**, 463–472.

Sabatini, S., Heidstra, R., Wildwater, M., and Scheres, B. (2003). SCARECROW is involved in positioning the stem cell niche in the *Arabidopsis* root meristem. *Genes Dev.* **17**, 354–358.

Scherer, G. F. (2002). Secondary messengers and phospholipase A2 in auxin signal transduction. *Plant Mol. Biol.* **49**, 357–372.

Scherer, G. F., Zahn, M., Callis, J., and Jones, A. M. (2007). A role for phospholipase A in auxin-regulated gene expression. *FEBS Lett.* **581**, 4205–4211.

Schmid, M., Davison, T. S., Henz, S. R., Pape, U. J., Demar, M., et al. (2005). A gene expression map of *Arabidopsis thaliana* development. *Nat. Genet.* **37**, 501–506.

Shin, R., Burch, A. Y., Huppert, K. A., Tiwari, S. B., Murphy, A. S., et al. (2007). The *Arabidopsis* transcription factor MYB77 modulates auxin signal transduction. *Plant Cell* **19**, 2440–2453.

Strader, L. C., Monroe-Augustus, M., and Bartel, B. (2008). The IBR5 phosphatase promotes *Arabidopsis* auxin responses through a novel mechanism distinct from TIR1-mediated repressor degradation. *BMC Plant Biol.* **8**, 41.

Tan, X., Calderon-Villalobos, L. I., Sharon, M., Zheng, C., Robinson, C. V., et al. (2007). Mechanism of auxin perception by the TIR1 ubiquitin ligase. *Nature* **446**, 640–645.

Tanaka, H., Dhonukshe, P., Brewer, P. B., and Friml, J. (2006). Spatiotemporal asymmetric auxin distribution: a means to coordinate plant development. *Cell Mol. Life Sci.* **63**, 2738–2754.

Tao, L. Z., Cheung, A. Y., and Wu, H. M. (2002). Plant Rac-like GTPases are activated by auxin and mediate auxin-responsive gene expression. *Plant Cell* **14**, 2745–2760.

Tao, L. Z., Cheung, A. Y., Nibau, C., and Wu, H. M. (2005). RAC GTPases in tobacco and *Arabidopsis* mediate auxin-induced formation of proteolytically active nuclear protein bodies that contain Aux/IAA proteins. *Plant Cell* **17**, 2369–2383.

Ueda, M., Matsui, K., Ishiguro, S., Sano, R., Wada, T., et al. (2004). The halted root gene encoding the 26S proteasome subunit RPT2a is essential for the maintenance of *Arabidopsis* meristems. *Development* **131**, 2101–2111.

Ulmasov, T., Murfett, J., Hagen, G., and Guilfoyle, T. J. (1997). Aux/IAA proteins repress expression of reporter genes containing natural and highly active synthetic auxin response elements. *Plant Cell* **9**, 1963–1971.

Vanneste, S. and Friml, J. (2009). Auxin: A trigger for change in plant development. *Cell* **136**, 1005–1016.

Wildwater, M., Campilho, A., Perez-Perez, J. M., Heidstra, R., Blilou, I., et al. (2005). The retinoblastoma-related gene regulates stem cell maintenance in *Arabidopsis* roots. *Cell* **123**, 1337–1349.

Winter D., Vinegar B., Nahal H., Ammar R., Wilson G. V., et al. (2007). An "electronic fluorescent pictograph" browser for exploring and analyzing large-scale biological data sets. *PLoS ONE* **2**, e718.

Index

Milton Keynes UK
Ingram Content Group UK Ltd.
UKHW031145141024
449569UK00024B/1064